Milton A. Anderson

GLP QUALITY AUDIT MANUAL

THIRD EDITION

CRC Press
Taylor & Francis Group
Boca Raton London New York

CRC Press is an imprint of the
Taylor & Francis Group, an **informa** business

GLP QUALITY AUDIT MANUAL
Third Edition

Milton A. Anderson

CRC Press
Taylor & Francis Group
6000 Broken Sound Parkway NW, Suite 300
Boca Raton, FL 33487-2742

First issued in paperback 2019

© 2000, 2001, 2002 by Taylor & Francis Group, LLC
CRC Press is an imprint of Taylor & Francis Group, an Informa business

No claim to original U.S. Government works

ISBN-13: 978-1-57491-106-0 (hbk)
ISBN-13: 978-0-367-39843-9 (pbk)

Library of Congress Cataloging-in-Publication Data

Anderson, Milton A.

GLP quality audit manual/Milton A. Anderson.-3rd ed.
    p.   cm.
    Includes bibliographical references.
    ISBN 1-57491-106-6
    1. Medical laboratories-quality control. I. Title.
    R860.A53   2000
    616.07'56'0685—dc21   00-056738

Visit the Taylor & Francis Web site at
http://www.taylorandfrancis.com

and the CRC Press Web site at
http://www.crcpress.com

CONTENTS

Section II—Quality Audit Checklists

Section III—Document/Report Examples

Section IV—References/Regulatory Texts

INTRODUCTION

As we enter the 21st century and the new millennium, our experience in the Good Laboratory Practice regulations and principles has extended over more than twenty years. Initially implemented reluctantly by many, these rules are now overwhelmingly accepted by the worldwide community. They have through the years given excellent direction to developing a consistent and efficient pathway for assuring the quality and integrity of research data. This mandate has provided the multinational approving authorities a high degree of confidence in the decisions they must provide regarding the scientific research data that sponsors have submitted.

This inherent system of providing "perpetual" data certification has also reduced the need for replicating studies, resulting in extensive research cost reductions and the saving of animal lives. Additionally, and perhaps most importantly, this confirmation of data quality and integrity has expedited the accessibility by consumers of deserving products, potentially benefiting and saving lives of the ultimate users—all of us.

Beginning in the last years of the 20th century, we also have been assisted by the ability to rapidly access information from around the world via the Internet. No longer is it necessary to wait weeks for a mailed response from a colleague overseas or days for a voluminous regulatory document to be shipped. Now these can be immediately accessed, reviewed, and downloaded, and printed instantaneously over the Internet. This convenience is a major change that will enhance the timeline of research, and, within the quality assurance functions, will provide immeasurable improvements in inspections, auditing, archiving, and administration. The future looks bright.

Milton Anderson

June 2000

ACKNOWLEDGMENTS

I would like to thank my many friends and colleagues at Abbott Laboratories, the Society of Quality Assurance, the American Society for Quality, and other professional organizations for their mutual interest in furthering the progress and understanding of the Good Laboratory Practices. For those who have been engaged in the development of the quality assurance role in GLP, I want to thank you. You have contributed to my overall GLP experience, knowledge, and encouragement to write this manual on Good Laboratory Practices.

I would also like to express my thanks to those individuals who recently urged me to proceed with a third edition of the GLP Quality Audit Manual. This manual should continue as a benefit to those of you engaged in the Good Laboratory Practices or the OECD Principles of Good Laboratory Practice. This version includes a new chapter on Internet Resources, which should be enlightening to many of you who have access to the Internet. The information from this global resource is endless, and its timeliness should result in improved quality assurance operations worldwide.

My enduring appreciation to my late wife, Dorothy, for her inspiration in my writing the initial publication, and the continued support and encouragement from my family and friends—Edward, Vickie, John, Brendan, Erin, Alexandria, Amanda, Peter, Marge, and Karen—to complete this third edition.

HOW TO USE THIS MANUAL

This manual is written as four separate sections:

Section I—Good Laboratory Practices

The reader is advised to read this section as background to understanding the development and the need for Good Laboratory Practice auditing in the healthcare and chemical industries. These chapters also provide a foundation of information in designing and initiating a Good Laboratory Practice quality assurance program.

Section II—Quality Audit Checklists

There are available in this section two types of checklists. First is the complete questionnaire checklist, which addresses all the relevant regulatory requirements. Secondly, are the abbreviated, condensed form, addressing preplacement audit and specific phases in auditing.

[The publisher grants permission to the original purchaser of this book to photocopy the Audit Checklists in reasonable quantities for use specifically by the purchaser in performing his/her job function at his/her job location. Permission is not granted for companies to copy the Audit Checklists for use by other personnel at other geographic locations or other distinct corporate units at the same location.]

Section III—Appendixes

The Quality Assurance Unit (QAU) described throughout the manual has some basic document requirements; those illustrated in this section may be helpful in achieving a better understanding of the QAU function.

Section IV—Regulatory Texts

For reader convenience, several helpful texts are provided in this section, including the full texts of both the U.S. Food and Drug Administration's (FDA) Good Laboratory Practice Regulations and the

---

Organization for Economic Co-operation and Development's (OECD) Principles of Good Laboratory Practice. To provide convenient references to the FDA's rationale for including or excluding certain regulations, the Federal Register comments sections for the June 20, 1979, and September 4, 1987, issues are furnished in this manual.

SECTION I
GOOD LABORATORY PRACTICES

CHAPTER 1

DEVELOPMENT OF REGULATIONS AND GUIDELINES

A. International Development

Although the policy of "good laboratory practice" has been observed to various degrees by private and public laboratories for years, implementation by governments to control laboratories did not begin until the early 1970s. Among the first countries to formally address the issue was New Zealand, with the promulgation in 1972 of the "Testing Laboratory" Registration Act.[1] This act defined the testing laboratory to include "the equipment, facilities, staff, records, procedures, and places used in testing." Additionally, it defined *test* as meaning "determining in whole or in part, the composition or physical properties of a substance or product, calibrating a piece of equipment, or determining the ability of any substance, product, or piece of equipment to satisfy particular requirements; . . ." The act also established a Testing Laboratory Registration Council, with functions and powers "to promote the development and maintenance of good laboratory practice in testing . . . ."

In March 1973, Denmark approved legislation that addressed laboratory practices. The "Law About the State's Technical Trial Board"[2,3] defined the duties of the Board as advancing, promoting, coordinating, and keeping control of authorized technical experiments for the purpose of bringing about safety and quality control. The research addressed includes that related to a material's physical, chemical, structural, and quality characteristics and how the material performs under various environmental conditions. The Board, the law explained, monitors the progress of authorized research activities, both public and private, including the auditing of documentation of issued research results. Approved research laboratories must satisfy certain criteria relating to the knowledge and competence of the staff, the capacity of the organization, and the suitability of research equipment.

Prior to this legislation there were many private and public laboratories worldwide that had established specific means to confirm the credibility, integrity, and quality of research. These voluntary policies were intensely enforced in some, but not all, organizations. In many countries, government was becoming more involved in both the awarding of financial grant programs for research and the requirement of government approval prior to marketing materials and products that affect the well-being and safety of many people. These government bureaus recognized that any decision to approve or disapprove the marketing of a product was only as good as the reliability of the research and data submitted to them. The significance of the New Zealand and Danish approaches was that these countries, for the first time, had legislated the beginning elements for prescribing laboratory research and reliability, through the concept of "good laboratory practices."

The need for major improvements in the authenticity and reliability of nonclinical laboratory studies in the United States was recognized in the mid-1970s. In 1975 in an existing "for-cause" compliance program, the Food and Drug Administration (FDA) noticed deficiencies during the inspections of some firms that seriously impugned the integrity of reported data. As a consequence of these findings and subsequent congressional action, the Good Laboratory Practice (GLP) regulations were implemented in June 1979.[4]

Because the GLP regulations as conceived in the United States had much broader implications than previous acts, they aroused interest in other countries and among international organizations such as the World Health Organization (WHO) and the Organization for Economic Cooperation and Development (OECD). In April 1978 the 22 member countries for the OECD, meeting in Stockholm, arrived at a consensus to assign the responsibility of developing the GLP internationally to the OECD. Beginning with an OECD meeting in Washington in April 1979, and subsequent deliberations at meetings in Brussels, Paris, and Berlin, the OECD Council approved the "OECD Principles of Good Laboratory Practice" on May 12, 1981. After more than 15 years of use, "Member countries considered that there was a need to review and update the Principles of GLP to account for scientific and technical progress in the field of safety testing and the fact that safety testing was currently required in many more areas . . . ." In 1995 an Expert Group was established, led by Germany with experts from Australia, Austria, Belgium, Canada, the Czech Republic, Denmark, Finland, France, Germany, Greece, Hungary, Ireland, Italy, Japan, Korea, the Netherlands, Norway, Poland, Portugal, the Slovak Republic, Spain, Sweden, Switzerland, the United Kingdom, the United States, and the International Organisation for Standardisation. Their work was completed in 1996, and the Revised OECD Principles of GLP were reviewed by the relevant policy bodies of the Organization and were adopted by Council on November 26, 1997.[5] These principles of GLP, although similar to the U.S. Food and Drug Administration GLP, differ in scope, being applicable to "all non-clinical testing of chemicals" whereas the FDA GLP applies to food and color additives, animal food additives, human and animal drugs, medical devices for human use, and biological and electronic products." Another agency in the United States, the Environmental Protection Agency (EPA), has responsibility for nonpharmaceutical chemicals as reflected through two distinct standards,[6,7] one for pesticides and the other for toxic chemical substances.

Although in the United States pharmacology studies have not routinely required compliance to GLPs, in many European countries their regulatory agencies have required that these studies also be conducted in compliance. Since many U.S. sponsors do have an interest in providing submissions to a variety of global regulatory agencies, sponsors in the United States are now conducting the appropriate safety-related pharmacology studies in compliance with the GLP regulations.

B. United States Development

The need for standard GLP criteria was well established in the United States in the mid-1970s. Although it has been said that these regulations were arbitrarily promulgated to extend the agency's enforcement base, most now concur that these regulations have been purposeful and beneficial. History proves an opportunity to evaluate the surveillance teams' reported conditions and to judge whether the GLP regulations were necessary. A review of these findings provides important historical perspective and an instructive first step in recognizing undesirable conditions that may still exist in some laboratories not working to GLP standards. Such a review may help in initiating action to prevent the continuance or future restoration of these conditions.

Some of the laboratory conditions existing during the mid-1970s were not necessarily the norm, but were found by the FDA and EPA during their review of industry operating procedures. The review was part of an extensive surveillance of the pharmaceutical and chemical industry and was initially reported in testimony to the United States Senate Subcommittee on Health in 1975 and 1976 and was subsequently publicized in newspapers and scientific journals. The most prominent evidence supporting the need for regulations was the flagrant research discrepancies noted by the FDA at a contract research laboratory. The laboratory, Industrial Biotest Corporation (IBT) of Northbrook, Illinois, was a leader in the contract testing field, performing over 20,000 studies to support the safety and efficacy data generated for hundreds of drugs and pesticides.[8] The alleged irregularities in the data were numerous. They included the falsification of laboratory work; replacement of animals that died under test with fresh animals without documenting the substitution; fabricating test results; and excluding test results if the results were not considered favorable by top company officials. Three company officials were found guilty of defrauding the government by falsifying drug and food additive research data. Their appeal to the Supreme Court was denied in 1986, and all were given lengthy jail sentences.[9]

Governmental surveys of nonclinical studies performed at various types of institutions revealed vast differences in the means of documenting study reliability. At the time of the compliance program survey, the institutions audited were not aware of the evaluation criteria. The agency had contended that the study data submitted in support of a regulated product were based on appropriate experimental procedures by a testing facility. Industry, contract laboratories, and academia were subsequently audited in the absence of published regulations, contributing widely to "compliance" disparities.

It is important to be aware of these governmental surveillance observations and conclusions,[10] particularly where regulations or guidelines such as the GLP regulations have not been implemented in a company performing laboratory studies. These observations and conclusions provide the rationale for the need and subsequent design and development of the GLP regulations or guidelines. The FDA's observations and conclusions follow.

Observations

1. Original autopsy records for certain studies were either unavailable or apparently transcribed to new records several years after the autopsies.

2. Pathology reports submitted to the agency were inconsistent with the original autopsy records.

3. Microscopic examinations of tissue slides were conducted by more than one pathologist, each of whom came to different conclusions; yet only the conclusions favorable to the drug were submitted to the agency.

4. Records of laboratory observations were neither dated nor signed.

5. Employees were unable to account for discrepancies between raw data and final reports submitted to the agency.

6. Animals were observed and recorded as normal for a variety of factors, including appearance, appetite, and thirst, when in fact the animals were dead.

7. Drugs under study were administered to animals in a manner that made it impossible to determine how much, if any, of the required dosage was actually ingested by the animal.

8. At one firm, the reproduction and teratology studies were conducted and laboratory personnel overseen by a senior scientist who did not have the proper qualifications or background to be conducting and supervising these critical studies.

9. In another case, necropsies were being performed by people without the proper training, as was recognized by a senior scientist who reviewed the work.

10. Treatment and control animals were not properly identified.

11. Weighings of the animals were not accurately recorded.

12. Animals were fixed in toto and not necropsied for several months.

13. A laboratory was sprayed and fogged with pesticides while the animals were in the laboratory.

14. In one study done by a contract laboratory, although the management had serious questions about the conduct of the study, they never questioned or exercised any control over the operating investigator.

15. In another study done by a contract laboratory, the FDA was told that animal tissue had been examined histopathologically when a review of the contract laboratory's original records indicated that these tissue samples were never even collected.

16. Significant discrepancies were found between gross observations on pathology sheets when compared with the individual pathology summaries submitted to the agency.

17. One firm submitted a study utilizing the wrong data and the wrong animal identification numbers, which were easily discovered by the agency, indicating that management did not check the data used.

Conclusions

1. Experiments were poorly conceived, carelessly executed, or inaccurately analyzed or reported.

2. Technical personnel were unaware of the importance of protocol adherence, accurate observations, accurate administration of the test substance, and accurate record keeping and record transcription.

3. Management did not assure critical review of data or proper supervision of personnel.

4. Assurance could not be given for the scientific qualifications and adequate training of personnel involved in the research study.

5. There was a disregard for the need to observe proper laboratory, animal care, and data management procedures.

6. Sponsors failed to monitor adequately the studies performed by contract testing laboratories.

7. Firms failed to systematically verify the accuracy and completeness of scientific data in reports of nonclinical laboratory studies before submission to the FDA.

These observations, together with other criteria, were used in the development of the November 1976 proposed GLP regulations.[10] The FDA, however, also initiated another surveillance program directed toward uninspected testing facilities to determine if there were "proposal" problems they had not anticipated.

Those in the health care field were exposed early to the governmental "compliance" survey. This survey,[11] conducted for approximately 3 months, consisted of 39 nonclinical laboratories, 23 sponsor laboratories, 11 contract laboratories, and 5 university laboratories. These audits were used in acquiring information for fine-tuning the final governmental regulations. Results of this program indicated some disparity in "compliance" between industry and academia. Sponsor laboratories were in compliance 69 percent; contract laboratories 56 percent; university laboratories 46 percent.

Primary noncompliance with the proposed regulations were with facilities failing to have a Quality Assurance Unit (QAU), failing to test each batch of test article-carrier mixture, and lacking standard operating procedures (SOP). The negative results regarding the QAU should not be surprising, since this was a relatively new concept, and in many operations the writing of formalized SOPs was a major task not yet completed.

As a result of the FDA Bioresearch Monitoring Program, including the pilot inspection program, published commentary, and public hearings, the U.S. FDA published the Good Laboratory Practice Regulations, which became effective on June 20, 1979.

Changes to the U.S. GLP were proposed in the October 29, 1984 (49FR43530), issue of The Federal Register. The proposals primarily involved changes to the provisions on quality assurance, protocol preparation, test and control article characterization, and retention of specimens and samples. These changes were subsequently issued in The Federal Register of September 4, 1987 (52FR33768).[12] Changes identified as significant by the FDA included provisions respecting quality assurance, protocol preparation, test and control article characterization, and retention of specimens and samples. Another Final Rule (effective September 13, 1991) was published in The Federal Register of July 15, 1991 (56FR32087), removing the examples of methods of animal identification.[13]

CHAPTER 2

THE QUALITY ASSURANCE UNIT

A. Personnel

The Quality Assurance Unit (QAU) may reside in one of several areas of an organization. It may consist of one or more persons, but the unit must be kept independent of management directly associated with the nonclinical study. Typically in large companies this is a separate group consisting of several auditors, a department manager, and a director (see Figure 2-1).

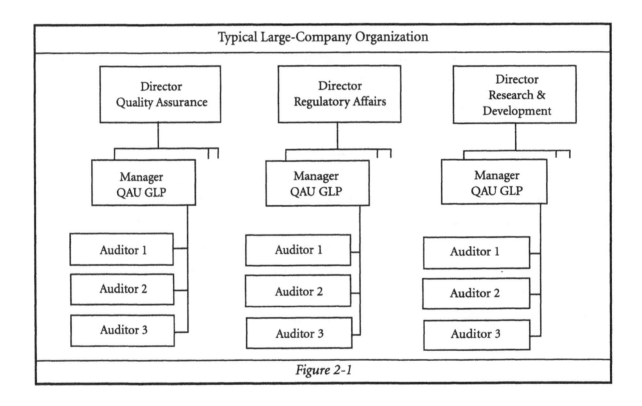

Figure 2-1

Section 58.35 of the GLP states that the testing facility shall have a QAU that shall be responsible for monitoring each study to assure management that the facilities, equipment, personnel, methods, practices, records, and controls conform with the regulations. The regulations are flexible and do allow for economies. In a small organization, study directors and QAU personnel could switch roles as long as the functions remain independent for each study. Management may assign toxicologist "A," for example, as study director of Study "A," and assign toxicologist "B" to perform the QAU functions of Study "A." A company, however, that has single functionaries, for example, a toxicologist and pathologist who participate professionally in each study, could not function as QAU personnel, for example, as study inspectors or auditors.

The QAU is not specifically designated by the GLPs to address the technical aspects of the study, but rather to assure conformity with the procedural and administrative requirements. The personnel engaged in inspecting the study should direct their attention toward observing the professionals' or technicians' compliance with the Standard Operating Procedures (SOPs) and toward the proper documentation of the event and the ensuing measurements as raw data. In auditing the scientific report, the auditors should confirm that it accurately reflects the raw data.

It is advantageous that personnel engaged with the QAU's inspection and auditing responsibilities have experience and education in science. Technically this type of background may not always be necessary; however, it is a definite asset. Scientific reports typically include many scientific and measurement terms, various descriptions of toxicological and pathological findings, summaries, conclusions, and many related tables along with statistical results. In some organizations, the scope of the QAU's responsibilities has been expanded to include the scientific aspects as well.

B. Responsibilities

The responsibilities of the QAU are not always limited to those listed in the GLP. Typically a QAU may also perform diversified functions associated with archiving, calibration, computer validation, animal welfare compliance,[14] regulatory affairs, and Good Clinical Practices. Allowances for engaging in these supplementary functions may depend on how demanding the nonclinical inspection and audit requirements are in a specific organization. Inclusion of some of these responsibilities should ease budgetary constraints in establishing and maintaining a QAU in your unique circumstances.

One of the QAU's primary functions is monitoring each regulated study to assure management that the facilities, equipment, personnel methods, practices, records, and controls conform with the GLP. Study protocols, or plans, are usually issued by the department responsible for administering the nonclinical study; in the case of a drug sponsor, the Drug Safety Evaluation Division (DSE); for contracted studies, the contractor. Typically the contract laboratory initiates the protocol, which is finalized after review with the sponsor. Copies of all applicable regulated nonclinical study protocols are submitted by the study administrative unit for retention. Two copies are normally received, one "date-stamped," filed according to drug and study identities, and the other copy assigned to the QAU auditor as a working reference. As an administrative aid, a protocol log should be maintained, referred to as the Protocol Record. This is indexed by study drug and may also include headings such as Study Identity, Type of Study, Duration, Route, Species, Study Director, Start Date, Receipt Date, and Auditor.

After the GLP study protocol has been issued, it should be included on the organization's Master Schedule. This schedule should reflect the activities of the various study directors engaged in nonclinical studies and provide a projected schedule of inspections for the QAU. The regulatory responsibility of the QAU is to maintain a copy of this schedule, not necessarily issue it. Frequently this document is issued by the QAU, but departments responsible for scheduling and administering the studies have also published it. An example of the Master Schedule for Northern Research, Inc. (NRI) is shown in Section III.

Some important GLP Master Schedule requirements usually included in an evaluation by external auditing agencies are:

- Indexing by test article

- Identities of the test system

- Nature of study

- Date study was initiated

- Current status of each study

- Identity of the sponsor

- Name of the study director

Compliance agencies typically use this schedule as one means of judging the workloads of various study directors. Keyed to the schedule are specific projected events, protocol issuance, inspections, report completion, and audits as reflected through the protocol or protocol schedule. Reissuance of the master schedule is normally done monthly, but the frequency may vary depending on study accession rates. Some companies may utilize a computer system to provide a continuing master schedule database, periodically, for example, monthly, providing an official hard copy for retention and reference of the official Master Schedule Sheet. This document is considered raw data and, therefore, the computer system maintaining and generating this document must be validated.

It is recommended that sponsors utilizing contract laboratories also maintain a master schedule of studies they have contracted. In addition to the preceding requirements, the identity of the contractor should be included and the extent of his involvement, for example, wet tissue cutting, preparing blocks and histology slides, or conducting the complete study. The contractor laboratory should, of course, be advised that the service or analysis performed be conducted in compliance to the GLP (58.10). In those instances in which a contract laboratory will not issue a final report, it is important that the sponsor request a QAU Statement, identifying the specific service the contractor performed and the dates inspections were conducted.

Documentation of each periodic inspection is also maintained by the QAU. Properly signed records include the date of inspection, the study inspected, the phase or segment of the study inspected, the auditor, findings and problems, corrective actions recommended, and, if appropriate, the scheduled date of reinspection. Examples of these types of documents, the Study Inspection Record, and the Inspection Report, can be found in Section III-1. Two separate reports are needed because one, the Study Inspection Record, may legally be requested and reviewed by the agency during their inspection. This record includes the study phases, inspection dates, auditors' signatures, and dates. The GLP specifies that this shall be made available to authorized individuals of the agency. The Inspection Report, however, containing findings, problems, and corrective actions, is a confidential record exempt from routine agency inspections.

If significant problems that could affect the study's integrity are encountered during a QAU inspection, these should be reported to the Study Director or to management immediately. If initially handled verbally, problems of this nature should subsequently be documented in an Inspection Report.

You should also be aware that management must be fully informed of any perceived serious adverse drug experience observed in testing laboratory animals. The sponsor has a responsibility of notifying the FDA and all participating investigators in a written Investigational New Drug (IND) safety report of any adverse experience associated with the use of the drug that is both serious and unexpected. A serious adverse drug experience includes any experience suggesting a significant risk for human subjects, including and finding of mutagenicity, teratogenicity, or carcinogenicity. Unexpected adverse experience means any adverse experience that is not identified in nature, severity, or frequency in the current investigator brochure; or, if an investigator brochure is not required, that is not identified in nature, severity, or frequency in the risk information described in the general investigational plan or elsewhere in the application. The notification would be required within 15 calendar days after the sponsor's initial receipt of the information (21 CFR 312.32). The complete details of this procedure should be described in the appropriate sponsor's management SOP.

All Inspection Reports, including those routine in nature, are submitted to the study director for a coordinated response. Depending on the observation, it may be necessary for several respondents to provide an integrated corrective action. The QAU auditor, for example, who has made an observation involving the statistics of a clinical chemistry table or summary, must include the study director as a respondent on the Inspection Report; but additionally it may be appropriate to require responses from the pathology monitor and the study statistician.

Management may be informed of these observations by submitting a copy of the Inspection Reports to their attention. After corrective actions have been recommended and completed, the Inspection Reports are submitted to management for their review and comment. Written status reports should also be submitted periodically on each study to the study director and management, including appropriately any problems and corrective actions. Both the Study Inspection Record and the Inspection Report are to be maintained by the QAU in accordance with current regulations.

These records along with the Master Schedule and copies of protocols must be maintained by the QAU for the period of time specified in section 58.195. This retention is basically for the shortest duration of time specified, as follows:

1. A period of two years following the date on which an application for a research or marketing permit is approved by the FDA.

2. A period of at least five years following the date on which the results of the nonclinical laboratory study are submitted to the FDA in support of an application for a research or marketing permit.

3. In instances in which the nonclinical laboratory study does not result in the submission of the study in support of an application for a research or marketing permit, a period of at least two years following the date on which the study is completed, terminated, or discontinued.

During the inspection the auditor should have various documents, including the GLPs, the protocol, laboratory manuals, and the applicable SOPs, available for immediate reference. Compliance with these requirements should continually be evaluated during the inspections. If variations are observed, documentation of authorized protocol changes or deviations should be ascertained. Protocol changes would include those events one is aware of before they take place, or prospectively, and are scheduled to occur routinely or systematically. For example, just prior to the start of the study it might be decided that an additional dosage group be included in the protocol or the dosage levels changed. Because of a clinical sign, it may be decided to schedule a supplemental series of blood collections and tests as part of the study. These prospective changes would be officially documented and authorized with a protocol change.

Deviations are variations from the study requirements as reflected in the protocol or SOPs. Deviations are usually a one-time, unanticipated occurrence. These events must be documented initially in the raw data and then in the report. For example, a temporary power failure caused the room temperature to drop below the protocol required range of 67°F to 77°F for three hours; a technician inadvertently misdosed the T1 group with T2 drug. These would first be mentioned in the appropriate raw data records, for example, Daily Journal, and then in the final report. Confirmation of the study director's timely documented acknowledgment of these events should also be part of the audit.

The culmination of all the study participants' efforts is best reflected in a scientific report. The scientists and technicians have persevered for months or years, carefully contributing their expertise to a study. Although they may have diligently tried to maintain compliance, there may be events that were not documented or records that differ. The QAU auditor's responsibility is to carefully review these various documents, assure that the report accurately reflects the raw data, and describe the methods and standard operating procedures.

After the report has been audited, the QAU manager (or equivalent) will complete what is commonly referred to as the QAU Statement as shown in Section III-1. This record, which becomes part of the scientific report, includes the dates inspections were made and dates findings were reported to management and the study director. All these data are readily available from the Study Inspection Record.

As part of the application for a research and marketing permit, the sponsor is also required to furnish a conforming statement relative to each of the nonclinical laboratory studies submitted to the FDA. In many companies this is a Regulatory Affairs responsibility, since they are typically the liaison between the company and the regulatory agency. The statement furnished should describe each study that is conducted in compliance with the GLPs, or if it was not conducted in compliance, a statement describing in detail the variations from these requirements. Noncompliant studies may be submitted, but the sponsor may have to justify that the noncompliance did not affect the integrity of the data (21 CFR 314.101).

The preceding QAU responsibilities, along with others that management may so designate, are documented in the SOPs. Included in these procedures should be SOPs that describe the following records:

*   Protocol Record

*   Master Schedule

*   Inspection Report

*   Inspection Record

*   QAU Statement

Regulations require that these records be maintained by the QAU in one specified location of the facility. This is particularly necessary in an organization that may have operations and individuals located in several facilities. As a precautionary measure some QAUs provide as a part of their "disaster planning" a policy of periodically photocopying or microfiching these unique records for storage in a remote secondary site.

C. Standard Operating Procedures

One of the more important concepts established by the Good Laboratory Practices and Principles is the need to conduct a nonclinical study accurately and with consistency. This can be accomplished only if the SOPs are written adequately to insure the quality and integrity of the data, and if relevant files are maintained of procedural revisions and deviations. These requirements apply to the QAU as well. Inspections, audits, maintenance of records, and issuing of reports must all be compliant with the GLPs and follow the management-authorized SOP instructions. As methods are modified, and revisions issued, historical files of the SOPs reflecting these changes must also be maintained.

The QAU SOP requirements are specified in part 58.35(c). None is identified by title, but in the regulations it specifies that the responsibilities and procedures applicable to the QAU, the records maintained by the QAU, and the method of indexing such records shall be in writing and shall be maintained. Some SOP titles that have been applied include the following:

1. Records Maintained by the Quality Assurance Unit

2. Maintenance of the Master Schedule

3. Maintenance of Study Protocols

4. Monitoring Nonclinical Studies

5. Systems Auditing

6. Scientific Report Auditing

7. Preparation of the Quality Assurance Unit Statement

8. Reports to Management and Study Director

Within these titles, the SOPs may be expanded to include subheadings that may more appropriately describe the QAU requirements. For example, the SOP, Monitoring Nonclinical Studies, would designate those study segments of concern, and those that are implied in the GLP as impacting the integrity of the study. These may include subheadings such as those that follow:

SOP: Monitoring Nonclinical Studies

    A.   Initiation Phase Inspection

        1.   Protocol Requirements

        2.   Baseline Procedures

    B.   In-Process Phase Inspection

        1.   Animal Weighing

        2.   Diet Preparation

        3.   Drug Preparation and Storage

        4.   Drug Administration

        5.   Facilities and Equipment

        6.   Data Collection

        7.   Special Procedures

    C.   Completion Phase Inspection

        1.   Necropsy

        2.   Histology

        3.   Clinical Pathology

        4.   Pathology Data Collection

        5.   Pathology Laboratory Facilities and Equipment

    D.   Reporting Phase

        1.   Protocol Review

        2.   Toxicology Data Review

        3.   Pathology Data Review

        4.   Final Report Review

    E.   QAU Reporting

        1.   Report Preparation

        2.   Report Distributions

        3.   Report Responses

        4.   QAU Inspection Statement

Like the laboratory disciplines requiring retention of SOPs, it is necessary for the QAU to maintain a historical file of their SOPs also, including all revisions. You might find it convenient as an annual routine to assemble in a single binder the above records along with those QAU SOPs that were in effect during the calendar year. Should an auditing agency come in to audit a study conducted during a previous year(s), the appropriate binder(s) containing the QAU SOPs and records could conveniently be retrieved and referenced.

D. Training

Currently, QAU personnel engaged in inspection, auditing, and administrative responsibilities of the GLPs come from a variety of career experiences and educational backgrounds. Perhaps all have had scientific training, baccalaureate, masters, or doctoral degrees. When the GLPs were first proposed, one of industry's immediate concerns was that there would be a shortage of personnel qualified to fulfill these responsibilities. Management felt that to meet their perceived qualifications it would be necessary to assign scientists already productive and familiar with their particular area of science. The final regulations, however, elaborated further and clearly defined the function of the QAU as one assuring compliance with procedural and administrative requirements rather than overseeing the technical aspects of study conduct. QAU personnel need not be limited to professionals or scientists, but these certainly remain major assets in this field and are recommended prerequisites.

The GLP defines the QAU's regulatory range of duties, but some management may prefer increasing the scope of the assignment, including auditing of the technical aspects of the study. Because of associated company functions that may be assigned, for example, archiving, calibration, regulatory affairs, and "good clinical practices," personnel qualifications may vary considerably between different organizations. Within a QAU, for example, there may be three or four levels of job responsibilities, some including only inspection, others inspection, auditing, or administration. A formal written job description is usually the way of communicating your specific needs and would increase the likelihood of efficiently acquiring the kind of individual required. Shortly after being assigned to the QAU, qualified individuals should proceed with pursuing their training needs. Depth of training may vary depending on job level, responsibilities, and prospective career paths. Attending external seminars on GLP topics as well as other subjects that enhance the auditor's understanding of the client group's nonclinical studies would be beneficial. For GLP inspection or auditing, in-depth scientific knowledge may not be necessary, but familiarity is essential. Internal seminars can also be organized by requesting brief presentations from your own organization's professionals in their field of expertise. A series of one- or two-hour sessions is common, and by providing video or audio records of these seminars, new employees may readily access the information for an efficient, continuing training program.

Specific training needs for each organization are unique, but the following syllabus may be a helpful reference to topics with which a new auditor should become familiar.

A. Quality Assurance Unit

   1. Organization and Personnel

   2. Management and Auditor Roles

   3. Good Laboratory Practice Regulations Overview

B. Investigational Facilities

   1. Animal Facilities

   2. Guide for the Care and Use of Laboratory Animals[14]

   3. Test Article Storage

   4. Formulating and Dispensing

   5. Clinical Pathology Laboratories

   6. Histology Laboratories

   7. Archives

C. Department (QAU) Standard Operating Procedures

D. Good Laboratory Practice Regulations/Guidelines

   1. Food and Drug Administration (FDA) (21 CFR Part 58)

   2. Environmental Protection Agency (EPA) (40 CFR Part 160)

   3. Environmental Protection Agency (EPA) (40 CFR Part 792)

   4. The Organisation for Economic Co-operation and Development (OECD) Principles of Good Laboratory Practice

E. Client Group Standard Operating Procedures

   1. Toxicology

   2. Pathology

   3. Pharmacology

   4. Animal Services

   5. Analytical Services

   6. Archives

   7. Nonclinical Statistics

F. GLP Inspections and Audit

    1. Initiation Phase

        a. Protocol

        b. Baseline

    2. In-Process Phase

        a. Documentation

        b. Proper Data Entries

        c. Drug Dispensing

        d. Dosing

    3. Completion Phase

        a. Necropsy

        b. Specimens

        c. Clinical Pathology

        d. Histology

    4. Reporting Phase

        a. Archives

        b. Raw Data

        c. Statistics

        d. QAU Statement

G. Metrology

    1. Calibration Program

H. Computer Systems

    1. Auditing Techniques

    2. Validation

    3. Internet Resources

Many of these topics are important and should preferably be covered in a proper sequence. The best learning arrangement is to have the prospective auditor accompany a senior auditor on several audits, addressing repeatedly several of these topics. Informational resources should be addressed early on, since the individual can access these at his or her own pace. The reference information is voluminous, and, therefore, it is suggested that a set of goals or schedule covering familiarity of these be established.

Documentation supporting satisfactory completion of this training should also be maintained. The document should include confirmation by management with signatures and dates indicating review and subsequently, if appropriate, acceptance of the individual as a qualified GLP auditor.

E. GLP Audit Checklists

Checklists have been found to be a convenient way to evaluate a facility, its operations, or a nonclinical study for compliance to Good Laboratory Practice Regulations or OECD Principles. It is particularly helpful for those individuals just becoming familiar with this activity, but also, for the more experienced, it provides a means of maintaining a consistency in the performance of their inspections and audits.

The wording in the officially published U.S. Food and Drug Administration Good Laboratory Practice Regulations was carefully developed and chosen by its authors. As with most laws and regulations semantics is important. In converting these regulatory statements from the usual imperative mood to a questionnaire checklist, I have tried to retain the intended meaning of the original provisions. Once familiarity with this relatively thorough GLP Audit Checklist—Complete is gained, the auditor can refer to the Abbreviated GLP Audit Checklist in Section II. These are more convenient checklists, paraphrasing the original in a shorter, more practical form. Additionally, these are unique to a specific inspection or audit phase, that is, Initiation, In-process, Completion, or Reporting. All checklists have references to the appropriate part of the Code of Federal Regulation (21 CFR) for example, 58.29 or 58.105, etc., the checklists and alphanumeric designations are also consistent, allowing one to readily switch from the abbreviated checklists to the specific questions of the full GLP Audit Checklist. The U.S. Environmental Protection Agency Standards[6,7] are also published in the Code of Federal Regulations, under different titles and parts, but have conveniently retained the same decimal identities after the part number. For example, in the section addressing the QAU:

Product	Agency	CFR Title	Part	Section
Drugs	FDA	21	58	58.35
Pesticides	EPA	40	160	160.35
Toxic Substances	EPA	40	792	792.35

There are similarities between these FDA and EPA documents, but they focus on distinctly different regulatory areas of responsibility and should be referred to separately for specific concerns.

The checklists are worded in such a manner that a positive response indicates regulatory compliance, and a negative response noncompliance. For some studies, the question may not be applicable, soliciting a response that may be indicated by checking the "N/A" box. A quality rating system was not considered a prudent checklist feature for use in GLP evaluations, for example, identifying varying degrees of compliance. The auditor should provide the client group with an explicit decision regarding compliance or noncompliance. If any questionable conditions exist, they should be identified and discussed and an action plan established to achieve satisfactory compliance.

An additional reference is made in the GLP Audit Checklist to the OECD Principles of Good Laboratory Practice, revised by the OECD Council in November 1997.[5] Both the FDA and the OECD references associated with the same question should be viewed as recommending or requiring similar conformance to the GLP, but they are not necessarily equivalent. The 1997 version of the OECD Principles has been a major improvement, but neither it nor the FDA GLPs have identical entries or coverage as is evident by the lack of associated OECD references for some specific regulations in the GLP Audit Checklist. Similarly, the GLP regulations do not include all the recommendations of the OECD Principles. The health and safety issues are not being ignored; however, in the United States, these responsibilities are reflected in regulations addressed by other federal agencies.

The Initiation Phase—Abbreviated GLP Audit Checklist, illustrated in the following, is a condensed version of the full text, GLP Audit Checklist—Complete, shown in Section II. To illustrate how these checklists are conveniently arranged in this manual, if the abbreviated checklist has an item in which, due to paraphrasing, an ambiguity is perceived, the complete checklists may be referenced. For example, refer to the column below identified as "Reference, 21 CFR." The reference shown, 58.120, is the numerical section of the regulation. If a protocol is being inspected using the abbreviated checklist and the protocol section "T," item "2.4," is not fully understood, refer to the same numeric section 58.120, in the questionnaire GLP Audit Checklist—Complete to find the full text of the question listed in the alphanumeric part, "T" item "2.4."

This example should be referred to for clarification:

| Initiation Phase—Abbreviated GLP Audit Checklist | | | | References | |
				21 CFR	OECD
T. **Protocol for and Conduct of a Nonclinical Laboratory Study**				**58.120**	
2. Protocol contains:					
2.4. Number, body weight, sex, source, species, strain, age	❑	❑	❑		8, 8.2(5)(b)

	21 CFR	OECD

T. Protocol for and Conduct of a
Nonclinical Laboratory Study   **58.120**

    2. Does the protocol contain, as applicable,
       the following information:

       2.4.   The number, body weight, range,
           sex, source of supply, species,
           strain, substrain, and age of
           test system?   ❏ ❏ ❏   8, 8.2(5)(b)

The abbreviated checklists are designed to be more convenient and efficient, including only those topics addressing a specific inspection phase, for example, Initiation, In-process, Completion, or Reporting phases. Most important they establish a degree of audit consistency. However, these are generic, and, therefore, you may require a checklist customized for your specific requirements. The new auditor should become familiar with the full text of the questions in the GLP Audit Checklist—Complete before using the shortened form, the Abbreviated GLP Audit Checklists. After making an assessment that a specific protocol requirement has in fact been included, excluded, or is not applicable, mark the appropriate box. If a detailed observation is warranted, reference to a supplementary set of inspection commentary on another sheet may be made.

Evaluation of the other phases, In-Process, Completion, and Reporting, may be facilitated similarly with the use of a checklist. These may not necessarily be complete regarding all the conditions or situations that are to be inspected or audited, but they can be used as guidelines and should furnish a base to establish your unique audit program.

Some sections between the abbreviated checklists are repeated. Although this may seem redundant, it should be understood that these checklists were designed with each phase relatively independent of the other. Any overlap provides additional coverage assurance should another individual be assigned subsequent phases of the same study.

Company policies regarding personnel assignments vary, some preferring a single assigned QAU person to the entire study, initiation through reporting phases. However, it is quite common to include in the QAU organization two classifications of GLP employees, inspectors and auditors. The inspectors typically perform the scheduled study inspections, and the usually more experienced individuals, the auditors, perform the study report audits and system audits, write SOPs, and provide GLP training.

CHAPTER 3

QUALITY INSPECTIONS AND AUDIT

A. Assuring Integrity

The QAU responsibilities described in the regulations [58.35(b)(3)] state that the QAU shall inspect each nonclinical laboratory study at intervals adequate to assure the integrity of the study. Formerly (21 CFR Part 58, June 20, 1979), this responsibility was expressed as "to inspect at periodic intervals each phase of a nonclinical laboratory study." The preamble of the amended regulation (FR33768, September 4, 1987) explains that each phase was applied to stress the need for repeated surveillance by the QAU, so that it observes each critical operation at least once during the course of the study.[12] The term *periodic* was used to emphasize the need for more than one inspection of certain repetitive, continuing operations. However, the FDA has determined during its inspection program that the quality of toxicology testing has improved since their surveillance was initiated in 1976. They have also concluded that more in-depth inspections (System Audits) of study processes (i.e., all operations required to accomplish a study phase) are more effective than quick spot checks of individual operations of a study.

The FDA has further defined the term *adequate*, explaining that "each study, no matter how short, needs to be inspected in-process at least once." They also emphasize that the QAU, across a series of studies, should inspect all phases in order to assure the integrity of the studies.

"Phase" has been a convenient reference to classifying the stage of a study, and is continuing in use. For example, some nonclinical laboratories are using the categories as shown in Table 3-1. Others have a more detailed interpretation, establishing a list of phases closely correlated with major or critical procedures scheduled during a specific type of study, such as those shown in Table 3-2.

The procedural identities are more specific, however; for each type of study they may not be entirely applicable. For example, in a Teratology, Fertility, and Reproductive study (TFR), blood and urine collection typically would not be needed, but other supplementary critical procedures would have to be introduced. With the broader identity definition the phases remain consistent. However, inspections should be directed toward those events most likely to reveal problems before the quality of the data are jeopardized. Therefore, a more favorable approach prospectively, allowing for some discretion in the choice of inspecting critical procedures within each phase of a study, is preferred.

In Section III-1, there are two illustrations of Study Inspection Records that should be examined. One, the Study Inspection Record (Version 1) for study M99-257, demonstrates the documentation of inspections following the broader definition of phases, that is, Initiation, In-Process, Completion, and Reporting. Similarly, a Study Inspection Record could also be developed in a more customized format for each type of study by referring to the major procedures inspected, for example, Protocol, or Animal Weighings, etc. Some sponsors, or contract laboratories, apply a compromise between the two systems by identifying the phase, and, within that phase, document the procedures that were inspected, such as illustrated in Section III-1, The Study Inspection Record (Version 2) for Study 99-258.

PHASE	IDENTITY
I	Initiation
II	In-Process
III	Completion
IV	Reporting
Table 3-1	

PHASE	PROCEDURE*	ABBREVIATION
I	Protocol	PRO
II	Animal Weighings	AWS
	Food Consumption	FCN
	Formulation	FOR
	Dosing	DOS
	Observations	OBS
	Palpitations	PAL
	Blood Collection	BCN
	Urine Collection	UCN
III	Necropsy	NEC
	Histology	HIS
	Clinical Chemistry	CCH
	Hematology	HEM
	Microscopy	MIC
IV	Reporting	RPT
*May occur in more than one phase		
Table 3-2		

There are four critical phases in the more generic interpretation, Initiation, In-Process, Completion, and Reporting. Normally the first three are referred to as *inspections,* and the fourth, Reporting Phase, is referred to as an *audit. Inspection* and *audit* as terms in the regulations have distinctly different intended meanings. Inspection is purposely applied, since it conveys a need for the QAU to examine and observe the facilities and operations of a study, whereas audit could connote merely a review of the records. Although we are aware of these differences, it is common to refer to QAU activities in both areas as "audits" and to the participants as "auditors."

B.   Initiation Phase

The first of these phases, Initiation, is centered primarily on the protocol, or study plan requirements and baseline activities. Auditors must keep in mind that the auditing concept, whether in accounting or in a quality assurance activity, normally requires a standard to be used for comparisons or reference of the condition or event. Our standard for protocol requirements in this instance is primarily the GLP regulations or the OECD Principles, but in addition, SOPs unique to your own organization must also be used as references.

As an aid in our audit, we will use a checklist, Initiation Phase—Abbreviated GLP Audit Checklist, furnished in Section II. The abbreviated checklists were developed from the longer questionnaire checklist, GLP Audit Checklist—Complete, which addresses most, if not all, of the pertinent questions associated with the GLP regulations. The abbreviated checklists include only those items considered applicable for each phase, such as in the Initiation Phase. Since they are specifically phase oriented, we will use these in the explanation of the phase inspections and audits that follow.

In a study conducted at a sponsor's facility, the Drug Safety Evaluation (DSE) group is usually responsible for developing the protocol. After the protocol has been properly approved, a copy is immediately submitted to the QAU, where it is registered or logged in and assigned directly to an auditor. On occasion, the expeditious development of some protocols may shorten the duration between protocol receipt and study initiation; therefore, it is important that the auditor first check the study initiation and start dates and establish communications with the study director. In some companies, "study start" may be defined as the first day of dosing, but preceding this may be an extended "baseline" evaluation for screening purposes. This is all part of the Initiation Phase and should also be inspected. Normally, the study protocol is reviewed first for confirmation of the GLP requirements, for example, inclusion of the procedure for identification of the test system, each dosage level, type and frequency of tests, chemical analyses, measurements to be made, and a statement of the proposed statistical methods to be used. For the new auditor, copies of some of the previous protocols may be reviewed for GLP compliance as a training project.

The auditor should then proceed with the remainder of the checklist. Are there SOPs, for example, describing all the procedural requirements of the protocol? Perhaps in this study a unique dosing method is going to be applied; is there an SOP describing this technique? Check with the study director; has he or she received the analytical results on the strength and purity of the test compound, and has the stability been determined? Are those responsible for the chemistry associated with the drug analyses, including drug and metabolite levels in various media, using validated test methods? Although the actual development of an analytical laboratory test method does not specifically fall within the scope of GLPs, confirming that a method applicable to a nonclinical study is documented and validated is a QAU, GLP responsibility. Review the specific test method validation reports and files. Have they followed their validation SOP? Did they include appropriate requirements of accuracy, precision, linearity, detection limit, system suitability, and specificity? For the data that may already be generated in the study baseline, are procedures being followed according to SOPs? Are specimen containers properly labeled, matching the specific animal identification? Are data being recorded directly, promptly, and legibly in ink?

During the protocol review, it might have been determined that two of the GLP protocol requirements were excluded, and in auditing the baseline the number of animals to be selected was increased. The study director should be contacted regarding the protocol oversight, and an inquiry made as to the reason for an increase in the number of study animals. The study director might have been unaware of the protocol oversights, but aware that a protocol change adding another dosage group, T4, is in the approval process. Under these conditions, it is apparent that while the necessary action has been initiated to amend the protocol with inclusion of another dosage group, the study director was unaware of the two protocol oversights, exclusions of the procedure for identification of the test system and a statement of the proposed statistical methods to be used. As a courtesy, the study director should be advised that an Inspection Report addressing the protocol oversights will be issued. An illustration of the Inspection Report with the audit observation and the response from the study director is shown in Section III-2, Inspection Report 00-019. A record of this inspection along with the auditor and identity inspection date should also be documented in the Study Inspection Record.

Inspection of the Initiation Phase may be done in just a few hours, or, if it has a prolonged baseline, it may cover several days or weeks. At the completion of the Initiation Phase, the checklist should be completed and kept on file. The Study Inspection Record should be carefully maintained, since it is one of the documents explicitly covered in the GLPs for agency surveillance and QAU retention.

C. In-Process Phase

In a nonclinical laboratory study, the In-Process phase is also commonly referred to as the "In-Life phase." It primarily addresses toxicological issues, heavily involving the toxicologist, but additionally unscheduled autopsies of dead animals may have to be performed by the pathologist, and blood and urine specimens may be scheduled for clinical pathologic analyses. It is a phase that generally includes several routine procedures, for example, formulating, dosing, clinical signs, animal weight, and food consumption, but also some studies may call for more sophisticated tests, for example, electrocardiograph (ECG) or electroretinograph (ERG).

The personnel involved in this phase must be well organized and methodical. Important also are the qualifications of these individuals; improper or poor technique could have serious implications. Proper responses to personnel checklist requirements are very important. Examine the curriculum vitae (CV) of the staff, the qualification records of the toxicology technicians, training records, and any procedural requirements regarding medical examinations, particularly those associated with primate studies.

Facility cleanliness should be evident throughout, not just in the study rooms, but in the corridors, supply rooms, drug dispensing room, and around any animal disposal operation, for example, the incinerator. Are the studies separated and identified properly, precluding dosing errors? For those compounds requiring special environmental conditions, are they adequate to preserve the strength and purity, or the stability of the test and control article (TCA) mixtures? This phase utilizes a variety of instruments, hythergraphs (recording temperature/humidity), balances, scales, volumetric infusion devices, graduates, refrigerators, incubators, thermometers, chronometers, etc. The importance of the precision and accuracy of these instruments cannot be underestimated. Reliable weighing of drug, perhaps in milligram quantities, and the automated infusion of drug in microliter quantities at specified rates are critically important. These measurement devices all require calibrations or standardizations traceable to some recognized standard, such as the National Bureau of Standards (NBS). Is there a formal calibration program at this facility, one that is creditable, that can trace retained documentation of instrument accuracy back to the time of the study, perhaps five or more years? Are they using a calibration service that is creditable, one that maintains documentation, and also has traceability to NBS or equivalent?

Numerous SOPs exist in toxicology and pathology, though perhaps many of them have not been reviewed or updated during the past year. Procedural changes may occur that should be reflected in the revision of the SOP. An approach utilized by some DSE groups is to have a policy of reviewing SOPs periodically, documenting the process, and issuing revised SOPs where necessary. This review of SOPs by department members may also be used in support of training documentation, for example, by recording the specific SOP revision letter, the date read, and the reviewee's signature.

One should be able to replicate the study with reference to the protocol and SOPs. As an auditor, had you observed that the technician was dosing with a 20-gauge needle, rather than a 23-gauge, or that the technician injected the drug over a 1-minute period rather than 2 minutes as required in the SOPs? Important? If management directed repetition of the study 3 years from now, would these changes impart differences in results by following both the protocol and the SOP requirements as they were written using a 23-gauge needle and a slower delivery rate? For some test articles it may; for others, perhaps not. However, it is not the auditor's prerogative to make that judgment; a change in the procedure should be viewed as significant. An audit observation should be made, allowing the study director along with other scientists to evaluate the deviation and the response to the Inspection Report, perhaps as illustrated in Section III-2, Inspection Report 00-006.

At some facilities, much of these data, such as animal weighings, water and food consumptions, and clinical signs, may be generated as direct computer input. These computer systems, of course, must be validated,[15] approved by management, and their supporting documentation packages complete before use of the system in a GLP study. Some basic requirements that should be queried include, Does the system require the toxicologist or technician to have unique entry codes for logging on, for accessing the data, and, if necessary, for modifying the data? Are these changes in computer entries traceable, the reason for change documented, along with the date and identity of the person who initiated the change?

Similarly, the inspection documentation procedure applied to this phase is applicable to the other phases, the auditor recording in the Study Inspection Record the inspections and dates, along with his or her signature and date. If an Inspection Report is issued, the appropriate information regarding it should be documented, such as illustrated in the Study Inspection Record in Section III-1.

D. Completion Phase

Now that the nonclinical study animals have been kept in controlled environmental conditions, have been properly fed and watered, and have been exposed to the proper dosage of test and control articles for the prescribed time, it is primarily the pathologist's expertise that prevails. His or her interest will be from a gross examination of the animal, both externally and internally at necropsy, and subsequently his microscopic analysis of the tissues may represent some of the most important data in the study. Also, in this phase the specimens of blood and urine obtained at necropsy, or shortly before, will furnish valuable data with regard to hematology, urinalysis, and clinical chemistry.

Most of the requirements of this phase are very similar to those of the previous phase. Personnel, for example, should have CVs reflecting proper credentials; the histology technicians should have education and experience supporting these very important functions. This area also requires personnel who are methodical and meticulous, but additionally skillful, with almost artistic qualities.

Are the necropsies performed in a separate area? Was the proper procedure used in anesthetizing the animal? Is a pathologist present if this is a requirement of the sponsor's SOP? Is the organ-weighing balance on a scheduled calibration program? Has the calibration date expired? Are the prosectors qualified, and has this been documented? Is there a continuation of the proper study and animal identity throughout the histology process? Are the urine and blood samples labeled and maintained in the proper sequence during urinalyses, hematologic analyses, and clinical chemistry analyses? Have load lists been retained, properly reflecting correct animal identities? If direct computer input is employed, is the person responsible for direct input identified at the time of input, and are changes properly documented?

If an observation is noted, it should be documented as in the previous phases. For example, at necropsy the auditor observed that spinal fluid was collected, but no SOP existed describing this procedure. The observation and response may appear similar to those shown in Section III-2, Inspection Report 99-112.

A record of the auditor's inspections should be continued in the Study Inspection Record, and if any observations warrant issuance of a formal Inspection Report, this should also be recorded, along with inclusion of the Corrective Action Acceptance Date.

E. Reporting Phase

The last phase, Reporting, is an auditing requirement of the QAU section of the GLP [58.35(b)(6)]. It states that the QAU shall review the final report to assure that it accurately describes the methods and standard operating procedures and that it accurately reflects the raw data of the nonclinical laboratory study. Each phase is significant, but this phase has very special implications. The report is the finished product of many contributors; it reflects information that will be carefully evaluated by the responsible agency, helping them reach a decision about the compound's safety based on scientific accuracy and integrity.

The scientific report, like the protocol, has certain GLP requirements regarding content. It shall include the dates on which the study was initiated and completed; the objectives and procedures as stated in the protocol, along with any changes; the dosage and regimen, route of administration, and duration, etc. These regulatory requirements can carefully be audited with reference to the checklist. As a training aid for the new auditor, earlier reports may be audited for confirmation of these GLP requirements.

This is also a phase to reexamine the study personnel CVs and training records and to confirm their proper archival disposition. Has the study director also had all the raw data, documentation, and specimens transferred to the archives? Has the QAU received copies of all protocol changes? Did the study director include in the final report all the study deviations? Did the calibrator properly tag the instrument, including an expiration date? Were instrument or equipment repairs documented? Were all procedures written and included as appropriate in the protocol or SOPs? Were the periodic visits of the pesticide service properly documented throughout the study? Did the dispensing scientist submit the various test article dosage formulations properly and as scheduled? Was the continued credibility of the computer program used in calculating dosages verified, using a test data set, and documented appropriately? Have stability analyses been completed and reported? Have file samples been submitted for retention? Has the disposition of any remaining test and control articles been documented?

As auditors, it is important to carefully read the scientific report. Checking for typographical and pagination errors should be included in the audit as well as examining report continuity and agreement between various sections of reports. Does the appended Statistical Report agree with the statements in the Scientific Report? Are all the tables of clinical signs, gross and microscopic observations, hematology, and clinical chemistry, etc., in agreement with the raw data? Are derived results, relative organ weights, food consumption, and summary tables correct? If a consultant was employed, is his or her report included, and is it in agreement with the final scientific report?

Reporting Phase audit observations should also be reviewed with the study director independently or in a joint meeting with other personnel who may be involved. A Study Inspection Report should be issued. If the observation involves other departments, for example, Nonclinical Statistics, the responsible manager should be included in the sign-off, confirming agreement with any ensuing corrective actions. A Study Inspection Report illustrating this situation is shown in Section III-2, Inspection Report 00-027.

At the completion of the Reporting Phase, it is important to confirm that any corrections or amendments to the Scientific Report are done accurately. These changes should once again be audited by the auditor. After the final report has been considered satisfactory, the QAU may issue, as required by the GLP, the QAU Statement, which includes the dates inspections were made and findings reported to management and the study director. The QAU Statement design and wording may vary, but the previously mentioned elements must be included. A QAU Statement is illustrated in Section III-1.

F. Statistical Auditing

One of the more important QAU responsibilities designated by the GLP is that of reviewing the final scientific report. These scientific reports for each sponsor or contract laboratory typically consist of a standard format, for example, Title Page, Table of Contents, Introduction, Materials and Methods, Conclusions, and perhaps supporting tables and appendixes. Some studies, particularly those of just a few weeks, such as "acute," have reports that are short. These may have the same reporting components of the more lengthy studies, but the body of the report is brief and the volume of data generated is usually very limited. Conversely, the longer, chronic, or carcinogenic studies, which may have an in-life phase of several months, or years, generate extensive numbers of data points, from several thousand to two, or three million. One of the QAU responsibilities, as you recall, is that it shall review the final study report to assure "that the reported results accurately reflect the raw data of the nonclinical laboratory study." In the extended study the evaluation of the expanded narrative and the voluminous amount of data can be very time consuming for an auditor. There are ways, fortunately, in which a report may be evaluated for accuracy without doing a 100 percent inspectional comparison with the raw data. This is not to imply that the sampling method to be described is a complete endorsement, but rather that it along with other techniques may be employed as the conditions warrant. The sampling procedure that is chosen should be fully described in an SOP. For internal audits, for example, an auditor has a great deal of familiarity with the technicians and professionals who are participating in a study. He may be acquainted through previous study audits with the accuracy of their data entries and the reliability of their reporting. Basically he or she has a quality assurance history of the individual technicians. In such instances the auditor's experience is an important discretionary factor in deciding on whether to proceed with a statistically formulated audit or one providing a more commonsense approach.

In the body of a scientific report the statements are usually completely evaluated for continuity and accuracy, but the numerous summaries and tables of data may be audited using a more efficient sampling technique. One of the most common sampling references is that of MIL-STD-105E, Sampling Procedures and Tables for Inspection by Attributes.[16] Another standard similar to this is the American National Standard ANSI/ASQZ1.4 published by the American Society for Quality. The development of these plans is based on probability, that is, the probability of finding a defective unit or data error.

Proceeding with this type of audit, for example, in the study, "Three-Month Intravenous Toxicity Study of BD-149 in Dogs," TABLE 1, Body Weight—Male Dogs, there are 5 dosage groups, consisting of 10 dogs in each group, or 50 dogs. Additionally, during this 3-month study, there are 2 baseline weighings, and during the study 1 weighing each week for 12 consecutive weeks, or 14 total weighings. The number of data points, therefore, is 5 X 10 X 14 or 700. A 100 percent inspection policy obviously would require having the auditor check each data report placement with that of the actual technician's entry. Statistically however, 100 percent inspection has been proven to have inherent human fallibilities, and, therefore, the statistical sampling approach may have merit. Please refer to a copy of MIL-STD-105E. The number of samples to be selected from this grid of 700 entries has to be chosen as indicated by the Inspection Level: Level II is normally used; however, if less discrimination is desired, Level I may be applied, or if more discrimination is preferred, Level III may be used, that is, a larger sample may be taken for an improved degree of confidence in the result. Referring to MIL-STD-105E Table I-Sample size code letters, for an Inspection Level II, and a matrix size, or lot size of 700, a sample size code letter of "J" is specified. In this instance our interest is in single sampling plans for normal inspection (Master table). Before consulting the table further, it must also be decided what type of acceptable quality level (AQL) should be targeted. The regulatory agencies have not established an acceptable level of accuracy, but they do recognize statistical sampling as a technique. A proper AQL, of course, involves the criticality of the measurement data. Body weight, food consumption, water consumption, etc., would not be considered relatively critical, and therefore, an AQL of 1 percent, for most organizations would be considered appropriate. However, a table consisting of the incidence of tumor observations should be more carefully audited, allowing fewer defects, or errors in the sample, for example, an AQL of 0.15 percent. A summary of this sampling plan is shown in Table 3-3.

Summary of Sampling Plan				
Code Letter	Sample Size	AQL %	Number	
			Accept	Reject
J	80	1.0	2	3
J	80	0.15	0	1
Table 3-3				

Next, which 80 of the 700 data points must be audited? In the matrix for body weights across the horizontal axis there are 14 entries, and along the vertical axis 50. Arbitrarily you could assign 14 columns, consisting of 50 rows each. The first column, for example, would consist of data points 1 to 50, the second, 51 to 100, and fourteenth column, 651 to 700. Now, select at random 80 data points in this arrangement. There are numerous means of randomly selecting these points; one is with the use of a randomization table. From this table, 80 random numbers between 001 and 700 must be selected. Circle these in the report table that you are auditing. Now compare these typed entries with the actual raw data entries. Does the selected random number nine, designating dog 0003, for the baseline date 9/6/99, with a body weight of 8.3 kg, match the raw data entry? Does the selected random number 667, designating dog 4009, for the baseline date 9/9/99, with a body weight of 7.8 kg, match the raw data?

As indicated above for the 1 percent AQL, 2 rejects in the random sample of 80 are acceptable in this table of 700 entries; however, if 3 data errors are encountered, the entire study table is rejected. If, for example, 4 errors or inconsistencies had been noted in the scientific report, a listing of discrepancies could be written in the inspection report as shown in Table 3-4.

Audit Observations Body Weight Errors (4/80)			
Dog No.	Date—1999	Report kg	Raw Data kg
1005	09/25	7.6	7.8
4009	10/22	8.6	8.2
2005	11/13	8.7	8.5
0001	12/05	7.9	8.9
Table 3-4			

This obviously indicates that other errors in the remaining 620 weight entries may exist, and, therefore, the study director should have the individual responsible for these entries examine this table thoroughly for additional errors before the table is issued again. The response in the Inspection Report should address this accordingly.

G. Computer Applications

Computer applications have made a significant contribution to the Drug Safety Evaluation (DSE) process since the inception of the GLPs. These computer systems are applicable to all phases of DSE and the submission of scientific reports to the regulatory agencies. Before these systems may contribute to any regulatory submission, however, they must be validated. The evaluation of a final computer system, confirming that it meets the user's prescribed criteria, is validation. In the validation process, the user, in conjunction with systems development, initially establishes the role of the computer and the criteria that it must meet. The FDA, industry, and academia have all made major contributions in developing criteria for computer systems used in pharmaceutical research. In late 1987, the Office of Regulatory Affairs of the FDA, the National Center for Toxicological Research (NCTR), and the NCTR Associated Universities invited chosen participants from the FDA, industry, and academia to a workshop held in Heber Springs, Arkansas. The purpose of the workshop was to initiate the development of a needed validation reference for establishing and ensuring the quality of computerized data systems. The final document, Computerized Data Systems for Nonclinical Safety Assessment,[15] has been a helpful reference for covering a variety of validation issues and includes chapters on system development, system verification and validation, quality considerations in a regulatory environment, and system integration.

Depending on the complexity of the computer system, the documentation supporting the validation package can be quite extensive. A validation package should exist for each regulatory (GLP) associated computer system. The QAU, of course, is involved in much of the initial validation, frequently in a consulting role regarding compliance to the regulations, but they may also fulfill a continuing role of auditing the validation documentation and systems for compliance.

In nonclinical research, the GLPs refer to specific regulations associated with computer equipment. As a QAU responsibility these requirements are applicable to all phases and must be inspected and audited for compliance. GLP regulations that are associated directly with the computer systems include the following:

58.3   Definitions

Raw data may include photographs, microfilm or microfiche copies, computer printouts, magnetic media, including dictated observations, and recorded data from automated instruments.

58.29   Personnel

Each individual engaged in the conduct of or responsible for the supervision of a nonclinical laboratory study shall have education, training, and experience, or combination thereof, to enable that individual to perform the assigned functions.

Each testing facility shall maintain a current summary of training and experience and job description for each individual engaged in or supervising the conduct of a nonclincal laboratory study.

58.61   Equipment Design

Equipment used in the generation, measurement, or assessment of data and equipment used for facility environmental control shall be of appropriate design and adequate capacity to function according to the protocol and shall be suitably located for operation, inspection, cleaning, and maintenance.

58.63 Maintenance and calibration of equipment

Equipment shall be adequately inspected, cleaned, and maintained. Equipment used for the generation, measurement, or assessment of data shall be adequately tested, calibrated and/or standardized.

The written standard operating procedures required under 58.81(b)(11) shall set forth in sufficient detail the methods, materials, and schedules to be used in the routine inspection, cleaning, maintenance, testing, calibration, and/or standardization of equipment, and shall specify, when appropriate, remedial action to be taken in the event of failure or malfunction of equipment.

The written standard operating procedures shall designate the person responsible for the performance of each operation.

Written records shall be maintained of all inspection, maintenance, testing, calibrating and/or standardizing operations. These records, containing the date of the operation, shall describe whether the maintenance operations were routine and followed the written standard operating procedures. Written records shall be kept of nonroutine repairs performed on equipment as a result of failure and malfunction. Such records shall document the nature of the defect, how and when the defect was discovered, and any remedial action taken in response to the defect.

58.81 Standard Operating Procedures

A testing facility shall have standard operating procedures in writing setting forth nonclinical laboratory study methods that management is satisfied are adequate to insure the quality and intregity of the data generated in the course of a study.

Standard operating procedures shall be established for, but not limited to, the following: . . . data handling, storage, and retrieval.

58.120 Protocol

Each study shall have an approved written protocol that clearly indicates the objectives and all methods for the conduct of the study.

58.130   Conduct of a nonclinical laboratory study

All data generated during the conduct of a nonclinical laboratory study, except those that are generated by automated data collection systems, shall be recorded directly, promptly, and legibly in ink.

Any change in entries shall be made so as not to obscure the original entry, shall indicate the reason for the change, and shall be dated and signed or identified at the time of the change.

58.190   Storage and retrieval of records and data

All raw data, documentation, protocols, final reports, and specimens . . . generated as a result of a nonclinical laboratory study shall be retained.

An individual shall be identified as responsible for the archives.

Only authorized personnel shall enter the archives.

Material retained or referred to in the archives shall be indexed to permit expedient retrieval.

58.195   Retention of records

. . . (R)aw data and specimens pertaining to a nonclinical laboratory study and required to be made by this part shall be retained in the archive(s) for whichever of the following periods is shortest:

A period of at least 2 years following the date on which an application for a research or marketing permit, in support of which the results of the nonclinical laboratory study were submitted, is approved by the Food and Drug Administration.

A period of at least 5 years following the date on which the results of the nonclinical laboratory study are submitted to the Food and Drug Administration in support of an application for a research or marketing permit.

In other situations . . . a period of at least 2 years following the date on which the study is completed, terminated, or discontinued.

These FDA retention durations are the minimum requirements specified by this agency. The EPA standards specify longer retention periods. In the current international environment of submitting applications for research or marketing permits to foreign regulatory agencies, the retention periods may be more extensive, some of them for as long as the test substance is used or marketed. The OECD retention statement, which is applicable to many countries, states that the raw data and other study records should be retained for the period specified by the appropriate authorities. It has been my experience that many sponsors prudently retain these data, and associated study records, indefinitely. To facilitate this requirement, the FDA has allowed records to be retained as originals, or true copies such as photocopies, microfilm, microfiche, or other accurate reproductions of original records. It is suggested, before any of these reproduction systems are initiated, that the appropriate agency be contacted for confirmation of their agreement.

CHAPTER 4

CONTRACT LABORATORY AUDITS

A. Preplacement Evaluation

At some companies it is policy to contract out all of nonclinical safety studies; at others, contracting studies may be done only when the need arises. The reasons for contracting varies. Animal facilities may be near or at capacity; a temporary shortage of professional help may be contemplated; the study may be unique, requiring special techniques, equipment, or environments; or the biohazards of the compound or positive control may make it desirable to test it away from the sponsor's facilities.

For studies that are planned with contract laboratories, the DSE group should inform the QAU of their intended study plan and request a GLP evaluation of the contractor. Usually someone from the DSE staff is aware of the scientific expertise of the contract organization, or they will also schedule a visit to the company and appraise it from a scientific standpoint.

The QAU auditor should coordinate his or her activities with the DSE director or with the sponsor's assigned scientific study monitor. Arrange a meeting with the monitor, who usually has or will acquire the pertinent information regarding the contract laboratory. Perhaps the monitor has on file a brochure describing its operations, specific study personnel titles, and individuals who may be contacted. As soon as possible, obtain a copy of the draft protocol and study schedule.

As an aid in this type of audit, a checklist is also helpful. Prospectively, the Preplacement and Maintenance Abbreviated GLP Audit Checklist shown in Section II may be reviewed prior to a visit, identifying areas where special inquiries might be directed or where specific operations will need an in-depth review. If it is a relatively unique study, or of a type not previously audited, it might be advisable to review some of the special equipment, dosing, and calibration features with a toxicologist in your organization who is familiar with the methodology. If the facility has been audited by the FDA or EPA, there may be available through Freedom of Information (FOI) copies of Establishment Inspection Reports (EIR) and inspectional observations, commonly referred to as "483s" (FDA form 483), which may reveal from a regulatory perspective that the company is relatively compliant, or that it has major violations. In fairness to the audited organization, include in any FOI document request copies of the sponsor's or contract laboratory's responses. These are sometimes enlightening, for example, revealing misunderstandings that originally occurred during the audit; a company rationale adamantly supporting that the animal identities were not compromised as concluded by the agency auditor; or statements expressing agreement, appreciation of the auditor's comments, and that his or her suggestions have been implemented.

Contact the contract laboratory's QAU manager, and advise him or her of the contemplated GLP study at their company. Schedule a visit, clarifying that it will be an evaluation audit of their facilities, operations, SOPs, and the QAU. If a brief "kickoff" meeting with their management is desired prior to the audit, advise the QAU manager accordingly. These are normally brief, informative exchanges that can be utilized to establish a cooperative, friendly, but professional tone.

A kickoff meeting provides management with a better understanding of the scope of your audit, and allows them to recommend changes in your agenda to facilitate a more efficient or comprehensive audit. If your interest is in drug dispensing and accounting, for example, they may suggest observing the operation within the next hour, rather than at the later scheduled time; or perhaps most technicians leave at 3:00 P.M., and your agenda included observations of some of their procedures through 5:00 P.M. Also at this introductory meeting, mention your intention to have a closeout meeting. If it is a small company, usually the president or the facility manager along with the QAU manager and a few of the professionals wish to know what your impressions and observations were. At larger companies, possibly just the QAU manager will note and relay any adverse conditions that have been observed. Normally, most observations are mentioned to the group, often allowing them to explain that corrective action had already been scheduled, or that they were not aware of the condition, but that it will be corrected. Some observations might be reserved for review with your QAU manager, the sponsor monitor assigned to the study, or the director of DSE.

After this preliminary "kickoff" meeting, meet with the QAU manager and the tentative study director. Advise them of the type of documentation to be reviewed. This might include the CVs of any professional personnel engaged in the study, along with those of supervision and study technicians, the SOPs associated with the facilities operations, and those SOPs addressing the specific type of study being contemplated and required by the GLPs. A thorough review should also be done of the QAU's SOPs and its personnel's CVs. This is extremely important, since in your absence their QAU will be delegated the responsibilities for the study inspection and audits.

While the QAU staff is gathering this documentation, it is typical initially to tour the facility. Unless requested otherwise, the QAU manager will arrange a tour only in those areas associated with your study interests. Familiarity with the checklist will be advantageous, allowing questions to be asked regarding the specific site or operation without having to return later. Are these studies being conducted in separate rooms? If biohazardous materials are being tested, are the air-handling systems separate, properly filtered, using 100 percent makeup air? Are the supply facilities for feed and bedding separate from the test systems and protected against infestation? Talk to the laboratory supervisor responsible for the inventory of test and control articles. Ask him or her to explain their system of receipt, storage, and accountability. How will they handle any protocol requirements regarding periodic analyses of test mixtures? Is the test methodology validated, the documentation available and compliant? Are the analytical balances in a scheduled calibration program? Can the SOP associated with this balance be reviewed, together with the documentation of its last three calibrations? Are the calibration standards traceable to the National Bureau of Standards or some other recognized standards organization? How are the various containers and mixing devices cleaned? Is the adequacy of cleaning determined, validated? Are the specimen archives secured, well maintained, and properly indexed? Are wet tissues, blocks, and slides separated? Is the autopsy room designed for multiple necropsies, with adequate space, providing separation; is it adequately equipped?

Continue with the tour, randomly examining several animal rooms for appearance, cleanliness, lighting, cage suitability, and conditions. Are the facilities certified by the United States Department of Agricultural Association (USDA) or accredited by the American Association for Accreditation of Laboratory Animal Care (AAALAC)? Are they using certified feeds, and is the water analyzed periodically? Ask whether the results of analysis for contaminants of the rodent chow lot used for one of the studies being conducted may be reviewed; and may the most recent report on water analysis be examined? In the clinical pathology area, ask about their overall calibration and standardization program. Do they have equipment logs containing cleaning and maintenance records? Are their reagents properly identified? Are reagents and specimens requiring refrigeration or freezing maintained in a monitored unit with calibrated monitoring devices? Are their methods described in properly approved SOPs? Ask the laboratory supervisor to explain their overall processes of urinalyses, hematologic analyses, and clinical chemistry analyses. Examine some of the controls and standards; are the expiration dates compliant? How do they correlate animal baseline identities with those subsequently used in the study? Is it adequate, assuring prevention of mixups?

Are the documentation archives secure, alarmed, adequately protected from fire? Who has access to the archives? Are documents allowed to be checked out? If so, what is the status of the checkout record? Are the computer backup tapes stored in a separately secured area? Is there an SOP describing archiving procedures?

Tours of this type are informative, not just from a facilities and operations viewpoint, but for impressions regarding key personnel. Now that the tour has been completed, can the previously requested documentation be reviewed? Examine the CV of the study director tentatively assigned. If there are questions about his or her credentials, experience, or length of employment, ask, and clear up your doubts. It might be prudent to request copies of CVs of key study personnel for review and supporting endorsement by your company's director of DSE.

Examine the SOPs required by the GLPs and associated with the specific study being contemplated at this nonclinical laboratory. Are the relevant SOPs included, and are they written to ensure data quality and integrity? When were they last modified or reviewed, and have they been approved by the individuals currently responsible for their implementation? Although the latter point would not necessarily be a requirement, it may give a measure of the attentiveness of management.

B. Evaluation of the QAU

One of the more significant aspects of this evaluation centers on the QAU. Your perception of its integrity and professionalism should be of paramount importance, since they will have the responsibility of performing the inspections for all phases, including the final audit. Although the contract laboratory has responsibility for the study, the sponsor submitting the study to the governmental agency still bears the responsibility for the work performed; therefore, careful and in-depth evaluation of this unit is extremely important. Have they had previous governmental agency audits? If so, are copies of the previous two or three EIRs (Establishment Inspection Report), "483s," and the contract laboratory's responses available? Are there any current issues that have not been resolved between the agency and the laboratory?

Review the Master Schedule. Besides evaluating the study director's workload, examine the QAU's activities. Judging by your experience, can their QAU personnel effectively manage this schedule? Ask for an organizational chart. Is there a conflict of interest, a reporting relationship that could bias the study (by the unit's not being independent of the personnel engaged in the direction and conduct of the study)? Is the study director, for example, also the intended study's pathology monitor? Arrangements of this type are not too likely, but in a smaller firm an individual may wear many hats.

Review the QAU's SOPs. How have they defined "phase," and will their activities include inspections during procedures of pivotal importance? Is their Record of Inspection report satisfactory? Ask for copies of some of their completed Inspection Reports. Are the observations pertinent, suitably written, and conscientiously addressed by the respondent group? Clear up a very important point: Should this nonclinical laboratory receive the contract, will their management allow the sponsor's QAU to review the Inspection Reports? The regulations specifically do not allow the agencies to inspect or copy QAU records of findings and problems or actions recommended and taken. However, as the sponsor's QAU representative, be aware of the significance of GLP violations, their resolution, and how they bear on the overall study submission. Accessibility to these records should be granted as a sponsor's option.

Has the QAU designated one location for study protocols, Master Schedules, Study Inspection Records, Inspection Reports, and Quality Assurance Unit Statements? Are these files properly indexed and adequately secured? Have backup copies of these documents been stored off-site?

Review their auditing procedures followed during the reporting phase. How are reports audited if raw data consist of electronic media, for example, disks, tapes, or output consisting of printouts and microfiche? Have these systems been validated? Can the supporting validation documentation be reviewed to confirm the reliability of the system?

At the conclusion of the evaluation, set aside some time for reviewing your notes and observations before meeting with any contract laboratory personnel. At the meeting, express your appreciation of their cooperation in facilitating this evaluation, and advise them of your findings, and the significance of any observations. As a courtesy contingent on their acceptance, inform the QAU manager of any subsequent inspections, for example, every month, once a quarter, or only near the end of the study. As a sponsor policy, the contractor should be advised that just before or shortly after the study report has been issued, a follow-up audit would normally be included by the sponsor's QAU.

After returning, schedule a meeting with the director of DSE and the study monitor. Review your findings, pointing out the significance of some of your concerns. For example, the study director has only been with the contractor for a month and already has responsibility for six other GLP studies on the Master Schedule, or in the animal studies no "check-weighings" were done before body weights or food consumption measurements were taken. It might be worth noting that at the close of the meeting they agreed to incorporate the "check-weighing" process into the appropriate SOPs. The DSE director may also be concerned about the new study director's assignment, possibly resulting in a readjustment of his or her schedule, or perhaps a replacement with a more experienced study director. In an audit evaluation of Southwestern Laboratories, Inc., there was serious concern about the lack of secondary computer backup tapes. SWL had backup tapes stored in their computer facility, but none was stored at a secure, off-site facility. Under these circumstances, in a worst-case scenario, fire or water damage at SWL's facility could destroy all their electromagnetic media data bases, without recovery provisions. Write a confidential report on the evaluation and complete a formalized audit or evaluation record, similar to the one illustrated in Section III-2, Audit Report—Contract Laboratory, 00-030. If you recommend this contract laboratory, you should be confident their QAU will perform the study inspection and audits diligently, contributing to the integrity, accuracy, and overall quality of a compliant, well-conducted study.

C. Study Audit

This is referred to as a study audit activity rather than inspection, primarily because the contract laboratory's QAU responsibility is to inspect the various phases of the study. In some instances, it might be warranted to proceed with an inspection similar to that conducted at your own facilities; however, there is usually a constraint on time. Ask for copies of the QAU's SOPs so a checklist may be developed prospectively. And if it involves an audit of the final report, acquire a copy in advance, again for the purpose of efficiently and effectively evaluating the report.

Although some of their procedures and operations have been observed, much of your time will be devoted to evaluation of the documentation that already exists, determining if they have followed both the study protocol and their SOPs. Normally, the Initiation, In-Process, and Completion checklists can readily be adapted for use in your own inspection operations; if not, proceed with developing a checklist associated with their specific QAU SOPs. Remember, however, the primary responsibility of the QAU functions rests with the QAU at this contract laboratory; therefore, it is important to evaluate their effectiveness. How frequently have they audited the study? What types of observations have been made? Has the scientific staff responded appropriately and in a timely manner to the observations? Have the protocol changes been properly approved and communicated to the sponsor's monitor? Check the test article dosing and dispensing records. Does an accounting of the test article usage support the resultant amount of compound remaining?

Audits of the final report should be conducted similarly to those conducted at your own facilities. Most contract laboratories will be fully aware of all your documentation needs and usually will have these immediately available in a room designated specifically for your audit. The study protocol, schedules, SOPs, and any interim and final reports will be readily available along with a QAU person to answer any questions that may arise. The Reporting Phase Abbreviated GLP Checklist should be applicable, but there may be some specific QAU SOPs that will have to be referenced to determine compliance. The comments made in Chapter 3 on Reporting Phase are also applicable for those audits conducted at a contract laboratory and should be reviewed.

As part of the application for a research and marketing permit, the sponsor is also required to furnish a conforming statement relative to each of the nonclinical laboratory studies submitted to the FDA. This statement should describe each study that is conducted in compliance with the GLPs, or if it was not conducted in compliance, a statement describing in detail the variations from these requirements. Noncompliant studies may be submitted, but the sponsor may have to justify that the noncompliance did not affect the integrity of the data (21 CFR 314.101).

Similarly, a contract laboratory should furnish a conforming statement to the sponsor for each nonclinical laboratory study they conducted, indicating for their delegated responsibilities compliance, or a statement identifying those GLP requirements that were noncompliant. From this supportive information, the sponsor may then prepare the appropriate conforming statement for submission to the regulatory agency.

CHAPTER 5

SYSTEM AUDITS

A. Effectiveness

In the publication of the original GLP regulation, considerable emphasis was placed on directing the inspection activities toward the specific study: "inspect each phase of a nonclinical study." Little, if any, was said of the benefits of performing the inspections of the associated systems. In the preamble of the revised FDA GLP, published in The Federal Register (52FR33772), considerable merit has been extended to the benefits of the systems, or maintenance type audits. "Contemporary concepts of quality assurance emphasize the effectiveness of thorough, in-depth inspections of study processes (i.e., all operations required to accomplish a study phase) in place of quick spot checks of individual operations with a study. Thorough examination of personnel, facilities, equipment, standard operating procedures, data collection procedures, raw data books, and other features associated with a study phase can achieve more effective quality assurance than does a more superficial observation of the conduct of the same study phase in a series of studies."

Much of the inspection and auditing activities of the GLP is routinely oriented in association with the study protocol of the nonclinical study. Because of the nature of the specific studies being conducted, some aspects of the GLPs may seldom be challenged or evaluated, therefore providing the need for a systems or maintenance audit. The following compliance issues are just some that may surface more readily through system audits of departments or functional groups:

- The reagents within a clinical chemistry laboratory may be used infrequently for some analysis. Have those with expired dating been used? Are there any unused with expired dating?

- A newly installed unit for diet mixing is being contemplated for use in several studies. Have they documented and validated both the mixing procedure and the equipment cleaning procedures?

- Several new animal rooms have been added to the facilities. Has the air-handling equipment been correspondingly upgraded to provide the proper climatic conditions and to meet air exchanges, temperature, humidity, and monitoring requirements?

- In the various laboratories, have the equipment or instrument maintenance logs been utilized, and are they on schedule, identifying dates of repair and cleaning?

- Are the historical SOP files being kept by each department or by a central function, for example, the designated archive? Are they complete?

- Have the CVs and training records of appropriate personnel been updated periodically?

- Have the histology slide or wet tissue storage areas been evaluated recently? Have slides or tissues been checked out recently? Are they following SOPs?

- Have all the nonclinical statistical computer programs been validated? Is there supporting documentation?

This type of audit may proceed similarly to that done for a preplacement audit. The same checklist may be followed, the Preplacement and Maintenance Abbreviated GLP Checklist shown in Section II, but the scope is more limited. Time constraint is not usually a factor in an audit of this type, and, therefore, various areas of the operations may be focused on during the year, duplicating to some extent the inspections associated with the individual studies, but proceeding in much greater depth.

B. Scheduling Convenience

These audits may be scheduled in advance, assigning an auditor to be chairperson for each quarter such as designated in Table 5-1. The system audits shown in Table 5-1 are just some of the system audits that may typically be identified. Other areas may include anatomical pathology, clinical pathology, toxicology, the archives, computer operations, and special analyses, for example, microbiological, pharmacokinetic, and metabolic. A specialty laboratory will, of course, require appropriate system audit assignments for their specific operations.

The functions in Table 5-1 may also be examined during a study inspection, but if they were not directly involved in a study recently, they may be overlooked or not examined in depth for an extended period.

C. Preparing

In conducting system audits, the audit chairperson would initially meet with audit team members, providing information about the department to be audited, the audit schedule, the scope, and the subject assignments. If this department had been audited previously, the observations and responses associated with the last audit should be reviewed, and any follow-up auditing required may be assigned to team members. Additionally, the entire QAU group should be queried regarding any significant observations noted during previous study inspections, which may justify further in-depth investigation.

   GLP Audit Manual ©2000, Interpharm Press

	Quality Assurance System Auditing Schedule—2001			
Qtr.	Audit Chairperson	Function	Audit	Subject
1st	Smith	Animal facilities	Standard Operating Procedures	Environmental air Analyses of feed Quarantine of Animals USDA accreditation Animal Welfare compliance Historical Retention
			Personnel	Curriculum Vitae
			Training	Documentation
			Medical	Required examinations
			Pesticide Application	Contract/GLP compliance
2nd	Brown	Analytical chemistry	Standard Operating Procedures	Test method validation Instrument calibration Standards/reagent labeling Historical retention
			Personnel	Curriculum Vitae
3rd	Jones	Formulating of Test Article	Standard Operating Procedures	Validated mixing procedures Validated cleaning procedures Documented accounting test article/control Retention samples Historical retention
			Personnel	Curriculum Vitae Qualified backup
			Training	Documentation
			Job description	Record

4th	Green	Histology	Standard Operating Procedures	Accession identity
				Processing
				Slide identities
				Retrievability
				   Wet tissues
				   Slides
				Historical retention

			Personnel	Curriculum vitae
			Training	Documentation
		Table 5-1		

It is sometimes helpful to supplement the auditing team by inviting a member of another department as a participant. Guest auditors, for example, may include a statistician for an audit of the data processing department, or a guest auditor of analytical chemistry could include the dispensing pharmacist. This cross-training provides a better understanding of auditing, and generally an improved acceptance of these regulations within his or her department. If you are contemplating auditing a function in which the QAU may not have adequate expertise, a consultant could also be contracted as an advisor in preparation of the audit or as a participating team member.

There could be, of course, the need to audit a department quickly, for example, in anticipation of an imminent regulatory agency audit; normally, however, audits are scheduled months in advance, as suggested in Table 5-1 by scheduling 1st Qtr, 2nd Qtr, etc., with a designated audit chairperson. The chairperson should schedule a kick-off meeting with his or her team members and the department manager and appropriate staff scheduled for the audit. Explain the scope of the audit and the areas within the manager's department that you intend to audit. Open it up to questions and suggestions on their part. Perhaps the manager has noted some concerns that need further in-depth evaluation, or staff members may want to know if they will need laboratory notebooks prior to 1995, which are stored off-site, requiring up to 24 hours to retrieve.

D. Conducting

Request copies of the department's SOPs, or if numerous, obtain a copy of the index and select SOPs that are considered critical for nonclinical studies. If, for example, this department is a supportive function, such as biometrics, important SOPs would include computer programs associated with generating randomization schedules or statistical results. Select several studies of different types: an acute, subchronic, chronic, and a TFR (Teratogenic, Fertility, and Reproductive) study. By referring to the appropriate SOP and the specific scientific reports, determine if the randomization and statistical software programs used were properly validated. Does timely documentation exist supporting the verification of the computer program versions referred to in the scientific reports? Do test sets of data exist? Have the results been retained, and do these match the expected values? If third-party computer programs are used, are they appropriately licensed and are they being supported? Are changes in software controlled, documented, and approved appropriately according to the SOPs?

SOPs in other departments can also be examined for compliance. Reagents within a clinical chemistry laboratory may be used infrequently for some analyses. Some reagents, although not used, are perhaps expired, and should be discarded. Is the department following the SOP, maintaining the availability of currently accepted reagents and appropriately discarding those that have expired?

The equipment used for diet mixing may have recently been received. Has it been properly grounded, preventing undue static electrical separations of components? Have the mixing procedures been validated? Also, have cleaning procedures been validated? Is the equipment in a maintenance program, and is there a corresponding equipment log showing installation, repair, and maintenance dates? Is the air-handling equipment in the facilities functioning properly? Is plant engineering continuing to maintain it, replacing filters as scheduled or when pressure differentials warrant changes? Are they in compliance with their SOPs?

Audit the accountability of the histology slides, blocks, and wet tissues of several current studies and a few that had been completed three to five years previously. Has the integrity of these materials been maintained?

In the departments audited, are historical SOP files being kept by each department, or by some central archive? Select several SOPs of multiple versions, and determine their traceability back to the originals. Have the CVs and training records of department personnel been updated periodically? Do these CV files now include all the newly hired technicians and professionals? Are job descriptions maintained, correctly describing the functions of various study personnel?

These audits are perceived by many managers as management tools for determining their department's degree of regulatory compliance. It's always encouraging to audit such a group, but occasionally a manager perceives your presence as adversarial, usually complicating and delaying the auditing process. Maintain a positive, even temperament among your auditing team members, and provide guidance for maintaining the auditing objectives: determination of this department's compliance with study protocols, Standard Operating Procedures, and the Good Laboratory Practice regulations or principles. If problems do exist, report them accordingly; best that these be brought to the surface by an internal auditing group in which a corrective action plan may be implemented than by a regulatory agency resulting, perhaps, in embarrassment and extensive explanation and documentation or, more tragic, rejection of the study.

E. Reporting

After completing an audit it is always helpful for the chairperson to schedule a meeting with the audit team and review their observations. Review the QAU SOPs that describe the audit procedures and report, for example, System Audit Report, 99-105, illustrated in Section III-2. Some members may perceive their observations as insignificant, but in a group discussion it may be apparent that the occurrence is frequent and, therefore, this observation could be elevated to an area of concern. One that comes to mind is the use of "whiteout" by new personnel to cover a mistake in data entry. A single instance should be mentioned at a closeout meeting, but if several occurrences are recognized, these should definitely be included in the written report. Similarly, there may be observations that the auditing team may have that would more appropriately be mentioned in the closeout meeting, but would not necessarily be documented. Identify for the group those observations that should be drafted into the audit report. When these have been identified, reviewed, and confirmed, then proceed with finalizing the report.

The system audit report should be titled; identify the department audited, the scope of the audit, the audit chairperson and auditors, the dates of the audit, its distribution, and the expected duration and date for completing responses to the observations. Additionally, it should include an audit summary and, of course, as illustrated, the audit observations, client's responses, targeted corrective action date commitments, completion date, and management's plan with accompanying approval.

It is then the chairperson's responsibility to schedule a closeout meeting with the client group, including management and those individuals directly involved with resolving the issues. Copies of the audit report should be limited, and, if used, copies should be controlled and returned at the end of the session with the exception of the original, which will be used by the audited department manager for documenting his or her responses. Normally, responses should be returned within 30 days; however, completion of a corrective action may conceivably take several months as indicated on the audit report. This should, however, not delay the return of the report to the QAU. The QAU should review the responses and decide as to their acceptance or rejection. Those that are recognized as significant observations with accompanying corrective actions should be scheduled for an audit follow-up and subsequently documented, if appropriate, as completed.

 GLP Audit Manual ©2000, Interpharm Press

CHAPTER 6

INTERNET RESOURCES

The Internet has undoubtedly become one of the most useful tools in research. Many of the governmental agencies throughout the world are now providing convenient Internet access to their regulations, guidelines, and other supplementary information associated with the Good Laboratory Practices. This immediate access of timely information should enable the researcher to provide a more expeditious and accurate processing of studies.

Listed here are several Internet sites that may provide helpful references. Besides the regulatory agencies' Uniform Resource Locators (URL) addresses, other relevant organizations, societies, and academic institution addresses are included. Descriptions preceded with an asterisk have a printed copy of their respective Internet first pages furnished in this section.

Many of the sites have been created in a hypertext format, allowing the user to visit a vast number of sites in a relatively short time. The government agency sites are organized particularly well, along with those of the Society of Quality Assurance (SQA) and the University of Georgia (UGA). The hypertext is extremely beneficial, and should allow the researcher to expeditiously acquire answers to a vast variety of regulatory questions.

A. Organizations

CDC (Centers for Disease Control and Prevention)

http://www.cdc.gov

*This Homepage provides a general introduction to the CDC. It includes several hyperlinks addressing the following topics: About CDC, In the News, Travelers' Health, Health Information, Publications, Software & Products, Data & Statistics, Training & Employment, Funding.

EPA (Environmental Protection Agency)

http://www.epa.gov

*This Homepage serves as a launching site for various titles, including: About EPA; News & Events; Offices, Labs & Regions; Projects & Programs; Laws & Registrations; Publications; Databases & Software; Money Matters.

http://www.ovpr.uga.edu/qau/epaglp_a.html

> This is an index provided by the USG's Quality Assurance Unit to the FIFRA GLP (40 CFR 160), formatted conveniently in hypertext, linking individuals to the standard of their choice on the Internet.

http://www.ovpr.uga.edu/qau/pr865.html

> This is an index provided by the USG's Quality Assurance Unit to the TSCA GLP (40 CFR 792), formatted conveniently in hypertext, linking individuals to the standard of their choice on the Internet.

http://www.epa.gov/irmpoli8/irm_galp

> The document is titled EPA Directive 2185, Good Automated Laboratory Practices (GALP). It includes hypertext titles: Introduction, GALP Overview, and GALP Implementation Assistance.

http://www.epa.gov/epahome/rules.html

> *Included in these documents are titles, Regulations & Proposed Rules, Codified Regulations, Current Legislation, and Laws. These subjects along with many other helpful entries are provided in hypertext for immediate access to other Internet sites.

http://www.ovpr.uga.edu/qau/advisor2.html

> This is titled: Index to the EPA GLP Advisories, which are numbered and may be readily accessed through hypertext linkage.

FDA (Food and Drug Administration)

http://www.fda.gov

> *This is the FDA's Homepage, providing an extensive range of subjects, all with hypertext entries to the Internet. Included are Biologics, Animal Drugs, Cosmetics, Medical Devices/Radiological Health, Freedom of Information, Field Operations, Regulations & Information Toxicology, and Medical Products Reporting.

http://www.fda.gov/cder/index.html

> *This is the Homepage for the Center for Drug Evaluation and Research. It provides hypertext links with the Internet regarding information about CDER, Drug Information, and Regulatory Guidance.

http://www.fda.gov/cvm

This is an introductory document to the Center for Veterinary Medicine (CVM). It describes the scope of the CVM's responsibilities and includes links to the Internet on other topics, including Important News, On-Line Library, The Green Book, Links, and About CVM.

http://www.ovpr.uga.edu/qau/prefda1.html

GLP Preamble; a hypertext index is provided for GLP (21 CFR 58) covering the full range of subjects with numbered comments.

http://www.ovpr.uga.edu/qau/fdaglp_a.html

This is an index provided by USG's Quality Assurance Unit to the GLP (21 CFR Part 58). Each Section, e.g., 58.31 Testing facility management, is formatted in hypertext, allowing convenient access to the specific Internet site.

http://www.cgmp.com/htmfiles/211.htm

This is the hypertext index for the current Good Manufacturing Practice (cGMP) regulations, finished pharmaceuticals (21 CFR Part 211).

http://www.mcs.com/__/kamm/gmpnew.html

This document includes the index and test of the Medical Devices cGMP (21 CFR Part 820). It is not provided in a hypertext Internet format.

http://www.fda.gov/cder/dmpq/cgmpnotes.htm

This is an internal FDA document providing access to cGMP issues in a timely manner, and includes updates on cGMP projects, and provides help in clarifying existing policy.

http://www.fda.gov/opacom/hpnews.html

*FDA News and Publications is a source of information about Recent Press Releases and Talk Papers. It also has topics on Current Issues of Other FDA Publications. These are all linked with the Internet. References include an Index and Archives.

http://www.access.gpo.gov/cgi-bin/cfrassemble.cgi?title=21

Title 21, Code of Federal Regulation. This provides access to all volumes and parts of Title 21.

Interpharm Press

http://www.interpharm.com

*This publisher is a premier provider of technical and regulatory information for manufacturers of biotechnology, bulk pharmaceutical chemicals, medical device and diagnostic products, and pharmaceuticals worldwide. Several hyperlinks are available including: Keyword Search, Interpharm Online Catalog of Books, Software, Videos and Regulatory Documents, Information for Authors, and Free Productivity Software.

The Library of Congress

http://lcweb.loc.gov/homepage/lchp.html

This is their Homepage, allowing through hypertext convenient access to many topics. Perhaps of primary interest would be the *Library Services*, which includes resources for libraries, information professionals, and researchers. These include Acquisitions, Cataloging, Preservation, Research, Special Programs, Standards, and access to the catalogs of the Library of Congress and other libraries.

National Archives and Records Administration

http://www.access.gpo.gov/nara/cfr/index.html

*This site provides access to the latest published Code of Federal Regulations. Additionally, it has other important references formatted in hypertext, allowing convenient access to the specific Internet sites. These other references include Federal Register, Privacy Act Issuances, Public Laws, United States Government manual, Weekly Compilation of Presidential Documents, U.S. Congress Information and GPO (Government Printing Office) Access Search Page.

National Institutes of Health (NIH)

http:www.nih.gov

This is the Homepage of the National Institutes of Health (NIH). In hypertext it provides an overview to NIH, Welcome to NIH, and additional topics including News and Events, Health Information, Funding Opportunities, Scientific Resources, and Institutes and Offices.

Organisation for Economic Co-operation and Development

http://www.oecd.org

This is the OECD's Homepage, providing a platform for accessing titles, which include Activities, About OECD, Statistics, News & Events, Recruitment, Bookshop, Free Documents, and Search OECD Centres.

The Society of Quality Assurance (SQA)

http://www.sqa.org

This is the Homepage of SQA, dedicated " . . . to promote the quality assurance profession through the discussion and exchange of ideas . . . ." The professionals in the Society are focused primarily on quality assurance–related requirements associated with the Good Laboratory Practices and Good Clinical Practices worldwide. More than 1,200 members of SQA participate in National Conferences or Regional Chapter Meetings. Besides those registered members from the United States, internationally listed are members from Australia, Canada, Denmark, England, France, Germany, Italy, Japan, Luxembourg, Mexico, Sweden, Switzerland, and the United Kingdom. This site provides hyperlinks to numerous regulatory related documents of interest to all those involved in the Good Laboratory Practices.

USDA (United States Department of Agriculture)

http://www.usda.gov

This is the USDA's Homepage, with links to sites including About USDA, News & Information, Opportunities, and Agencies.

National Agriculture Library

http://www.nalusda.gov

Agriculture Research Service

http://www.ars.usda.gov

U.S. House of Representatives

http://www.house.gov

This site, United States House of Representatives, has numerous hypertext entries that could be very helpful for your interests, including Member Offices, Committee Offices,

---

Leadership Offices, Other House Organizations, Commissions and Task Forces, Annual Congressional Schedule, The Legislative Process, Roll Call Votes, and House Committee Hearing Schedules and Oversight Plans.

http://www.house.gov/science

The United States House of Representatives Committee on Science provides a good example of the government's involvement in various aspects of scientific development. Hypertext topics currently include Technology Transfer Challenges and Partnerships; Report on Genetically-Modified Plants; Report on the Assessment of the Benefits, Safety, and Oversight of Plant Genomics and Agricultural Biotechnology; Science Policy Study; and Science Education.

U.S. Senate

http://www.senate.gov

This United States Senate site includes a platform for linking to a variety of interests: Learning About the Senate, Legislative Activities, Committees, Senators, Committee Hearing Schedule, Bill Search, and Today's Schedule.

B. Internet Terms

Internet Services Providers (ISP)

This is an organization that provides access to the Internet by dialing in. In order to access the Internet it is necessary to connect through an ISP. Before choosing an ISP it would be appropriate to have an experienced Internet user provide a recommendation. There are more than 6,000 at this time. Through the use of a computer online, you may find a listing of some ISPs at Web address: http://thelist.iworld.com/

Homepage

This is the introductory page of a Web site, generally formatted with hyperlinks connecting the user to other related sites. It should be noted that in my research of the Internet this word is spelled as a single word, Homepage, or as two words, Home Page.

Hyperlink

A programmed connection between Web pages, usually highlighted, or underlined. By clicking on it, the Web browser is activated, seeking out the requested page from another location of the Internet.

Hypertext Markup Language (HTML)

The standard programming language used for creating and formatting WWW pages.

Search Engine

These are useful tools utilized for seeking or searching for information on the Web. They each have different parameters in their search criteria and, therefore, provide a variety of source references. Some examples of search engines are Alta Vista (http://www.altavista.com), Excite (http://www.excite.com), Infoseek (http://infoseek.com), Lycos (http://www.lycos.com), Netscape Netcenter (http://www.netscape.com), and Yahoo (http://www.yahoo.com).

Universal Resource Locator (URL)

This is the addressing system used in locating Web sites on the Internet.

Web Browser

Software that provides users with an interface in accessing and navigating the World Wide Web.

http://www.cdc.gov

CDC
Centers for Disease Control and Prevention

| CDC Home | Search | Health Topics A-Z |

Centers for Disease Control and Prevention

| About CDC | Announcements | Funding | Publications | Contact Us |

SPOTLIGHTS

▶ Women and Heart Disease: An Atlas of Racial and Ethnic Disparities in Mortality
This a comprehensive document with national and state maps depicting county-level heart disease death rates for women for the five major racial and ethnic groups.

▶ Healthy People 2010 - Cancer
Has the focus changed? The health objectives to be achieved by the year 2010 are now available.

▶ Register now for PREVENTION 2000
The Association of Teachers of Preventive Medicine & the American College of Preventive Medicine Conference.

▶ Child Passenger Safety Week
Learn how to protect child passengers and about the effectiveness of child safety seats and booster seats.

▶ False Internet Report About Bananas

▶ About CDC
Information about CDC's organization, facilities, people, budget, and mission.

▶ Announcements
A calendar of events, current topics, and recent reports and publications.

▶ Data and Statistics
CDC health data standards, scientific data, surveillance, health statistics reports and laboratory information.

▶ Funding
Information about grant and cooperative agreement funding opportunities.

▶ Health Topics A-Z
Fact sheets, disease prevention and health information from A to Z (e.g., Anthrax, Cancer, Drownings, Zoster).

▶ In the News
Press releases and current health news.

▶ Other Sites
CDC information networks, public health partners, state and local health departments, and web resources.

▶ Publications, Software, Products
Order and download brochures, catalogs, publications, software, slides, and videos.

▶ Subscriptions
Sign up to receive CDC and ATSDR health publications, software, and other products by email.

▶ Training and Employment

Contents
▶ In the News
▶ Travelers' Health
▶ Health Topics A-Z
▶ Publications, Software & Products
▶ Data & Statistics
▶ Training & Employment
▶ Subscriptions
▶ Other Sites
▶ Visitor Survey

Highlighted Resources
▶ MMWR
▶ EID Journal
▶ CDC FOUNDATION
▶ CDC Prevention Guidelines
▶ CDC Y2K Compliance

Search
[Search] [Clear]

Public Inquiries
(404) 639-3534
(800) 311-3435

Centers for Disease Control
and Prevention
1600 Clifton Rd.
Atlanta, GA 30333
U.S.A
(404) 639-3311

Department of Health and
Human Services - USA

http://www.epa.gov

Text Version
Comments

In the News

Topics

Your Community

Audience Groups

Laws & Regulations

Programs

Information Sources

About EPA

Search

For KIDS

REMEMBER THE PAST
Regional Reports
PROTECT THE FUTURE

≋EPA **United States**
Environmental Protection Agency

"...to protect human health and to safeguard the natural environment..."

"Americans now will have the best picture ever of the actual amounts of toxic pollution being emitted by industry into local communities."
—Administrator Carol M. Browner, May 11, 2000

Headlines

New Toxics Data Enhance Public Right to Know
Toxic emissions figures for seven major industrial sectors are available for the first time through EPA's internet accessible Toxic Release Inventory.

Administrator Carol M. Browner's Toxic Release Inventory Announcement
"Today, we are making public critical information on discharges of toxic pollution from seven major industries in America."

Other Stories

Profiting from Pollution Prevention Planning
An efficient production process and effective resource utilization reduces costs and maximizes profits.

EPA Headquarters Press Releases

Local Stories

Click on a region for local news stories

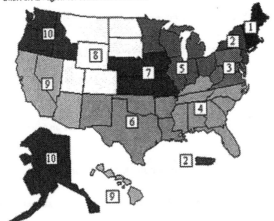

Announcements

-Activities Update
-What's New on the EPA Web

Features

Speeches & Testimony
-Speeches and Testimony

Education Matters
-Design for the Environment Program for Small Businesses

Emerging Issues
-The Small Business Regulatory Enforcement Fairness Act of 1996

-Energy Star Small Business

People & Profiles
-Karen Brown, EPA's Small Business Ombudsman

Business News
-EPA's Small Business Gateway

-EPA Small Business Assistance Program

Public Participation
-2000 Small Business Ombudsman/Small Business Assistance Program National Conference

Money Matters & Jobs
-Small Business Innovation Research

-Environmental Finance Program

Environmental IQ
-Small Business facts

http://www.epa.gov/epahome/rules.html

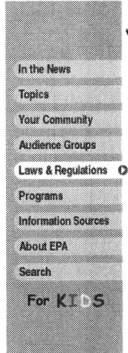

⊜EPA United States
Environmental Protection Agency

In the News

Topics

Your Community

Audience Groups

Laws & Regulations ◯

Programs

Information Sources

About EPA

Search

For KIDS

Laws and Regulations

Laws and regulations are a major tool in protecting the environment.
Find out about:

Regulations & Proposed Rules
New regulations, proposed rules, important notices and the regulatory agenda of future regulations.

Current Legislation
Current legislation before the U.S. Congress, Congressional Committees, and uncompiled Public Laws.

Codified Regulations
Federal regulations codified in the Code of Federal Regulations and additional material related to Title 40: Protection of Environment.

Laws
Public Laws passed by the U.S. Congress and codified in the U.S. Code.

Regulations and Proposed Rules

Federal Register - Environmental Documents
Full text of all Federal Register documents issued by EPA, and of selected documents issued by other Departments and Agencies. Notices, meetings, proposed rules, and regulations are divided into twelve topical categories for easy access (eg. air, water, pesticides, toxics, waste).

Federal Register database **EXIT EPA**
The Government Printing Office (GPO) maintains the official electronic database that covers Federal Register documents beginning in January, 1994. Unlike the EPA Environmental Documents service, it provides access to all Federal Register documents from all Departments and Agencies.

The Semiannual Regulatory Agenda
All agencies publish semiannual regulatory agendas describing regulatory actions they are developing or have recently completed. These agendas are published in the Federal Register, usually during April and October each year, as part of the Unified Agenda of Federal Regulatory and Deregulatory Actions.

- EPA Unified Agenda Preamble **EXIT EPA** Explains the rulemaking process, EPA's regulatory priorities and the contents of the Unified Agenda. Made available by the Regulatory Information Service Center of the U.S. General Services Administration (GSA).
- EPA Semiannual Regulatory Agenda Table of Contents **EXIT EPA** provided by the Regulatory Information Service Center.
- EPA's Statement of Regulatory and Deregulatory Priorities **EXIT EPA** Made available by the Regulatory Information Service Center.
- Search the Unified Agenda This site, maintained by the Government Printing Office, offers instructions and tips on how to search for Unified Agenda information and upcoming actions. The site searches the entire Unified Agenda.

Dockets
Dockets contain information and supporting documentation related to the rulemaking process. Public comments received on rules and proposed rules are also maintained in the appropriate Docket.

Electronic Enhancement of the Regulatory Process - This site is a comprehensive source for information on EPA's current electronic regulatory

http://www.fda.gov

 **Welcome to Internet FDA**
The Nation's Foremost Consumer Protection Agency

U.S. Food and Drug Administration

WHAT'S NEW

Year 2000

Bioengineered Foods

FDA Modernization Act

INDEX

SEARCH

FAQs

DOCKETS

INTERNATIONAL

CONTACT US

International Year of Older Persons

 **Buying Medical Products Online?**
Shop Smart!

 Foods   *Human Drugs*

 Biologics   *Animal Drugs*   *Cosmetics*

 Medical Devices/ Radiological Health   *Freedom of Information*

 Field Operations   **CHILDREN &TOBACCO** *Regulations & Information*

 Toxicology Research   **MEDWATCH** *Medical Products Reporting & Safety Information*

 Department of Health and Human Services

Special Information for:
Consumers • Industry • Health Professionals • Patients • State & Local Officials • Women

 KIDS

Privacy Statement

http://www.fda.gov/cder/index.html

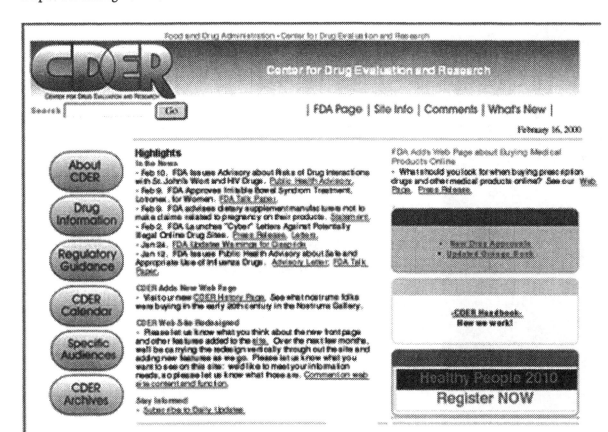

http://www.fda.gov/opacom/hpnews.html

FDA News and Publications

Some information previously found on this page is now located on the What's New page. Older press releases, talk papers, and publications can be found in the FDA Archives.

Recent Press Releases and Talk Papers

FDA statement Concerning Structure/Function Rule and Pregnancy Claims *(Feb. 9, 2000)*

FDA Approves Irritable Bowel Syndrome Treatment for Women *(Feb. 9, 2000)*

Next Year's Budget Request for FDA *(Feb. 7, 2000)*

FDA Launches "Cyber" Letters Against Potentially Illegal Foreign-Based Online Drug Sites *(Feb. 2, 2000)*

FDA Approves First Digital Mammography System *(Jan. 31, 2000)*

2000 FDA Science Forum: FDA and the Science of Safety: New Perspectives *(Jan. 28, 2000)*

FDA Issues Nationwide Warning on Senor Felix's, Trader Joe's, Delicioso, and the Carryout Café Brands of 5 Layer Dip Because of Possible Health Risk *(Jan. 27, 2000)*

Current Issues of Other FDA Publications

Latest FDA backgrounder: "HACCP: A State-of-the-Art Approach to Food Safety" (updated in August 1999)

Latest FDA Enforcement Report (Feb. 16, 2000)

Latest Issue of FDA Consumer magazine (January-February 2000)

Index | Search | Archives | Comments | Home
(Hypertext updated by db, 2000-FEB-15)

---

http://www.access.gpo.gov/nara/cfr/index.html

National Archives and
Records Administration

Attention: New Code of Federal Regulations Browse Feature

Code of Federal Regulations	Federal Register	Privacy Act Issuances	Public Laws	United States Government Manual	Weekly Compilation of Presidential Documents	Public Papers of the Presidents of the United States	U.S. Congress Information	GPO Access Search Page

About the CFR online
Establishing HTML links to GPO's CFR WAIS databases
Search the entire set of CFR databases by keyword (current data)
Retrieve CFR sections by citation (current and/or historical data)
Search or browse your choice of CFR titles and/or volumes (current and/or historical data)
LSA (List of CFR Sections Affected) (current and/or historical data)
Search the Federal Register for related documents (current and/or historical data)

PARALLEL TABLE OF AUTHORITIES AND RULES (TXT 992k) (PDF 400k)
(Extracted from the January 1, 1998, revision of the *CFR Index and Finding Aids* -- pp. 709-817)

An IHS Group Company

Who's New

Keyword Search

Per Publication Announcements

New Publications

Interpharm Online Catalog of Books, Software, Videos and Regulatory Documents

About Interpharm

Information for Authors

Contacting Interpharm

Interpharm Press
15 Inverness Way East
Englewood, CO
80112-5776
United States of America
Phone: +1 303 662 9101
Fax: +1 303 754 3993
Toll-free (U.S. & Canada)
1-877-4-INFORM

Impact the way you do business!
visit www.medicaldata.com/alt

YOUR ONE STOP GLOBAL RESOURCE FOR HEALTHCARE MANUFACTURING AND REGULATORY INFORMATION

We are the premier provider of technical and regulatory information for manufacturers of biotechnology, bulk pharmaceutical chemicals, and medical device and diagnostic products and pharmaceuticals worldwide.

Interpharm Press provides the world's largest selection of technical books, productivity software, regulatory documents, and audio-visual training tools for the pharmaceutical and chemical manufacturing industries.

Coming Soon!

Training Tracker III for Windows 95/98/NT is a software program designed to help you manage your training records and data, including requirements, scheduling, history and training costs for your employees.

The program provides an audit trail for validations, necessary in today's regulatory and compliance driven marketplace. For example, meeting 21 CFR Part 2 requirements, Training Tracker III will be available this spring. Send us your email information and we'll be happy to advise you when you can get your copy and when our online and CD-ROM demonstrations will be available.

Information Request

NEW PUBLICATIONS

Keep watching our site for more exciting announcements of new products and services designed for the pharmaceutical and biotechnical manufacturing industries.

©2000 Interpharm Press, Inc. U.S.A.

CHAPTER 7

CONTRIBUTIONS TO GLP UNDERSTANDING

A. FDA Management Briefing and Inquiry Publications

Although the GLP regulations have been in effect for more than twenty years, the need still exists for understanding and interpreting. In Section IV in this manual there are several sources of information that can be useful in acquiring further knowledge about the regulations and the rationale supporting certain interpretations. The first of these documents are

1. Good Laboratory Practice Regulations Management Briefings, 1979, Food and Drug Administration, and

2. Good Laboratory Practice Regulations, Questions and Answers, 1981, Dr. Paul Lepore, Bioresearch Monitoring Staff, Food and Drug Administration.

Those of us who had our beginnings in GLP in the late 1970s found that these references were extremely helpful, providing insight and understanding for interpreting the regulations. These publications are written in a question-and-answer format, typically arranged according to the specific sections of the GLP regulations, e.g., 58.1 Scope—58.15 Inspection of a Testing Facility. Some examples from the first reference illustrate the types of questions from the 1979 Management Briefings:

Subpart A—General Provisions

2. Does the Agency intend to audit analytical data collected on a test article?

   Yes, insofar as it contributes to the evaluation of a nonclinical laboratory study.

30. It is not clear whether a laboratory involved solely in chemical analysis support of a nonclinical laboratory study would be required to comply with the GLPs. Can this be clarified?

   Yes. Analytical laboratories must comply with the GLPs to the extent that they provide data which support the nonclinical laboratory study. Only those portions of the laboratory, those procedures and those personnel involved are required to be in compliance with the GLPs.

32. Does the term "nonclinical laboratory study" include animal laboratory studies that are designed for the explicit purpose of determining whether a test article has reasonable

promise of clinical effectiveness and in which observations bearing on clinical safety are only incidental or fragmentary, or, at most, clearly secondary?

No.

It should be emphasized that familiarity with the regulations and their evolving interpretations should be a continuing objective of all engaged in the GLP.

The second reference, "Questions and Answers," also has some interesting questions and answers that should improve your overall understanding of the regulations. Like the previous reference, this publication also has questions categorized, but the arrangement is more much specific to a regulatory section, for example:

Section 58.3 Definitions

1. Are animal cage cards considered to be raw data?

   Raw data is defined as "any laboratory worksheets, records memorandum, notes . . . that are the result of original observations and activities . . . and are necessary for the reconstruction and evaluation of the report of that study." Cage cards are not raw data if they contain information like animal number, study number, study dates, and cage number (information that is not the result of original observations and that is not necessary for study reconstruction). However, if an original observation is put on the cage cards, then all cards must be saved as raw data.

Section 58.130 Conduct of a Nonclinical Laboratory Study

7. Is it acceptable to manually transcribe raw data into notebooks if it is verified accurate by signature and date?

   Technically the GLPs do not preclude such an approach. It is not a preferred procedure, however, since the chance of transcription errors would exist. Accordingly, such an approach should be used only when necessary and in this event the raw data should also be retained.

B. Preambles to the GLP

Some of the more basic publications that can be used for broadening your knowledge of the Good Practice Regulations include not just the regulations, but rather their accompanying preambles. Two of these documents may also be found in Section IV, Regulatory Texts:

---

1. U.S. Food and Drug Administration, Good Practice Regulations, effective June 20, 1979, published in The Federal Register of December 22, 1978.

2. U.S. Food and Drug Administration, Good Practice Regulations, effective October 5, 1987, published in The Federal Register of September 4, 1987.

The preamble to these documents is important, providing a background of information associated with commentary that the FDA had solicited from industry and the public, and it describes in many cases the rationale applied in finalizing a regulation. The first of these is the preamble of the GLP Regulations issued June 20, 1979:

Study Director

65. Five comments objected to the proposed requirement that the study director assure that test and control articles or mixtures be appropriately tested. The comments argued that this was not a proper function of the study director.

The Commissioner agrees that this responsibility is more properly assigned to testing facility management. Therefore, the requirement has been transferred to 58.31(d).

This is the current wording in the regulation, confirming the above statement.

58.31—Testing facility management

(d) Assure that test and control articles or mixtures have been appropriately tested for identity, strength, purity, stability, and uniformity, as applicable.

Specimen and Data Storage Facilities

111. Several comments asked whether 58.51 applied to completed or ongoing studies. Concern was also expressed that limiting access to storage areas to authorized personnel was not feasible.

This section is amended to apply to archive storage of all raw data and specimens from completed studies. The commissioner cannot agree, however, that limiting access of the archives to authorized personnel only is not feasible. Prudence would dictate such limited access even in the absence of a requirement. The potential for misplaced data and specimens is too great to allow unlimited access to the archives.

The wording reflecting this requirement is expressed in the current regulation:

58.51—Specimen and data storage facilities

> Space shall be provided for archives, limited to access by authorized personnel only, for the storage and retrieval of all raw data and specimens from completed studies.

The second preamble associated with the GLPs is that published in the Federal Register of October 5, 1987.

Quality Assurance Unit

13. One comment recommended that FDA delete the word "sheet" from the term "master schedule sheet" in 58.35 (b)(1) on the ground that there are methods for maintaining a master schedule other than use of an actual "sheet."

> FDA acknowledges that current technology allows for various methods for maintaining a master schedule, ranging from sophisticated computerized procedures to procedures whereby such information is contained in written records. Regardless of the method utilized, however, the master schedule information is "raw data" within the meaning of the GLP regulations and copies of the master schedule are required to be retained in the study archives in accordance with 58.195(b). The agency is, therefore, retaining the term "master schedule sheet" to emphasize that the master schedule constitutes raw data subject to agency inspection and that the records must be retained.

Test and Control Article Characterization

38. Several comments recommended that FDA revise 58.105(c) to remove the current requirement that storage containers be assigned to a test article for the duration of a nonclinical laboratory study. The comments argued that the requirement is unnecessary and is inconsistent with the Organization for Economic Cooperation and Development principles of GLP. One comment alleged that the statement is vague and questioned whether the provision would permit a storage container that a laboratory emptied during the conduct of, but before completion of, a study to be destroyed or reused. The same comment also questioned whether, as the test article is depleted during the conduct of a study, the provision would permit a laboratory to transfer the test article into a container smaller than that originally assigned to the article.

FDA does not believe that it would be appropriate to eliminate the storage container provision in 58.105(c). FDA advises that the provision simply requires that each test article storage container be assigned to a test article at the beginning of a study and remain so assigned until the study is completed, terminated, or discontinued. The test article may not be transferred to different sized storage containers as a study progresses, nor may assigned storage containers be destroyed while a study is in progress.

This response continues for several more paragraphs, the FDA ultimately concluding that assignment of storage containers is necessary to ensure the integrity of a study. It is suggested that you become familiar with these documents and review them in Section IV, for complete reference and understanding.

C. Freedom of Information

In the United States the FDA has an extensive group of field investigators who are continually inspecting or auditing sponsor facilities for either Good Manufacturing Practices (GMPs), Good Clinical Practices (GCPs), or Good Laboratory Practices (GLPs). GLP training and laboratory inspection techniques for the FDA's field investigators are a continuing program, normally conducted at its National Center for Toxicological Research in Jefferson, Arkansas. With the varied background of these inspectors it is not always possible to predict how an inspection at your facility may progress. It is possible, however, to assess how other sponsor or contract laboratories have fared through FDA audits by requesting from the FDA through the Freedom of Information (FOI) a copy of the Establishment Inspection Report (EIR) and any accompanying FDA-483, the report describing any regulatory compliance violations as perceived by the inspector. Just write a request to one of the following addresses, for example:

As provided through the Freedom of Information Act, please send the EIR and FDA-483 for the November 19, 1990, inspection of Southwestern Laboratories, Ltd., located in Oklahoma City, Oklahoma:

Department of Health and Human Services
Food and Drug Administration
HFI-35, Room 12A
5600 Fishers Lane
Rockville, Maryland 20857

or

FOI Services, Inc.

12315 Wilkins Avenue

Rockville, Maryland 20852-1877

Phone: 301-881-0401

Fax: 301-881-0415

The FOI documents obtained directly from the FDA will be the most economical source, but not necessarily the most expeditious. If there is any litigation, or if for some reason the FDA has not closed out a specific inspection file, the information will not be released to the FOI archives, and, therefore, the arrival of the documentation may be delayed.

D. Compliance Program Guidance Manual—Good Laboratory Practice

Another document that you should be aware of is the FDA's Compliance Program Guidance Manual, titled Good Laboratory Practice (Nonclinical Laboratories), a recent copy of which is in Section IV.[17] This provides a step-by-step program for an investigator to follow. The instructions state: "This program provides for the inspection of public, private, and government nonclinical laboratories which may be performing tests on food and color additives, human and animal drugs, human medical devices, biological products, pesticides, disinfectants, and electronic products." This program should be helpful, since it provides information about the types of questions the investigator may pursue. Agency investigators do differ in their approach, but after reviewing a few hundred of the FDA GLP audit documents available through FOI, it is apparent many of them follow this program routinely. Here are some statements and questions from just a few sections of this compliance program document:

Establishment Inspections

The facility inspection should be guided by the GLPs. The following areas should be evaluated and described as appropriate:

1(b) Personnel—21 CFR 58.29 Review personnel records, policies, and operations to determine if:

- Summaries of training and job descriptions are maintained and current for selected employees engaged in the study;

- Personnel have been adequately trained to carry out the study functions that they perform;

The investigator should identify key laboratory and management personnel including any outside consultants used, and review their CVs, training records, and position descriptions. Review facility SOPs covering training and personnel practices.

If the firm has computerized operations, determine:

- Who was involved in the design and validation of the computer system;

- Who is responsible for the operation of the computer system, including inputs, processing, and output of the data; and

- If computer system personnel are trained in GLPs.

Continuing with some additional illustrations, the following is the program associated with the QAU:

2. Quality Assurance Unit (QAU)—21 CFR 58.35

   Purpose: To determine if the firm has an effective QAU which monitors overall facility operations, reviews records and reports, and assures management of GLP compliance.

   a. QAU Operations—21 CFR 58.35(b)–(d)

      Review QAU SOPs to ensure that they cover all methods and procedures for carrying out the required QAU functions, and confirm that they are being followed. Verify that SOPs exist for the following activities:

      - Maintenance of a master schedule sheet;

      - Maintenance of copies of all protocols and amendments;

      - Inspection of each nonclinical laboratory study at intervals adequate to assure the integrity of the study, and maintenance of records of each inspection;

      - Notification to the study director and management of any problems which are likely to affect the integrity of the study;

      - Submission of periodic status reports on each study to the study director and management;

- Review of the final study report;

- Preparation of a statement to be included in the final report which specifies the dates inspections were made and findings reported to management and to the study director; and

- Inspection of computer operations.

Indicate whether the QAU maintains the protocols and amendments.

Verify how the QAU schedules audits and in-process inspections, including equipment maintenance, and to whom, when, and how it reports findings and retains records.

NOTE: The investigator may request the firm's management to certify in writing that inspections are being implemented, performed, documented, and followed up in accordance with this section [see 58.35(d)].

There are numerous other GLP sections in the Compliance Program available for complete review in Section IV of this manual.

This information is provided primarily as a convenience for introducing you to some of the available information associated with your interest in GLP. Hopefully, you have been encouraged to seek out these publications for review and comprehension of this very interesting regulatory area.

If you are also interested in joining a professional society associated primarily with Good Laboratory Practices, there are several available, including:

Society of Quality Assurance (SQA)
515 King Street, Suite 420
Alexandria, VA 22314-3137, USA
Ph: 703-684-4050
Fax: 703-684-6048
Web site: www.sqa.org

Drug Information Association
501 Office Center Drive, Suite 450
Fort Washington, PA 19034-3211, USA
Ph: 215-628-2288
Fax: 215-641-1229
E-mail: dia@diahome.org

British Institute of Regulatory Affairs (BIRA)
34 Dover St.
London WIX 3RA, England, UK

International Society of Quality Assurance (ISQA)
1512 Wackena Rd.
Morrisville, NC 27560, USA
Ph: 919-319-1155
Fax: 919-467-7724
E-mail: isqahq@mindspring.com

American Society for Quality (ASQ)
611 E. Wisconsin Ave., P.O. Box 3005
Milwaukee, WI 53201-3005, USA
Ph: 414-272-8575
Fax: 414-272-1734
Email: asq@asq.org
Website: www.asq.org

GLOSSARY

These definitions are derived from the following referenced organizations' documents:

(a) Title 21 CFR Part 58, Food and Drug Administration, Good Laboratory Practice Regulations, published in the Federal Register December 22, 1978

(b) Title 21 CFR Part 58, Food and Drug Administration, Good Laboratory Practice Regulations, published in The Federal Register September 4, 1987

(c) OECD Principles of Good Laboratory Practice, OECD Council November 26, 1997 (OECD publication of 1998)

(d) Title 40, CFR Part 160, Environmental Protection Agency (EPA), Pesticide Programs, Good Laboratory Practices Standards, effective 5/2/84

(e) Title 40, CFR Part 792, EPA Toxic Substances Control, Good Laboratory Practice Standards, effective 12/29/83

Batch[a]

A specific quantity or lot of a test or control article that has been characterized according to part 58.105(a).

Batch[c]

A specific quantity or lot of a test item or reference item produced during a defined cycle of manufacture in such a way that it could be expected to be of a uniform character and should be designated as such.

Control Article[b]

Any food additive, color additive, drug, biological product, electronic product, medical device for human use, or any article other than a test article, feed or water that is administered to the test system in the course of a nonclinical laboratory study for the purpose of establishing a basis for comparison with the test article.

Control Substances[d, e]

Any chemical substance or mixture or any other material other than a test substance that is administered to the test system in the course of a study for the purpose of establishing a basis for comparison with the test substance.

Experimental Completion Date[c]

The last date on which data are collected from the study.

Experimental Starting Date[c]

The date on which the first study specific data are collected.

Good Laboratory Practice (GLP)[c]

A quality system concerned with the organisational process and the conditions under which non-clinical health and environmental safety studies are planned, performed, monitored, recorded, archived, and reported.

Master Schedule[c]

A compilation of information to assist in the assessment of workload and for the tracking of studies at a test facility.

Non-clinical Health and Environmental Safety Study ("Study")[c]

An experiment or set of experiments in which a test item is examined under laboratory conditions or in the environment to obtain data on its properties and/or its safety, intended for submission to appropriate regulatory authorities.

Nonclinical Laboratory Study[b]

Any in vivo or in vitro experiments in which test articles are studied prospectively in test systems under laboratory conditions to determine their safety. The term does not include studies utilizing human subjects or clinical studies or field trials in animals. The term does not include basic exploratory studies carried out to determine whether a test article has any potential utility or to determine physical or chemical characteristics of a test article.

Person[a]

Includes an individual, partnership, corporation, association, scientific or academic establishment, government agency, or organizational unit thereof, and any other legal entity.

Principal Investigator[c]

An individual who, for a multi-site study, acts on behalf of the Study Director and has defined responsibility for delegated phases of the study. The Study Director's responsibility for the overall conduct of the study cannot be delegated to the Principal Investigator(s); this includes approval of the study plan and its amendments, approval of the final report, and ensuring that all applicable Principles of Good Laboratory Practice are followed.

Quality Assurance Programme[c]

A defined system, including personnel, which is independent of study conduct and is designed to assure test facility management of compliance with these Principles of Good Laboratory Practice.

Quality Assurance Unit[a]

Any person or organizational element, except the study director, designated by testing facility management to perform the duties relating to quality assurance of nonclinical laboratory studies.

Raw Data[a]

Any laboratory worksheets, records, memoranda, notes, or exact copies thereof, that are the result of original observations and activities of a nonclinical laboratory study and are necessary for the reconstruction and evaluation of the report of the study. In the event that exact transcripts of raw data have been prepared (e.g., tapes which have been transcribed verbatim, dated, and verified accurate by signature), the exact copy or exact transcript may be substituted for the original source as raw data. "Raw data" may include photographs, microfilm or microfiche copies, computer printouts, magnetic media, including dictated observations, and recorded data from automated instruments.

Raw Data[c]

All original test facility records and documentation, or verified copies thereof, which are the result of the original observations and activities in a study. Raw data also may include, for example, photographs, microfilm or microfiche copies, computer readable media, dictated observations, recorded data from automated instruments, or any other data storage medium that has been recognized as capable of providing secure storage of information for a time period as stated in Section 2, Good Laboratory Practice Principles, part 10, Storage and Retention of Records and Materials.

Reference Item ("Control item")[c]

Any article used to provide a basis for comparison with the test item.

Short-term Study[c]

A study of short duration with widely used, routine techniques.

Specimen[a]

Any material derived from a test system for examination or analysis.

Specimen[c]

Any material derived from a test system for examination, analysis, or retention.

Sponsor[a]

(1) A person who initiates and supports, by provision of financial or other resources, a nonclinical laboratory study;

(2) A person who submits a nonclinical study to the Food and Drug Administration in support of an application for a research or marketing permit;

(3) A testing facility, if it both initiates and actually conducts the study.

Sponsor[c]

An entity which commissions, supports and/or submits a non-clinical health and environmental safety study.

Standard Operating Procedures (SOPs)[c]

Documented procedures which describe how to perform tests or activities normally not specified in detail in study plans or test guidelines.

Study Completion Date[b]

The date the final report is signed by the study director.

Study Completion Date[c]

The date the Study Director signs the final report.

Study Director[a]

The individual responsible for the overall conduct of a nonclinical laboratory study.

Study Director[c]

The individual responsible for the overall conduct of the nonclinical health and environmental safety study.

Study Initiation Date[b]

The date the protocol is signed by the study director.

Study Initiation Date[c]

The date the Study Director signs the study plan.

Study Plan[c]

A document which defines the objectives and experimental design or the conduct of the study, and includes any amendments.

Study Plan Amendment[c]

An intended change to the study plan after the study initiation date.

Study Plan Deviation[c]

An unintended departure from the study plan after the study initiation date.

Test Article[a]

Any food additive, color additive, drug, biological product, electronic product, medical device for human use, or any other article subject to regulation under the act.

Test Facility[c]

The persons, premises and operational unit(s) that are necessary for conducting the non-clinical health and environmental safety study. For multi-site studies, those which are conducted at more than one site, the test facility comprises the site at which the Study Director is located and all individual test sites, which individually or collectively can be considered to be test facilities.

Test Facility Management[c]

The person(s) who has the authority and formal responsibility for the organisation and functioning of the test facility according to these Principles of Good Laboratory Practice.

Test Facility[a]

A person who actually conducts a nonclinical laboratory study, i.e., actually uses the test article in a test system. It encompasses only those operational units that are being or have been used to conduct nonclinical laboratory studies.

Test Item[c]

An article that is the subject of a study.

Test Site[c]

The location(s) at which a phase(s) of a study is conducted.

Test Site Management[c]

If appointed, the person(s) responsible for ensuring that the phase(s) of the study, for which he is responsible, are conducted according to these Principles of Good Laboratory Practice.

Test Substance or Mixture[d]

A substance or mixture administered or added to a test system in a study, which substance or mixture

(1) Is the subject of an application for a research or marketing permit supported by the study, or is the comtemplated subject of such an application; or

(2) Is an ingredient, impurity, degradation product, metabolite, or radioactive isotope of a substance related to a substance described by that paragraph, which is used in the study to assist in characterizing the toxicity, metabolism, or other characteristics of a substance described by the paragraph.

Test Substance or Mixture[e]

A substance or mixture administered or added to a test system in a study, which substance or mixture is used to develop data to meet the requirements of Toxic Substances Control Act (TSCA) section 4(a) test rule and/or is developed under a negotiated testing agreement or section 5 rule/order to the extent the agreement or rule/order references this part.

GLP Audit Manual ©2000, Interpharm Press

Test System[a]

Any animal, plant, microorganism, or subparts thereof to which the test or control article is administered or added for study. It also includes appropriate groups or components of the system not treated with the test or control article.

Test System[c]

Any biological, chemical or physical system or a combination thereof used in a study.

Validation

(1) The process of evaluating a system at the end of the development process to assure compliance with user requirements.

(2) The process of evaluating software at the end of the software development process to ensure compliance with software requirements (FIPS PUB 132 May 1988) (ANSI/IEEE Std 1012-1986 and Std 729-1983).

Vehicle[c]

Any agent which serves as a carrier used to mix, disperse, or solubilise the test item or reference item to facilitate the administration/application to the test system.

Verification

The demonstration of consistency, completeness and correctness of the software at each state and between each stage of the development life cycle (FIPS PUB101 6 June 1983).

REFERENCES

The Good Laboratory Practice references included in this text are those associated with the United States Federal Regulations and the Organization for Economic Cooperation and Development. Other useful journal references are listed below:

1. New Zealand Testing Laboratory Registration Act 1972, No. 36, October 20, 1972.

2. Danish National Testing Board Act No. 144, March 21, 1973.

3. Law About the State's Technical Trial Board, effective April 1, 1973, translation from Danish, E. Napier.

4. U.S. Food and Drug Administration, Nonclinical Laboratory Studies, Good Laboratory Practice Regulations, Code of Federal Regulations, Title 21, Part 58, effective June 20, 1979.

5. The OECD Principles of Good Laboratory Practice, No. 1, revised 1997. Organization for Economic Cooperation and Development (OECD).

6. U.S. Environmental Protection Agency, Toxic Substances Control, Good Laboratory Practice Standards, Code of Federal Regulations, Title 40, Part 792, effective December 29, 1983.

7. U.S. Environmental Protection Agency, Pesticide Programs, Good Laboratory Practice Standards, Code of Federal Regulations, Title 40, Part 160, effective May 2, 1984.

8. Marshall, E., "The Murky World of Toxicity Testing," *Science,* Vol. 220, p. 1130, June 10, 1983.

9. FDA Prosecutions, Fall 1986, "Decisions on Long Standing Cases."

10. Federal Register, Vol. 41, No. 225, Friday, November 19, 1976, Department of Health, Education, and Welfare, Food and Drug Administration, Nonclinical Laboratories Studies, Proposed Regulations for Good Laboratory Practice, pp. 51207–51208.

11. Department of Health and Human Services, "Analysis of Sponsor Laboratories for Compliance with the GLP Regulations," July 24, 1975.

12. Federal Register, Vol. 52, No. 172, Friday, September 4, 1987, Department of Health and Human Services, Food and Drug Administration, 21 CFR Part 58, Good Laboratory Practice Regulations.

13. Federal Register Vol. 56, No. 135, Monday, July 15, 1991. Department of Health and Human Services, Food and Drug Administration, 21 CFR Part 58, Good Laboratory Practice Regulations.

14. Guide for Care and Use of Laboratory Animals, NIH Publication No. 85-23, Revised 1985.

15. Computerized Data Systems for Nonclinical Safety Assessment, Current Concepts and Quality Assurance Drug Information Association, Maple Glen, PA 19002, September 1988.

16. Sampling Procedures and Tables for Inspection by Attributes, Military Standard MIL-STD-105E, 10 May 89, Department of Defense, USA.

17. Food and Drug Administration Compliance Program Guidance Manual, Program 7348.808, Chapter 48, Bioresearch Monitoring, Good Laboratory Practice, Completion Date 09/03/93.

SECTION II
QUALITY AUDIT CHECKLISTS

GLP AUDIT CHECKLIST—COMPLETE

The *GLP Audit Checklist—Complete,* included here for your reference, includes all the regulatory parts listed below. In the abbreviated GLP Audit Checklists that follow in subsequent sections of this manual, some of these regulatory parts may appropriately be excluded. They are, however, included in other checklists that are considered more fitting for that inspection phase. In some instances, because of the type of study, or management's preference for increased redundancy, additional regulatory parts may be chosen from *GLP Audit Checklist—Complete* to form a more suitable abbreviated checklist, unique to a specific inspection or auditing requirement.

A. General Provisions

B. Personnel

C. Testing Facility Management

D. Study Director

E. Quality Assurance Unit

F. Facilities

G. Animal Care Facilities

H. Animal Supply Facilities

I. Facilities for Handling Test and Control Articles

J. Laboratory Operation Areas

K. Specimen and Data Storage Facilities

L. Equipment Design

M. Maintenance and Calibration of Equipment

N. Standard Operating Procedures

O. Reagents and Solutions

P. Animal Care

Q. Test and Control Article Characterization

R. Test and Control Article Handling

S. Mixture of Articles with Carriers

T. Protocol for and Conduct of a Nonclinical Laboratory Study

U. Conduct of a Nonclinical Laboratory Study

V. Reporting of Nonclinical Laboratory Study Results

W. Storage and Retrieval of Records and Data

X. Retention of Records

Checklists have been found to be a convenient way to evaluate a facility, its operations, or a nonclinical study for compliance with GLP regulations or OECD Principles. The wording in the officially published United States Food and Drug Administration GLP regulations was carefully developed and chosen by its authors. In the conversion of these regulatory statements from the usual imperative mood to a questionnaire checklist I have tried to retain the intended meaning of the general provisions. Once gaining familiarity with this relatively complete GLP Audit Checklist, the auditor can refer to the *Abbreviated GLP Audit Checklists,* also in this section of the manual. These are more convenient checklists, paraphrasing the original, resulting in a shorter, more practical form. All checklists have references to the appropriate Code of Federal Regulations, 21 CFR Part 58, Good Laboratory Practice Regulations, and are standardized in their alpha title designations for convenient cross-referencing. This checklist, *GLP Audit Checklist—Complete,* also has references to the OECD Principles of Good Laboratory Practice Number 1, revised 1997, which is also available in this manual, Section IV—References/Regulatory Texts. The checklist OECD references should aid the reader in locating the appropriate text in the actual OECD document.

INSPECTION/AUDITING HISTORY GLP AUDIT

Facility Location

Name: _____

Address: _____

Date(s) of Inspection Auditing: _____

Protocol

Title: _____

Number: _____ Date Issued: _____

Scientific Report

Number: _____ Date Issued: _____

Assigned Inspector/Auditor: _____

Title: _____

Alternate Inspectors/Auditors: _____

Name: _____

Inspection Reports

Number: _____ Date Issued: _____ Date Returned: _____

Number: _____ Date Issued: _____ Date Returned: _____

Number: _____ Date Issued: _____ Date Returned: _____

Quality Assurance Unit Statement

Issued: _____ Date: _____

Discussion Sessions

Attendees: _____   Date: _____

Study Director: _____

Comments

_____

_____

_____

_____

_____

_____

_____

_____

_____

_____

_____

_____

_____

_____

_____

_____

_____

_____

_____

_____

_____

_____

GLP Audit Manual ©2000, Interpharm Press

GLP AUDIT CHECKLIST	Response			References	
	Yes	No	N/A	21 CFR	OECD

A. General Provisions

1. Has the sponsor, in utilizing the services of a consulting laboratory, contractor, or grantee to perform an analysis or other service, notified them that the service is part of a nonclinical laboratory study and must be conducted in compliance with the provisions of this part? ☐ ☐ ☐ 58.10

2. Does the testing facility permit the FDA, at reasonable times and in a reasonable manner, to inspect the facility and all records and specimens required to be maintained? ☐ ☐ ☐ 58.15

B. Personnel 58.29

1. Does each individual engaged in the conduct of or supervision of the study have the education, training, and experience to perform the assignments? ☐ ☐ ☐   1, 1.1(2)(b)

2. Does the facility maintain a current summary of training, experience, and job descriptions for each person engaged in or supervising the study? ☐ ☐ ☐   1, 1.1(2)(c)

3. Are there sufficient personnel for the timely and proper conduct of the study according to the protocol? ☐ ☐ ☐   1, 1.1(2)(b)

4. Do personnel take sanitation and health precautions to avoid contamination of test and control articles and test systems? ☐ ☐ ☐   1, 1.4(4)

5. Do personnel engaged in the study wear appropriate clothing, changed at a frequency to prevent microbiological, radiological, or chemical contamination of test systems and test and control articles? ☐ ☐ ☐

6. Are personnel with an illness that may adversely affect the test systems, test or control article (TCA), and any other operation excluded from the study until corrected? ☐ ☐ ☐ 1, 1.4(4)

7. Are personnel instructed to report to their supervisor any health or medical conditions that may have an adverse effect on the study? ☐ ☐ ☐ 1, 1.4(4)

C. Testing Facility Management   58.31

Does testing facility management:

1. Designate a study director before study initiation? ☐ ☐ ☐ 1, 1.1(2)(g)

2. If necessary, replace the study director promptly? ☐ ☐ ☐ 1, 1.1(2)(g)

3. Assure there is a Quality Assurance Unit (QAU) established? ☐ ☐ ☐ 1, 1.1(2)(f)

4. Assure that test and control articles ormixtures are appropriately tested for identity, strength, purity, stability, and and uniformity, as applicable? ☐ ☐ ☐ 1, 1.1(2)(p)

5. Assure that personnel, resources, facilities, equipment, materials, and methodologies are available as scheduled? ☐ ☐ ☐ 1, 1.1(2)(a), (b), & (e)

6. Assure that personnel clearly understand the functions they are to perform? ☐ ☐ ☐ 1, 1.1(2)(d)

GLP AUDIT CHECKLIST	Response			References	
	Yes	No	N/A	21 CFR	OECD

7. Assure that any deviations from these regulations reported by the QAU are communicated to the study director and corrective actions are taken and documented? ☐ ☐ ☐   2, 2.2(1)(e)

D. Study Director   **58.33**

1. Does the study director have appropriate education, training, and experience? ☐ ☐ ☐   1, 1.1(2)(g)

2. Does the study director exercise overall responsibility for the technical conduct of the study, including interpretation, analysis, documentation, and reporting of results? ☐ ☐ ☐   1, 1.2(1)

3. Does the study director assure:

   3.1. That the protocol and changes are approved as provided by 58.120 and followed? ☐ ☐ ☐   1, 1.2(2)(a) & (e) 8, 8.1(1) & (2)(a)

   3.2. That all experimental data, including observations of unanticipated responses to the test system, are accurately recorded and verified? ☐ ☐ ☐   1, 1.2(2)(f)

   3.3. That unforeseen circumstances that may affect the quality and integrity of the study are noted when they occur, and corrective action is taken and documented? ☐ ☐ ☐   1, 1.2(2)(e)

3.4. That test systems are as specified in the protocol? ☐ ☐ ☐ | 1, 1.2(2)(e)

3.5. That all applicable GLP regulations are followed? ☐ ☐ ☐ | 1, 1.2(2)(h)

3.6. That all raw data, documentation, protocols, specimens, and final reports are transferred to the archives during or at the close of the study? ☐ ☐ ☐ | 1, 1.2(2)(i)

E. Quality Assurance Unit | 58.35

1. Is the QAU responsible for monitoring each study? ☐ ☐ ☐ | 2, 2.1(2)

2. Does the QAU assure management that the facilities, equipment, personnel methods, practices, records, and controls are in conformance with these regulations? ☐ ☐ ☐ | 2, 2.1(1) / 2, 2.2(1)(b)

3. For any given study, is the QAU separate from and independent of the personnel engaged in the direction and conduct of that study? ☐ ☐ ☐ | 2, 2.1(3)

4. Does the QAU:

4.1. Maintain a copy of a master schedule sheet of all nonclinical laboratory studies conducted at the testing facility indexed by test article and containing the test system, nature of study, date study was initiated, current status of each study, identity of the sponsor, and name of the study director? ☐ ☐ ☐ | 2, 2.2(1)(a)

GLP AUDIT CHECKLIST	Response			References	
	Yes	No	N/A	21 CFR	OECD
4.2. Maintain copies of all protocols pertaining to all studies for which the QAU is responsible?	❑	❑	❑		2, 2.2(1)(a)
4.3. Inspect each nonclinical laboratory study at intervals adequate to assure the integrity of the study and maintain written and properly signed records of each periodic inspection showing the date of the inspection, the study inspected, the phase or segment of the study inspected, the person performing the inspection, findings and problems, action recommended and taken to resolve existing problems, and any schedule date for reinspection?	❑	❑	❑		2, 2.2(1)(c)
4.4. Immediately inform the study director and management of any significant problems that are likely to affect study integrity?	❑	❑	❑		2, 2.2(1)(e)
4.5. Periodically submit to management and the study director written status reports on each study, noting any problems and the corrective actions taken?	❑	❑	❑		
4.6. Determine that no deviations from approved protocols or standard operating procedures were made without proper authorization and documentation?	❑	❑	❑		2, 2.1(1)(c)

4.7. Review the final study report to assure that it accurately describes the methods and standard operating procedures, and that the reported results accurately reflect the raw data of the study? ☐ ☐ ☐ 2, 2.2(1)(d)

4.8. Prepare and sign a statement to be included with the final study report that specifies the dates inspections were made and findings reported to management and to the study director? ☐ ☐ ☐ 2, 2.2(1)(f)

4.9. Maintain written documentation of the responsibilities and procedures applicable to the QAU, and the method of indexing such records? ☐ ☐ ☐ 7, 7.4(5)

4.10. Maintain a record including inspection dates, the study inspected, and the phase and inspector for review by authorized regulatory personnel? ☐ ☐ ☐ 2, 2.2(1)(f)

4.11. Assure that inspections are being implemented, performed, documented, and followed up in accordance with GLP? ☐ ☐ ☐ 2, 2.1(1)

4.12. Maintain the QAU records in one location at the testing facility? ☐ ☐ ☐

F. Facilities **58.41**

1. Is the testing facility of suitable size and construction to facilitate the proper conduct of the study? ☐ ☐ ☐ 3, 3.1(1)

GLP AUDIT CHECKLIST	Response			References	
	Yes	No	N/A	21 CFR	OECD

2. Is it designed to provide a degree of separation that will prevent any function or activity from having an adverse affect on the study?   ❐   ❐   ❐   3, 3.1(2)

G. Animal Care Facilities   **58.43**

1. Does the facility have a sufficient number of animal rooms or areas, as needed, to assure proper:

   1.1. Separation of species or test systems?   ❐   ❐   ❐   3, 3.2(1)

   1.2. Isolation of individual projects?   ❐   ❐   ❐   3, 3.2(1)

   1.3. Quarantine of animals?   ❐   ❐   ❐   3, 3.2(2)
                                                5, 5.2(2)

   1.4. Routine or specialized housing of animals?   ❐   ❐   ❐   3, 3.2(2)

2. Does the facility have a number of rooms separate from those above to ensure isolation of studies being done with test systems or test and control articles known to be biohazardous, including volatile substances, aerosols, radioactive materials, and infectious agents?   ❐   ❐   ❐   3, 3.2(1)

3. Are separate areas provided as appropriate for the diagnosis, treatment, and control of laboratory animal diseases?   ❐   ❐   ❐   3, 3.2(2)

4. Do the facilities provide for the collection and disposal of all animal waste and refuse or for safe sanitary storage of waste before removal from the facility?   ❐   ❐   ❐   3, 3.5

5. Are the disposal facilities provided and operated as to minimize vermin infestation, odors, disease hazards, and environmental contamination? ☐ ☐ ☐   3, 3.5

6. Are the facilities designed, constructed, and located so as to minimize disturbances that interfere with the study? ☐ ☐ ☐   3, 3.1(1)

H. Animal Supply Facilities   **58.45**

1. Are there storage areas, as needed, for feed, bedding, supplies, and equipment? ☐ ☐ ☐   3, 3.2(3)

2. Are the storage areas for feed and bedding separated from areas housing the test systems? ☐ ☐ ☐   3, 3.2(3)

3. Are these storage areas protected against infestation or contamination? ☐ ☐ ☐   3, 3.2(3)

4. Are perishable supplies preserved by appropriate means? ☐ ☐ ☐   3, 3.2(3)

I. Facilities for Handling Test and Control Articles   **58.47**

1. As necessary, to prevent contamination, mix-ups, are there separate areas for:

   1.1. Receipt and storage of test and control articles? ☐ ☐ ☐   3, 3.3(1)

   1.2. Mixing of the test and control articles with a carrier, e.g., feed? ☐ ☐ ☐   3, 3.3(1)

   1.3. Storage of the test and control article mixtures? ☐ ☐ ☐   3, 3.3(2)

		Yes	No	N/A	21 CFR	OECD
2.	Are storage areas for the test and/or control article and test and control mixtures separate from areas housing the test systems?	❑	❑	❑		3, 3.3(2)
3.	Are they adequate to preserve the identity, strength, purity, and stability of the articles and mixtures?	❑	❑	❑		3, 3.3(2)
J.	**Laboratory Operation Areas**				58.49	
1.	Is separate laboratory space provided for the performance of the routine and specialized procedures required by nonclinical laboratory studies?	❑	❑	❑		3, 3.2(1)
K.	**Specimen and Data Storage Facilities**				58.51	
1.	Is space provided for archives, limited to access by authorized personnel only, for the storage and retrieval of all raw data and specimens from completed studies?	❑	❑	❑		10, 10.3
L.	**Equipment Design**				58.61	
1.	Is equipment used in the generation, measurement, or assessment of data and equipment used for facility environmental control of appropriate design and adequate capacity to function according to the protocol?	❑	❑	❑		4(1) 5, 5.1(1)
2.	Is this equipment suitably located for operation, inspection, cleaning, and maintenance?	❑	❑	❑		4(1)

		Response			References	
GLP AUDIT CHECKLIST		Yes	No	N/A	21 CFR	OECD

M. Maintenance and Calibration of Equipment — 58.63

1. Is this equipment adequately inspected, cleaned, and maintained? □ □ □ — 4(1)

2. Is equipment used for the generation, measurement, or assessment of data adequately tested, calibrated, and/or standardized? □ □ □ — 4(2)

3. Do the Standard Operating Procedures (SOPs) required in 58.81(b)(11) set forth in sufficient detail the methods, materials, and schedules to be used in the routine inspection, cleaning, maintenance, testing, calibration, and/or standardization of equipment? □ □ □ — 4(2)

4. Do these SOPs specify, when appropriate, the remedial action to be taken in the event of failure or malfunction of equipment? □ □ □

5. Do these SOPs also designate the person responsible for the performance of each operation? □ □ □

6. Are written records maintained of all inspection, maintenance, testing, calibration and/or standardizing operations? □ □ □ — 4(2)

7. Do these records, containing the date of operation, describe whether the maintenance operations were routine and followed the written SOPs? □ □ □

8. Are written records kept of nonroutine repairs performed on equipment as a result of failure and malfunction? □ □ □

GLP Audit Manual ©2000, Interpharm Press

9. Do these records document the nature of the defect, how and when the defect was discovered, and any remedial action taken in response to the defect? ☐ ☐ ☐

| **N.** | **Standard Operating Procedures** | | | | 58.81 | |

1. Are the SOPs in writings setting forth study methods adequate to insure the quality and integrity of the data generated in the course of a study? ☐ ☐ ☐   1, 1.1(2)(e) 7, 7.1

2. Are all deviations in a study from SOPs authorized by the study director? ☐ ☐ ☐   1, 1.2(2)(e)

3. Are the deviations documented in the raw data? ☐ ☐ ☐   7, 7.3

4. Are significant changes in established SOPs properly authorized in writing by management? ☐ ☐ ☐   7, 7.1

5. Are SOPs established for, but not limited to, the following:

    5.1. Animal room preparation? ☐ ☐ ☐   7, 7.4(4)(a)

    5.2. Animal care? ☐ ☐ ☐   7, 7.4(4)(b)

    5.3. Receipt, identification, storage, handling, mixing, and method of sampling of the test and control articles? ☐ ☐ ☐   7, 7.4(1)

    5.4. Test systems observations? ☐ ☐ ☐   7, 7.4(4)(c)

    5.5. Laboratory Tests? ☐ ☐ ☐   7, 7.1

    5.6. Handling of animals found moribund or dead during study? ☐ ☐ ☐   7, 7.4(4)(d)

GLP AUDIT CHECKLIST	Yes	No	N/A	21 CFR	OECD
	Response			References	

| | Yes | No | N/A | 21 CFR | OECD |

	Response			References	
GLP AUDIT CHECKLIST	Yes	No	N/A	21 CFR	OECD
5.7. Necropsy of animals or postmortem examination of animals?	❑	❑	❑		7, 7.4(4)(d)
5.8. Collection and identification of specimens?	❑	❑	❑		7, 7.4(4)(e)
5.9. Histopathology?	❑	❑	❑		7, 7.4(4)(e)
5.10. Data handling, storage, and retrieval?	❑	❑	❑		7, 7.4(3)
5.11. Maintenance and calibration of equipment?	❑	❑	❑		7, 7.4(2)(a)
5.12. Transfer, proper placement, and identification of animals?	❑	❑	❑		7, 7.4(4)(b)
6. Does each laboratory area have immediately available manuals and SOPs relative to the laboratory procedures being performed?	❑	❑	❑		7, 7.2
7. If published literature is used, is it used as a supplement to SOPs and not in lieu of SOPs?	❑	❑	❑		7, 7.2
8. Has an historical file been established and maintained of SOPs and all revisions, including the dates of such revisions?	❑	❑	❑		1, 1.1(2)(k) 10, 10.1(f)
O. Reagents and Solutions				58.83	
1. Are all reagents and solutions in the laboratory areas labeled to indicate identity, titer or concentration, storage requirements, and expiration date?	❑	❑	❑		4(4) 7, 7.4(2)(c)
2. Are deteriorated or outdated reagents and solutions not used?	❑	❑	❑		4(4)

GLP Audit Manual ©2000, Interpharm Press

GLP AUDIT CHECKLIST	Response			References	
	Yes	No	N/A	21 CFR	OECD
P. Animal Care				58.90	
1. Is there an SOP for the housing, feeding, handling, and care of animals?	❑	❑	❑		7, 7.4(4)(b)
2. Are all newly received animals from outside sources isolated and their health status evaluated in accordance with acceptable veterinary medical practice?	❑	❑	❑		5, 5.2(2)
3. Are these evaluations in accordance with acceptable veterinary medical practice?	❑	❑	❑		5, 5.2(1)
4. At the initiation of the study are the animals free of any disease or condition that might interfere with the purpose or conduct of the study?	❑	❑	❑		5, 5.2(2)
5. In the course of a study are the animals that contract such a disease or condition isolated, if necessary?	❑	❑	❑		5, 5.2(2)
6. If these animals are treated for the disease or signs of the disease, does the treatment not interfere with the study?	❑	❑	❑		5, 5.2 (2)
7. Are the diagnosis, authorizations of treatment, and each date of treatment documented and retained?	❑	❑	❑		5, 5.2(2)
8. Do warm-blooded animals, excluding suckling rodents, used in laboratory procedures that require manipulations and observations over an extended period of time receive appropriate identification?	❑	❑	❑		5, 5.2(5)

GLP AUDIT CHECKLIST	Response			References	
	Yes	No	N/A	21 CFR	OECD
9. Do these aforementioned animals used in studies that require the animals to be removed from and returned to their home cages for any reason (e.g., cage cleaning, treatment) receive appropriate identification?	❒	❒	❒		5, 5.2(5)
10. Does all information needed to specifically identify each animal within an animal-housing unit appear on the outside of that unit?	❒	❒	❒		5, 5.2(5)
11. Are animals of different species housed in separate rooms when necessary?	❒	❒	❒		3, 3.2(1) 5, 5.2(1)
12. Are animals of the same species, but used in different studies, not ordinarily housed in the same room when inadvertent exposure to control or test articles or animal mix-up could affect the outcome of either study?	❒	❒	❒		3, 3.2(1) 5, 5.2(1)
13. If such mixed housing is necessary, is adequate differentiation by space and identification made?	❒	❒	❒		3, 3.2(1) 5, 5.2(1)
14. Are animal cages, racks, and accessory equipment cleaned and sanitized at appropriate intervals?	❒	❒	❒		5, 5.2(6)
15. Are feed and water used for the animals analyzed periodically to ensure that contaminants known to be capable of interfering with the study and reasonably expected to be present in such feed or water are not present at levels above those specified in the protocol?	❒	❒	❒		5, 5.2(6)

GLP Audit Manual ©2000, Interpharm Press

GLP AUDIT CHECKLIST	Response			References	
	Yes	No	N/A	21 CFR	OECD
16. Are such analyses maintained as raw data?	☐	☐	☐		8, 8.3(3)
17. Does the bedding used in animal cages or pens not interfere with the purpose or conduct of the study?	☐	☐	☐		5, 5.2(6)
18. Is the bedding changed as often as necessary to keep the animals dry and clean?	☐	☐	☐		5, 5.2(6)
19. If pest control materials are used, is their use documented?	☐	☐	☐		5, 5.2(6)
20. Are cleaning and pest control materials that interfere with the study not used?	☐	☐	☐		5, 5.2, 6
Q. Test and Control Article Characterization				**58.105**	
1. Are the identity, strength, purity, and composition or other characteristics that will appropriately define the test or control article (TCA) determined and documented for each batch?	☐	☐	☐		6, 6.2(1) & (2)
2. Are the methods of synthesis, fabrication, or derivation of the test and control articles documented by the sponsor or the testing facility?	☐	☐	☐		
3. Are marketed products used as control articles characterized by their labeling?	☐	☐	☐		
4. Is the stability of each TCA determined by the testing facility or by the sponsor either before initiation of a study or concomitantly according to written SOPs, which provide for periodic analysis of each batch?	☐	☐	☐		6, 6.2(4) & (5)

		Yes	No	N/A	21 CFR	OECD
5.	Is each storage container for a test or control article labeled by name, chemical abstract number or code number, batch number, expiration date, if any, and, where appropriate, storage conditions necessary to maintain the identity, strength, purity, and composition of the TCA?	☐	☐	☐		6, 6.1(3)
6.	Are storage containers assigned to a particular test article for the duration of the study?	☐	☐	☐		
7.	For studies lasting more than 4 weeks duration, are reserve samples from each batch of TCA retained for the period of time provided in 58.195?	☐	☐	☐		6, 6.2(6)

R. Test and Control Article Handling   **58.107**

Are procedures established for a system for the handling of the TCA articles to ensure that:

		Yes	No	N/A	21 CFR	OECD
1.	There is proper storage?	☐	☐	☐		6, 6.1(2)
2.	Distribution is made in a manner designed to preclude the possibility of contamination, deterioration, or damage?	☐	☐	☐		6, 6.1(2)
3.	Proper identification is maintained throughout the distribution process?	☐	☐	☐		6, 6.1(2)
4.	The receipt and distribution of each batch is documented, including the date and quantity of each batch distributed or returned?	☐	☐	☐		6, 6.1(1)

			Response			References	
GLP AUDIT CHECKLIST		Yes	No	N/A	**21 CFR**	**OECD**

S.	**Mixtures of Articles with Carriers**				**58.113**	
1.	For each TCA that is mixed with a carrier, are tests by appropriate analytical methods conducted:					
	1.1. To determine the uniformity of the mixture and to determine, periodically, the concentration of the TCA in the mixture?	❒	❒	❒		6, 6.2(2) & (5)
	1.2. To determine the stability of the TCA in the mixture?	❒	❒	❒		6, 6.2(2) & (5)
2.	Is stability of the TCA in the mixture as required by the conditions of the study determined either before initiation of the study, or concomitantly according to written SOPs that provide for periodic analysis of the TCA in the mixture?	❒	❒	❒		6, 6.2(5)
3.	Where any of the components of the TCA carrier mixture has an expiration date, is that date clearly shown on the container?	❒	❒	❒		6, 6.1(3)
4.	If more than one component has an expiration date, is the earliest date shown?	❒	❒	❒		6, 6.1(3)
T.	**Protocol for and Conduct of a Nonclinical Laboratory Study**				**58.120**	
1.	Does each study have an approved written protocol that clearly indicates the objectives and all methods for the conduct of the study?	❒	❒	❒		8, 8.1(1)
2.	Does the protocol contain, as applicable, the following information:					

GLP AUDIT CHECKLIST	Response			References	
	Yes	No	N/A	21 CFR	OECD
2.1. A descriptive title and statement of the purpose of the study?	❒	❒	❒		8, 8.2(1)(a) & (b)
2.2. Identification of the TCA by name, chemical abstracts number, or code number?	❒	❒	❒		8, 8.2(1)(c)
2.3. The name of the sponsor and the name and address of the testing facility at which the study is being conducted?	❒	❒	❒		8, 8.2(2)(a) & (b)
2.4. The number, body weight, range, sex, source of supply, species, strain, substrain, and age of test system?	❒	❒	❒		8, 8.2(5)(b)
2.5. The procedure for identification of the test system?	❒	❒	❒		8, 8.2(5)(b)
2.6. A description of the experimental design, including the methods for control of bias?	❒	❒	❒		8, 8.2(5)(e)
2.7. A description and/or identification of diet used in the study as well as solvent, emulsifiers, and/or other materials used to solubilize or suspend the TCA before mixing with the carrier?	❒	❒	❒		
2.8. A description including specifications for acceptable levels of contaminants that are reasonably expected to be present in the dietary materials and are known to be capable of interfering with the purpose or conduct of the study if present at levels greater than established by the specifications?	❒	❒	❒		8, 8.2(5)(e)

GLP Audit Manual ©2000, Interpharm Press

2.9. The reason for route-of-administration choice? ☐ ☐ ☐ | 8, 8.2(5)(c)

2.10. Each dosage level, expressed in milligrams per kilogram of body weight or other appropriate units, of the TCA to be administered and the method and frequency of administration? ☐ ☐ ☐ | 8, 8.2(5)(d)

2.11. The type and frequency of tests, analyses, and measurements to be made? ☐ ☐ ☐ | 8, 8.2(5)(e)

2.12. The records to be maintained? ☐ ☐ ☐ | 8, 8.2(6)

2.13. The date of approval of the protocol by the sponsor and the dated signature of the study director? ☐ ☐ ☐ | 8, 8.2(3)(a)

2.14. A statement of the proposed statistical methods to be used? ☐ ☐ ☐ | 8, 8.2(5)(e)

3. Are all the changes in or revision of an approved protocol and the reasons documented signed by the study director, dated, and maintained with the protocol? ☐ ☐ ☐ | 8, 8.1(2)(a)

U. Conduct of a Nonclinical Laboratory Study   58.130

1. Is the nonclinical laboratory study conducted in accordance with the protocol? ☐ ☐ ☐ | 8, 8.3(2)

2. Is the test system monitored in conformity with the protocol? ☐ ☐ ☐ | 1, 1.2(2)(e) \ 2, 2.2(1)(c)

3. Are specimens identified by test system, study, nature, and date of collection? ☐ ☐ ☐ | 8, 8.3(1)

GLP AUDIT CHECKLIST	Response			References	
	Yes	No	N/A	21 CFR	OECD
4. Is this information located on the specimen container or does it accompany the specimen in a manner that precludes error in the recording and storage of data?	☐	☐	☐		8, 8.3(1)
5. Are records of gross findings for a specimen from postmortem observations available to a pathologist when examining that specimen histopathologically?	☐	☐	☐		8, 8.3(3) 10, 10.1(a)
6. Are all data generated during the conduct of a nonclinical laboratory study, except those that are generated by automated data collection systems, recorded directly, promptly, and legibly, in ink?	☐	☐	☐		8, 8.3(3)
7. Are all data entries dated on the date of entry and signed or initialed by the person entering the data?	☐	☐	☐		8, 8.3(3)
8. Is any change in entries made so as not to obscure the original entry, indicate the reason for such change, and is dated and signed or identified at the time of change?	☐	☐	☐		8, 8.3(4)
9. In automated data collection systems, is the individual responsible for direct data input identified at the time of data input?	☐	☐	☐		8, 8.3(5)
10. Is any change in automated data entries made so as not to obscure the original entry, indicate the reasons for change, and is dated and the responsible individual identified?	☐	☐	☐		8, 8.3(5)

GLP Audit Manual ©2000, Interpharm Press

GLP AUDIT CHECKLIST	Response			References	
	Yes	No	N/A	21 CFR	OECD
V. Reporting of Nonclinical Laboratory Study Results				58.185	
1. Has a final report been prepared for each nonclinical laboratory study?	❐	❐	❐		9, 9.1(1)
2. Does the final report include, but is not necessarily limited to, the following:					
2.1. Name and address of the facility performing the study?	❐	❐	❐		9, 9.2(2)(a) & (b)
2.2. The dates on which the study was initiated and completed?	❐	❐	❐		9, 9.2(3)
2.3. Objectives and procedures stated in the approved protocol, including any changes in the original protocol?	❐	❐	❐		9, 9.2(5)(a) & (b)
2.4. Statistical methods employed for analyzing the data?	❐	❐	❐		9, 9.2(6)(c)
2.5. The TCA identified by name, chemical abstracts number or code number, strength, purity, and composition or appropriate characteristics?	❐	❐	❐		9, 9.2(1)(b)
2.6. Stability of the TCA under the conditions of administration?	❐	❐	❐		9, 9.2(1)(d)
2.7. A description of the methods used?	❐	❐	❐		9, 9.2(5)(a)
2.8. A description of the test system used?	❐	❐	❐		9, 9.2(5)(b)
2.9. Where applicable, the number of animals used, sex, body weight range, source of supply, species, strain and substrain, age?	❐	❐	❐		

GLP AUDIT CHECKLIST	Response			References	
	Yes	No	N/A	21 CFR	OECD
2.10. The procedure used for animal identification?	❒	❒	❒		
2.11. A description of the dosage, dosage regimen, route of administration, and duration?	❒	❒	❒		
2.12. A description of all circumstances that may have affected the quality or integrity of the study?	❒	❒	❒		
2.13. The name of the study director, the names of other scientists or professionals, and the names of all supervisory personnel involved in the study?	❒	❒	❒		9, 9.2(2)(c), (d), & (e)
2.14. A description of the transformations, calculations, or operations performed on the data?	❒	❒	❒		9, 9.2(6)(c)
2.15. A summary and analysis of the data and a statement of the conclusions drawn from the analysis?	❒	❒	❒		9, 9.2(6)(a), (c), & (d)
2.16. The signed and dated reports of each of the individual scientists or other professionals involved in the study?	❒	❒	❒		9, 9.1(2)
2.17. The locations where all specimens, raw data, and the final report are to be stored?	❒	❒	❒		9, 9.2(7)
2.18. The statement prepared and signed by the QAU as described in 58.35?	❒	❒	❒		9, 9.2(4)
3. Is the final report signed and dated by the study director?	❒	❒	❒		9, 9.1(3)

GLP Audit Manual ©2000, Interpharm Press

4. Are corrections or additions to a final report in the form of an amendment by the study director?   ❏   ❏   ❏   9, 9.1(4)

5. Does the amendment clearly identify that part of the final report that is being added to or corrected and the reasons for the correction or addition?   ❏   ❏   ❏

6. Is the amendment signed and dated by the person responsible?   ❏   ❏   ❏   9, 9.1(4)

W. Storage and Retrieval of Records and Data   58.190

1. Are all raw data, documentation, protocols, final reports, and specimens (except those specimens obtained from mutagenicity tests and wet specimens of blood, urine, feces, and biological fluids) generated as a result of a nonclinical laboratory study retained?   ❏   ❏   ❏   10, 10.1(a)

2. Is there an archive for orderly storage and expedient retrieval of all raw data, documentation, protocols, specimens, and interim and final reports?   ❏   ❏   ❏   3, 3.4
                       10, 10.2

3. Do the conditions of storage minimize deterioration of the documents or specimens in accordance with the requirements for the time period of their retention and the nature of the documents or specimens?   ❏   ❏   ❏   3, 3.4

4. If the facility has contracted with a commercial archive to provide a repository for all material to be retained, has specific reference been made in the archive to those other locations?   ❏   ❏   ❏

GLP AUDIT CHECKLIST	Response			References	
	Yes	No	N/A	21 CFR	OECD

5. Is an individual identified as responsible for the archives? ☐ ☐ ☐   1, 1.1(2)(l)

6. Do only authorized personnel enter the archive? ☐ ☐ ☐   10, 10.3

7. Is material retained or referred to in the archives indexed to permit expedient retrieval? ☐ ☐ ☐   10, 10.2

X. Retention of Records   58.195

1. Except for wet specimens, samples of TCA, and specially prepared material (e.g., histochemical, electron microscopic, blood mounts, teratological preparation, and uteri from dominant lethal mutagenesis tests), are documentation records, raw data, and specimens pertaining to a nonclinical laboratory study and required to be made by this part retained in the archive(s) for whichever of the following periods is shortest:

   1.1. A period of at least 2 years following the date on which an application for a research or marketing permit, in support of which the results of the nonclinical laboratory study were submitted, is approved by the FDA? ☐ ☐ ☐   10, 10.1(a)–(g)

   1.2. A period of at least 5 years following the date on which the results of the nonclinical laboratory study are submitted to the FDA in support of an application for a research or marketing permit? ☐ ☐ ☐   10, 10.1(a)–(g)

GLP Audit Manual ©2000, Interpharm Press

	Response			References	
GLP AUDIT CHECKLIST	Yes	No	N/A	21 CFR	OECD

1.3. In other situations (e.g., where the nonclinical laboratory study does not result in the submission of the study in support of an application for a research or marketing permit), a period of at least 2 years following date on which the study is completed, terminated, or discontinued? ☐ ☐ ☐   10, 10.1(a)–(g)

2. Are wet specimens (except those specimens obtained from mutagenicity tests and wet specimens of blood, urine, feces, and biological fluids), samples of test or control articles, and specially prepared material, which are relatively fragile and differ markedly in stability and quality during storage, retained only as long as the quality of the preparation affords evaluation? ☐ ☐ ☐   10, 10.1("h")

3. Are the master schedule sheet, copies of protocols, and records of quality assurance inspections, as required by 58.35, maintained by the QAU as an easily accessible system of records for the period of time specified in (1) previously? ☐ ☐ ☐   10, 10.1(b)

4. Are summaries of training and experience and job descriptions required to be maintained by 58.29 retained along with all other testing facility employment records for the length of time specified in (1) previously? ☐ ☐ ☐   10, 10.1(c)

5. Are records and reports of the maintenance, calibration, and inspection of equipment, as required by 58.63, retained for the length of time specified in (1) previously? ☐ ☐ ☐   10, 10.1(d)

6. Are records required by this part retained either as original records or as true copies such as photocopies, microfilm, microfiche, or other accurate reproduction of the original records?   ❏   ❏   ❏

7. If a facility conducting nonclinical testing goes out of business, are all raw data, documentation, and other material specified previously transferred to the archives of the sponsor of the study?   ❏   ❏   ❏   10, 10.4

8. If the above transfer occurs, is the FDA notified in writing?   ❏   ❏   ❏

Observations/Comments

_____

_____

_____

_____

_____

_____

_____

_____

_____

_____

_____

_____

Inspected/Audited by: _____   **Date:** _____

PREPLACEMENT AND MAINTENANCE
ABBREVIATED GLP AUDIT CHECKLIST

The *Preplacement and Maintenance Abbreviated GLP Audit Checklist* illustrated here includes the regulatory parts listed below, except for those designated with an "*," which are excluded. Those excluded from this checklist are included in others that are considered more appropriate. In some instances, because of the type of study, or management's preference for increased redundancy, additional regulatory parts may be amended to form a more suitable checklist, unique to a specific inspection or auditing requirement.

A. General Provisions

B. Personnel

C. Testing Facility Management

D. Study Director

E. Quality Assurance Unit

F. Facilities

G. Animal Care Facilities

H. Animal Supply Facilities

I. Facilities for Handling Test and Control Articles

J. Laboratory Operation Areas

K. Specimen and Data Storage Facilities

L. Equipment Design

M. Maintenance and Calibration of Equipment

N. Standard Operating Procedures

O. Reagents and Solutions

P. Animal Care

Q. Test and Control Article Characterization

R. Test and Control Article Handling

S. Mixture of Articles with Carriers

T. Protocol for the Conduct of a Nonclinical Laboratory Study*

U. Conduct of a Nonclinical Laboratory Study

V. Reporting of Nonclinical Laboratory Study Results*

W. Storage and Retrieval of Records and Data

X. Retention of Records

Before a nonclinical study is placed at a contractor's site, an evaluation of the facilities and operations should be conducted by the sponsor's Quality Assurance Unit (QAU). Special emphasis should be made in auditing their QAU, since normally the responsibility for the required inspection and auditing of contracted studies is placed with the contractor's QAU. If they do not have a functional QAU, or for some reason it is inadequate, the sponsor will have to provide for the periodic inspections and audits of each study. Additionally, periodic maintenance inspections and audits should be conducted at the sponsor's locations and at any contracted operations.

The checklists that follow have reference to the appropriate Code of Federal Regulation, 21 CFR, Part 58, Good Laboratory Practice, and the OECD Principles of Good Laboratory Practice, for cross-referencing, allowing one to switch from the abbreviated checklists to the specific questions of the *GLP Audit Checklist—Complete*, or to the official agency document itself.

The checklists are retained by some organizations; on each list space has been provided for the inspector/auditor signature and date. The regulations do require completion and retention of the Study Inspection Record and Inspection Report, examples of which are found in Section III. These documents provide the necessary signed records of each periodic inspection [58.35(b)(3)].

PREPLACEMENT/MAINTENANCE
ABBREVIATED GLP AUDIT CHECKLIST

AUDITING HISTORY

Facility Location

Name: _____

Address: _____

Date(s) of Inspection Auditing: _____

Protocol

Title: _____

Number: _____   Date Issued: _____

Scientific Report

Number: _____   Date Issued: _____

Assigned Inspector/Auditor: _____

Title: _____

Alternate Inspectors/Auditors: _____

Name: _____

Inspection Reports

Number: _____   Date Issued: _____   Date Returned: _____

Number: _____   Date Issued: _____   Date Returned: _____

Number: _____   Date Issued: _____   Date Returned: _____

Phase Study Inspection Record Completed: _____   Date:_____

Quality Assurance Unit Statement

Issued: _____   Date: _____

Discussion Sessions

Attendees: _____   Date: _____

Study Director: _____

Comments

_____

_____

_____

_____

_____

_____

_____

_____

_____

_____

_____

_____

_____

_____

_____

_____

_____

_____

_____

PREPLACEMENT/ MAINTENANCE	Response			References	
	Yes	No	N/A	21 CFR	OECD

A. General Provisions

1. Consultant laboratory, contractor, grantee notified of nonclinical study, conduct per GLP

	☐	☐	☐	58.10	

2. Laboratory permits inspections of facilities, records, specimens required

	☐	☐	☐	58.15	

B. Personnel 58.29

1. Education, training, experience for assignments

	☐	☐	☐		1, 1.1(2)(b)

2. Summaries, training, experience, job descriptions

	☐	☐	☐		1, 1.1(2)(c)

3. Staffed for timely, proper conduct per protocol

	☐	☐	☐		1, 1.1(2)(b)

4. Sanitation, health precautions

	☐	☐	☐		1, 1.4(4)

5. Appropriate clothing worn, proper changes made

	☐	☐	☐		

6. Illness adversely affecting test systems, TCA, excluded until corrected

	☐	☐	☐		1, 1.4(4)

7. Report health or medical conditions that may adversely affect study

	☐	☐	☐		1, 1.4(4)

C. Testing Facility Management 58.31

1. Designates study director before initiation

	☐	☐	☐		1, 1.1(2)(g)

2. If necessary, replaces study director promptly

	☐	☐	☐		1, 1.1(2)(g)

		Response			References	
PREPLACEMENT/ MAINTENANCE		Yes	No	N/A	**21 CFR**	**OECD**
3.	Assures there is a QAU	☐	☐	☐		1, 1.1(2)(f)
4.	Assures proper tests of TCA or TCA mixtures	☐	☐	☐		1, 1.1(2)(p)
5.	Assures resources and methodologies as scheduled	☐	☐	☐		1, 1.1(2)(b) & (e)
6.	Assures personnel understand functions	☐	☐	☐		1, 1.1(2)(d)
7.	Assures deviations communicated, documented	☐	☐	☐		2, 2.2(1)(e)
D.	**Study Director**				**58.33**	
Assures:						
1.	Protocol changes approved per protocol section	☐	☐	☐		1, 1.2(2)(a) & (e) 8, 8.1(1) & (2)(a)
2.	Experimental data, including unanticipated responses, accurately recorded and verified	☐	☐	☐		1, 1.2(2)(f)
3.	Unforeseen circumstances affecting quality and integrity noted; corrective action taken, documented	☐	☐	☐		1, 1.2(2)(e)
4.	Test systems as specified in protocol	☐	☐	☐		1, 1.2(2)(e)
5.	Applicable GLP regulations followed	☐	☐	☐		1, 1.2(2)(h)
6.	All raw data, documentation, specimens transferred to archives by close of study	☐	☐	☐		1, 1.2(2)(i)

PREPLACEMENT/ MAINTENANCE	Response			References	
	Yes	No	N/A	21 CFR	OECD
E. Quality Assurance Unit				**58.35**	
1. Responsible for monitoring each study	❏	❏	❏		2, 2.1(2)
2. Assures management facilities, equipment, personnel, methods, records, etc., in conformance	❏	❏	❏		2, 2.1(1)
3. Separate, independent of those directing, conducting	❏	❏	❏		2, 2.1(3)
4. Maintain copy of master schedule as required	❏	❏	❏		2, 2.2(1)(a)
5. Maintain copies of relevant protocols	❏	❏	❏		2, 2.2(1)(a)
6. Inspect at intervals to assure and document integrity of study	❏	❏	❏		2, 2.2(1)(c)
7. Immediately informs study director of significant problems likely to affect integrity	❏	❏	❏		2, 2.2(1)(e)
8. Periodically submit to management and study director reports, including corrective actions	❏	❏	❏		
9. Determine no deviation from protocols or SOPs without authorization and documentation	❏	❏	❏		2, 2.1(1)(c)
10. Reviews final report assuring it is accurate and reflects raw data	❏	❏	❏		2, 2.2(1)(d)
11. Issues QAU statement for inclusion in final report	❏	❏	❏		2, 2.2(1)(f)
12. Documented responsibilities, SOPs, records maintained, method of indexing	❏	❏	❏		7, 7.4(5)

PREPLACEMENT/ MAINTENANCE	Response			References	
	Yes	No	N/A	21 CFR	OECD

13. Maintains record with inspection dates, study, phase, inspector	❏	❏	❏		2, 2.2(1)(f)
14. Assures inspections done according to GLP	❏	❏	❏		2, 2.1(1)
15. Maintains QAU records in one location of facility	❏	❏	❏		

F. **Facilities**				**58.41**	
1. Suitable size, construction for proper conduct	❏	❏	❏		3, 3.1(1)
2. Provide separation, preventing adverse affect	❏	❏	❏		3, 3.1(2)

G. **Animal Care Facilities**				**58.43**	
1. Sufficient rooms, areas, to assure proper:					
1.1. Separation of species or test systems	❏	❏	❏		3, 3.2(1)
1.2. Isolation of projects	❏	❏	❏		3, 3.2(1)
1.3. Quarantine	❏	❏	❏		3, 3.2(2) 5, 5.2(2)
1.4. Routine or specialized housing	❏	❏	❏		3, 3.2(2)
2. Separate rooms ensuring isolation for studies with biohazards, radioactives, infectious agents	❏	❏	❏		3, 3.2(1)
3. Separate areas for diagnosis, treatment of diseases	❏	❏	❏		3, 3.2(2)
4. Proper collection, disposal of waste	❏	❏	❏		3, 3.5
5. Vermin, odors, disease hazards minimized	❏	❏	❏		3, 3.5

PREPLACEMENT/ MAINTENANCE	Response			References	
	Yes	No	N/A	21 CFR	OECD

6.	Designed, located to minimize interference	☐	☐	☐		3, 3.1(1)

H. Animal Supply Facilities — 58.45

1.	Storage areas, feed, bedding, supplies, equipment	☐	☐	☐		3, 3.2(3)
2.	Storage areas for feed, bedding separate from areas housing test system	☐	☐	☐		3, 3.2(3)
3.	Protected against infestation, contamination	☐	☐	☐		3, 3.2(3)
4.	Perishables preserved appropriately	☐	☐	☐		3, 3.2(3)

I. Facilities for Handling Test and Control Articles — 58.47

1.	To prevent contamination, mix-ups, separate areas for:					
	1.1. Receipt, storage of test/control articles (TCA)	☐	☐	☐		3, 3.3(1)
	1.2. Mixing of TCA with carrier	☐	☐	☐		3, 3.3(1)
	1.3. Storage of TCA mixtures	☐	☐	☐		3, 3.3(2)
2.	Storage TCA/mixtures separate from test system	☐	☐	☐		3, 3.3(2)
3.	Adequate to preserve identity, strength, purity, stability of TCA/mixtures	☐	☐	☐		3, 3.3(2)

J. Laboratory Operation Areas — 58.49

1.	Separate space provided for routine and specialized procedures	☐	☐	☐		3, 3.2(1)

		Response			References	
PREPLACEMENT/ MAINTENANCE		Yes	No	N/A	21 CFR	OECD

		Yes	No	N/A	21 CFR	OECD
K.	**Specimen and Data Storage Facilities**				**58.51**	
1.	Archives, limited to those authorized only, for storage, retrieval of raw data and specimens	❏	❏	❏		10, 10.3
L.	**Equipment Design**				**58.61**	
1.	Equipment used in generation, measurement, assessment, environmental control, appropriate design, capacity to function according to protocol	❏	❏	❏		4(1) 5, 5.1(1)
2.	Suitably located for operation, inspection, cleaning, maintenance	❏	❏	❏		4(1)
M.	**Maintenance and Calibration of Equipment**				**58.63**	
1.	Adequately inspected, cleaned, maintained	❏	❏	❏		4(1)
2.	Adequately tested, calibrated/standardized	❏	❏	❏		4(2)
3.	SOPs provide sufficient detail of methods, materials, schedules for routine maintenance, calibration	❏	❏	❏		4(2)
4.	SOPs address when to take appropriate remedial action, if malfunction	❏	❏	❏		
5.	SOPs identify responsible person	❏	❏	❏		
6.	Written records of inspections, maintenance, testing, calibration, and/or standardization	❏	❏	❏		4(2)
7.	Records contain whether routine, followed SOPs	❏	❏	❏		

 GLP Audit Manual ©2000, Interpharm Press

		Response			References	
PREPLACEMENT/ MAINTENANCE		Yes	No	N/A	21 CFR	OECD

		Yes	No	N/A	21 CFR	OECD
8.	Records maintained of nonroutine repairs	❏	❏	❏		
9.	Records document defect, discovery, remedy	❏	❏	❏		
N.	**Standard Operating Procedures**				**58.81**	
1.	SOPs written to ensure data quality, integrity	❏	❏	❏		7, 7.1
2.	Changes in SOPs authorized	❏	❏	❏		1, 1.2(2)(e)
3.	SOPs established, but not limited to:					
	3.1. Animal room preparation	❏	❏	❏		7, 7.4(4)(a)
	3.2. Animal care	❏	❏	❏		7, 7.4(4)(b)
	3.3. Receipt, identification, storage, handling, mixing, and method of sampling TCA	❏	❏	❏		7, 7.4(1)
	3.4. Test systems observations	❏	❏	❏		7, 7.4 (4)(c)
	3.5. Laboratory tests	❏	❏	❏		7, 7.1
	3.6. Handling of animals found moribund or dead	❏	❏	❏		7, 7.4(4)(d)
	3.7. Necropsy or postmortem examinations	❏	❏	❏		7, 7.4(4)(d)
	3.8. Collection and identification of specimens	❏	❏	❏		7, 7.4(4)(e)
	3.9. Histopathology	❏	❏	❏		7, 7.4(4)(e)
	3.10. Data handling, storage, and retrieval	❏	❏	❏		7, 7.4(3)
	3.11. Maintenance and calibration of equipment	❏	❏	❏		7, 7.4(2)(a)

		Response			References	
PREPLACEMENT/		Yes	No	N/A	**21 CFR**	**OECD**
MAINTENANCE						
	3.12. Transfer, placement, and identification of animals	❏	❏	❏		7, 7.4(4)(b)
4.	Relevant SOPs, manuals immediately available	❏	❏	❏		7, 7.2
5.	Literature used as supplement to SOPs, not in lieu of	❏	❏	❏		7, 7.2
6.	Historical file of SOPs and all revisions	❏	❏	❏		1, 1.1(2)(k) 10, 10.1(f)
O.	**Reagents and Solutions**				**58.83**	
1.	Labeled for identity, concentration, storage, expiration	❏	❏	❏		4(4) 7, 7.4(2)(c)
2.	Deteriorated, expired reagents not used	❏	❏	❏		4(4)
P.	**Animal Care**				**58.90**	
1.	SOPs for housing, feeding, handling, care of animals	❏	❏	❏		7, 7.4(4)(b)
2.	Newly received isolated until evaluated	❏	❏	❏		5, 5.2(2)
3.	Evaluation in accordance with veterinary medicine	❏	❏	❏		5, 5.2(2)
4.	At initiation, free of disease or condition that might interfere	❏	❏	❏		5, 5.2(2)
5.	During study, diseased animals isolated, if necessary	❏	❏	❏		5, 5.2(2)
6.	Treatment does not interfere	❏	❏	❏		5, 5.2(2)

GLP Audit Manual ©2000, Interpharm Press

PREPLACEMENT/ MAINTENANCE	Response			References	
	Yes	No	N/A	21 CFR	OECD
7. Diagnosis, treatment authorization documented	☐	☐	☐		5, 5.2(2)
8. Proper identification	☐	☐	☐		5, 5.2(5)
9. Required returns receive appropriate I.D.	☐	☐	☐		5, 5.2(5)
10. All specific I.D. information on outside of housing unit	☐	☐	☐		5, 5.2(5)
11. Different species in separate rooms, if necessary	☐	☐	☐		3, 3.2(1) 5, 5.2(1)
12. Different studies not housed in same room	☐	☐	☐		3, 3.2(1) 5, 5.2(1)
13. If mixed housing, adequate differentiation	☐	☐	☐		3, 3.2(1) 5, 5.2(1)
14. Adequate cleaning, sanitizing of cages, racks, etc.	☐	☐	☐		5, 5.2(6)
15. Periodic analyses of feed, water for contaminants	☐	☐	☐		5, 5.2(6)
16. Bedding changes adequate	☐	☐	☐		5, 5.2(6)
17. Pest control materials documented	☐	☐	☐		5, 5.2(6)
18. Cleaning, pest materials that interfere not used	☐	☐	☐		5, 5.2(6)

PREPLACEMENT/ MAINTENANCE	Response			References	
	Yes	No	N/A	21 CFR	OECD

Q. Test and Control Article Characterization — 58.105

1. Storage containers: name, abstract or code number, batch number, expiration date, storage conditions ☐ ☐ ☐ — 6, 6.1(3)

2. Storage containers assigned for study duration ☐ ☐ ☐

3. Reserve samples retained for proper period ☐ ☐ ☐ — 6, 6.2(6)

R. Test and Control Article Handling — 58.107

SOPs for handling TCA to ensure:

1. Proper storage ☐ ☐ ☐ — 6, 6.1(2)

2. Distribution that precludes change, damage ☐ ☐ ☐ — 6, 6.1(2)

3. Maintenance of identification during distribution ☐ ☐ ☐ — 6, 6.1(1)

4. Documentation of receipt, distribution returned, each batch ☐ ☐ ☐ — 6, 6.1(1)

S. Mixtures of Articles with Carriers — 58.113

1. For TCA mixed with carrier, analytical methods:

   1.1. Determine uniformity, and determine periodically, concentration of TCA in mixture ☐ ☐ ☐ — 6, 6.2(2) & (5)

   1.2. Determine stability of TCA in mixture ☐ ☐ ☐ — 6, 6.2(2) & (5)

PREPLACEMENT/ MAINTENANCE	Response			References	
	Yes	No	N/A	21 CFR	OECD
2. Stability of TCA in mixture determined either before initiation,	☐	☐	☐		6, 6.2(5)
or					
3. Concomitantly per SOPs providing for periodic analysis of TCA in the mixture	☐	☐	☐		6, 6.2(5)
T. Protocol for the Conduct of a Nonclinical Laboratory Study				**58.120**	
U. Conduct of a Nonclinical Laboratory Study				**58.130**	
1. Data recorded directly, promptly, legibly in ink	☐	☐	☐		8, 8.3(3)
2. Entry dated on date of entry, signed by same person	☐	☐	☐		8, 8.3(3)
3. Changes do not obscure, reason given, dated, and signed at time of change	☐	☐	☐		8, 8.3(4)
4. Individual responsible for automated data entries identified at time of data input	☐	☐	☐		8, 8.3(5)
5. Changes in automated entries do not obscure, reason given, dated, responsible individual identified	☐	☐	☐		8, 8.3(5)
V. Reporting of Nonclinical Laboratory Study Results				**58.185**	
W. Storage and Retrieval of Records and Data				**58.190**	
1. Raw data, documentation, protocols, specimens, and final reports retained	☐	☐	☐		10, 10.1(a)

		Response			References	
PREPLACEMENT/ MAINTENANCE		Yes	No	N/A	21 CFR	OECD

2.	Storage, retrieval of raw data, protocols, specimens, and reports	❏	❏	❏		3, 3.4 10, 10.2
3.	Deterioration minimized by conditions of storage	❏	❏	❏		3, 3.4
4.	Reference in archive to any contracted storage	❏	❏	❏		
5.	Individual identified as responsible for archive	❏	❏	❏		1, 1.1(2)(l)
6.	Entry to archive by authorized personnel only	❏	❏	❏		10, 10.3
7.	Archived material indexed to permit expedient retrieval	❏	❏	❏		10, 10.2

X.	**Retention of Records**				**58.195**	
1.	Except for wet specimens, samples of TCA, the documentation records, raw data, and specimens retained in archive are retained as required:					
	1.1. At least 2 years following NDA approval	❏	❏	❏		10, 10.1(a)–(g)
	1.2. At least 5 years following date on which study submitted in support of an NDA	❏	❏	❏		10, 10.1(a)–(g)
	1.3. Other situations at least 2 years following date study completed, terminated, or discontinued	❏	❏	❏		10, 10.1(a)–(g)

PREPLACEMENT/ MAINTENANCE	Response			References	
	Yes	No	N/A	21 CFR	OECD

2.	Wet specimens, samples of TCA retained as long as quality of preparation can be evaluated	❏	❏	❏		10, 10.1("h")
3.	Master schedule, protocols, QAU records retained by QAU as easily accessible for time applicable	❏	❏	❏		10, 10.1(b)
4.	Summaries of training, experience, job descriptions, testing facility employment records retained for time applicable	❏	❏	❏		10, 10.1(c)
5.	Records, reports of maintenance, calibration, and inspection of equipment retained for time applicable	❏	❏	❏		10, 10.1(d)
6.	Records of this part retained as original or true copies (e.g., photocopies, microfilm, microfiche) or other accurate reproduction of original records	❏	❏	❏		
7.	Materials specified above transferred to sponsor should contracted facility discontinue business	❏	❏	❏		10, 10.4
8.	FDA notified in writing should No. 7 occur	❏	❏	❏		

Observations/Comments

_____
_____
_____
_____
_____
_____
_____
_____
_____
_____
_____
_____
_____
_____
_____
_____
_____
_____
_____
_____
_____
_____
_____
_____
_____
_____
_____
_____
_____

Inspected/Audited by: _____ **Date:** _____

INITIATION PHASE ABBREVIATED GLP AUDIT CHECKLIST

The *Initiation Phase Abbreviated GLP Audit Checklist* illustrated here includes the regulatory parts listed below, except for those designated with an "*," which are excluded. Those excluded from this checklist are included in others that are considered more appropriate. In some instances, because of the type of study, or management's preference for increased redundancy, additional regulatory parts may be amended to form a more suitable checklist, unique to a specific inspection or auditing requirement.

    A. General Provisions

    B. Personnel*

    C. Testing Facility Management

    D. Study Director*

    E. Quality Assurance Unit*

    F. Facilities*

    G. Animal Care Facilities*

    H. Animal Supply Facilities*

    I. Facilities for Handling Test and Control Articles*

    J. Laboratory Operation Areas*

    K. Specimen and Data Storage Facilities*

    L. Equipment Design*

    M. Maintenance and Calibration of Equipment*

    N. Standard Operating Procedures

    O. Reagents and Solutions*

    P. Animal Care*

    Q. Test and Control Article Characterization

    R. Test and Control Article Handling

    S. Mixtures of Articles with Carriers

    T. Protocol for and Conduct of a Nonclinical Laboratory Study

    U. Conduct of a Nonclinical Laboratory Study—Baseline

    V. Reporting of Nonclinical Laboratory Study Results*

    W. Storage and Retrieval of Records and Data*

    X. Retention of Records*

This is the first of four critical phases designated by the Quality Assurance Unit (QAU) for inspection during a nonclinical study. All checklists have references to the appropriate part of the Code of Federal Regulation, 21 CFR, Part 58, Good Laboratory Practice, and the OECD Principles of Good Laboratory Practice, for cross-referencing, allowing one to switch from the abbreviated checklists to the specific questions of the *GLP Audit Checklist—Complete*, or to the official regulatory document itself.

The checklists are retained by some organizations; on each list, space has been provided for the inspector/auditor signature and date. The regulations do require completion and retention of the Study Inspection Record and Inspection Report, examples of which are found in Section III. These documents provide the necessary signed records of each periodic inspection [58.35(b)(3)].

GLP Audit Manual ©2000, Interpharm Press

INITIATION PHASE ABBREVIATED GLP AUDIT CHECKLIST

AUDITING HISTORY: PHASE I

Facility Location

Name: _____

Address: _____

Date(s) of Inspection Auditing: _____

Protocol

Title: _____

Number: _____ Date Issued: _____

Scientific Report

Number: _____ Date Issued: _____

Assigned Inspector/Auditor: _____

Title: _____

Alternate Inspectors/Auditors: _____

Name: _____

Inspection Reports

Number: _____ Date Issued: _____ Date Returned: _____

Number: _____ Date Issued: _____ Date Returned: _____

Number: _____ Date Issued: _____ Date Returned: _____

Phase Study Inspection Record Completed: _____ Date: _____

Quality Assurance Unit Statement

Issued: _____ Date: _____

Discussion Sessions

Attendees: _____ Date: _____

Study Director: _____

Comments

_____

_____

_____

_____

_____

_____

_____

_____

_____

_____

_____

_____

_____

_____

_____

_____

_____

_____

_____

_____

_____

_____

GLP Audit Manual ©2000, Interpharm Press

INITIATION PHASE	Response			References	
	Yes	No	N/A	21 CFR	OECD

A. General Provisions

1. Consultant laboratory, contractor, grantee notified of nonclinical study, conduct per GLP

| | ❒ | ❒ | ❒ | 58.10 | |

2. Laboratory permits inspections of facilities, records, specimens required

| | ❒ | ❒ | ❒ | 58.15 | |

B. Personnel   58.29

C. Testing Facility Management   58.31

1. Assure proper tests of TCA or TCA mixtures

| | ❒ | ❒ | ❒ | | 1, 1.1(2)(p) |

2. Assure resources as scheduled

| | ❒ | ❒ | ❒ | | 1, 1.1(2)(b) & (e) |

3. Assure deviations communicated, documented

| | ❒ | ❒ | ❒ | | 2, 2.2(1)(e) |

D. Study Director   58.33

E. Quality Assurance Unit   58.35

F. Facilities   58.41

G. Animal Care Facilities   58.43

H. Animal Supply Facilities   58.45

I. Facilities for Handling Test and Control Articles   58.47

J. Laboratory Operation Areas   58.49

K. Specimen and Data Storage Facilities   58.51

| INITIATION PHASE | Response | | | References | |
	Yes	No	N/A	21 CFR	OECD
L. **Equipment Design**				**58.61**	
M. **Maintenance and Calibration of Equipment**				**58.63**	
N. **Standard Operating Procedures**				**58.81**	
1. SOPs written to ensure data quality, integrity	❏	❏	❏		7, 7.1
2. Changes in SOPs authorized	❏	❏	❏		7, 7.1
O. **Reagents and Solutions**				**58.83**	
P. **Animal Care**				**58.90**	
Q. **Test and Control Article Characterization**				**58.105**	
1. Identity, strength, purity, etc., documented	❏	❏	❏		6, 6.2(1) & (2)
2. Methods of derivation documented	❏	❏	❏		
3. Marketed products characterized by labeling	❏	❏	❏		
4. Stabilizes before initiation,	❏	❏	❏		6, 6.2(4) & (5)
or					
5. Concomitantly per SOPs providing for periodic analysis of each batch	❏	❏	❏		6, 6.2(4) & (5)
6. Storage containers: name, abstract or code number, batch, expiration date, storage conditions	❏	❏	❏		6, 6.1(3)

		Response			References	
INITIATION PHASE		Yes	No	N/A	**21 CFR**	**OECD**

					21 CFR	**OECD**
R.	**Test and Control Article Handling**				**58.107**	
1.	SOPs ensure receipt, distribution, dates, quantity returned	❏	❏	❏		6, 6.1(1)
S.	**Mixtures of Articles with Carriers**				**58.113**	
1.	Stability of TCA in mixture determined either before initiation,	❏	❏	❏		6, 6.2(5)
	or					
2.	Concomitantly per SOPs providing for periodic analysis of TCA in the mixture	❏	❏	❏		6, 6.2(5)
3.	Component expirations clearly shown	❏	❏	❏		6, 6.1(3)
4.	Earliest expiration shown	❏	❏	❏		6, 6.1(3)
T.	**Protocol for and Conduct of a Nonclinical Laboratory Study**				**58.120**	
1.	Approved, indicating objectives and methods	❏	❏	❏		8, 8.1(1)
2.	Protocol contains:					
	2.1. Title and purpose	❏	❏	❏		8, 8.2(1)(a) & (b)
	2.2. TCA name, chemical abstracts number, code number	❏	❏	❏		8, 8.2(1)(c)
	2.3. Sponsor name, test facility name and address	❏	❏	❏		8. 8.2(2)(a) & (b)
	2.4. Number, body weight, sex, source, species, strain, age	❏	❏	❏		8, 8.2(5)(b)

INITIATION PHASE	Response			References	
	Yes	No	N/A	21 CFR	OECD
2.5. Procedure for I.D. of test system	☐	☐	☐		8, 8.2(5)(b)
2.6. Experimental design, method, control of bias	☐	☐	☐		8, 8.2(5)(e)
2.7. I.D. of diet, solvent, materials for suspending	☐	☐	☐		
2.8. Specs acceptable level of contaminants in diet	☐	☐	☐		8, 8.2(5)(e)
2.9. Reason for route choice	☐	☐	☐		8, 8.2(5)(c)
2.10. Dosage levels in units, method, frequency	☐	☐	☐		8, 8.2(5)d
2.11. Test, frequency, analyses, measurements	☐	☐	☐		8, 8.2(5)(e)
2.12. Records maintained	☐	☐	☐		8, 8.2(6)
2.13. Sponsor approval date, dated signature of study director	☐	☐	☐		8, 8.2(3)(a)
2.14. Proposed statistical methods	☐	☐	☐		8, 8.2(5)(e)
3. Changes, reasons dated, signed by study director	☐	☐	☐		8, 8.1(2)(a)
4. Protocol and study schedule match, including:					
4.1. Arrival, exams, isolation, acclimation	☐	☐	☐		
4.2. Baseline, exams, body weight, food consumption, selection, randomization observations, specimen collections, analyses, etc.	☐	☐	☐		

INITIATION PHASE

		Response			References	
		Yes	No	N/A	21 CFR	OECD

		Yes	No	N/A	21 CFR	OECD
4.3.	Dosing, observations, body weight, food consumption, exams, collections, analyses, necropsy, etc.	❐	❐	❐		
U.	**Conduct of a Nonclinical Laboratory Study—Baseline**				**58.130**	
1.	Conducted according to protocol	❐	❐	❐		8, 8.3(2)
2.	Monitored in conformity with protocol	❐	❐	❐		1, 1.2(2)(e), 2, 2.2(1)(c)
3.	Specimens identified by test system, study, nature, date of collection	❐	❐	❐		8, 8.3(1)
4.	Specimen I.D. secure	❐	❐	❐		8, 8.3(1)
5.	Data recorded direct, promptly, legibly, in ink	❐	❐	❐		8, 8.3(3)
6.	Entry dated on date of entry, signed by same person	❐	❐	❐		8, 8.3(3)
7.	Changes proper, reason, dated, initialed	❐	❐	❐		8, 8.3(4)
8.	Individual responsible for automated data entries identified at time of input	❐	❐	❐		8, 8.3(5)
9.	Changes in automated entries do not obscure reason given, dated, responsible person identified	❐	❐	❐		8, 8.3(5)
10.	Raw data documentation includes such items:					
10.1.	Animal purchase orders, transfers, health certificates, immunizations, isolation, sex verification, physicals, animal I.D., randomizations, drug lot approval, clinical pathology	❐	❐	❐		

| | | | ABBREVIATED
GLP AUDIT CHECKLIST | | |
| | | Response | | References | |
INITIATION PHASE	Yes	No	N/A	21 CFR	OECD

V.	**Reporting of Nonclinical Laboratory Study Results**				58.185	
W.	**Storage and Retrieval of Records and Data**				58.190	
X.	**Retention of Records**				58.195	

Observations/Comments

_____

_____

_____

_____

_____

_____

_____

_____

_____

_____

_____

_____

_____

_____

_____

_____

_____

_____

_____

Inspected/Audited by: _____ **Date:** _____

GLP Audit Manual ©2000, Interpharm Press

IN–PROCESS PHASE ABBREVIATED GLP AUDIT CHECKLIST

The *In-Process Phase Abbreviated GLP Audit Checklist* illustrated here includes the regulatory parts listed below, except for those designated with an "*," which are excluded. Those excluded from this checklist are included in others that are considered more appropriate. In some instances, because of the type of study, or management's preference for increased redundancy, additional regulatory parts may be amended to form a more suitable checklist, unique to a specific inspection or auditing requirement.

A. General Provisions

B. Personnel

C. Testing Facility Management

D. Study Director*

E. Quality Assurance Unit*

F. Facilities

G. Animal Care Facilities

H. Animal Supply Facilities*

I. Facilities for Handling Test and Control Articles

J. Laboratory Operation Areas*

K. Specimen and Data Storage Facilities*

L. Equipment Design

M. Maintenance and Calibration of Equipment

N. Standard Operating Procedures

O. Reagents and Solutions

P. Animal Care

Q. Test and Control Article Characterization

R. Test and Control Article Handling

S. Mixtures of Articles with Carriers

T. Protocol for the Conduct of a Nonclinical Laboratory Study

U. Conduct of a Nonclinical Laboratory Study

V. Reporting of Nonclinical Laboratory Study Results*

W. Storage and Retrieval of Records and Data*

X. Retention of Records*

This is the second of four critical phases designated by the Quality Assurance Unit (QAU) for inspection during a nonclinical study. All checklists have references to the appropriate part of the Code of Federal Regulation, 21 CFR, Part 58, Good Laboratory Practice, and the OECD Principles of Good Laboratory Practice, for cross-referencing, allowing one to switch from the abbreviated checklists to the specific questions of the *GLP Audit Checklist—Complete,* or to the official regulatory document itself.

The checklists are retained by some organizations; on each list, space has been provided for the inspector/auditor signature and date. The regulations do require completion and retention of the Study Inspection Record and Inspection Report, examples of which are found in Section III. These documents provide the necessary signed records of each periodic inspection [58.35(b)(3)].

IN–PROCESS PHASE ABBREVIATED GLP AUDIT CHECKLIST

INSPECTION HISTORY: PHASE II

Facility Location

Name: _____

Address: _____

Date(s) of Inspection Auditing: _____

Protocol

Title: _____

Number: _____ Date Issued: _____

Scientific Report

Number: _____ Date Issued: _____

Assigned Inspector/Auditor: _____

Title: _____

Alternate Inspectors/Auditors: _____

Name: _____

Inspection Reports

Number: _____ Date Issued: _____ Date Returned: _____

Number: _____ Date Issued: _____ Date Returned: _____

Number: _____ Date Issued: _____ Date Returned: _____

Phase Study Inspection Record Completed: _____ Date: _____

Quality Assurance Unit Statement

Issued: _____   Date: _____

Discussion Sessions

Attendees: _____   Date: _____

Study Director: _____

Comments

_____

_____

_____

_____

_____

_____

_____

_____

_____

_____

_____

_____

_____

_____

_____

_____

_____

_____

_____

				ABBREVIATED GLP AUDIT CHECKLIST	
	Response			References	
IN–PROCESS PHASE	Yes	No	N/A	**21 CFR**	**OECD**

A. General Provisions

1. Consultant laboratory, contractor, grantee, notified of nonclinical study, contract per GLP	❑	❑	❑	58.10	
2. Laboratory permits inspections of facilities, records, specimens required	❑	❑	❑	58.15	

B. Personnel — 58.29

1. Education, training, experience for assignments	❑	❑	❑		1, 1.1(2)(b)
2. Summaries, training, experience, job descriptions	❑	❑	❑		1, 1.1(2)(c)
3. Sanitation, health precautions	❑	❑	❑		1, 1.4(4)
4. Appropriate clothing worn, proper changes made	❑	❑	❑		
5. Illness adversely affecting test systems, TCA, excluded until corrected	❑	❑	❑		1, 1.4(4)
6. Report health or medical conditions that may adversely affect study	❑	❑	❑		1, 1.4(4)

C. Testing Facility Management — 58.31

1. Assures required tests	❑	❑	❑		1, 1.1(2)(p)
2. Assures resources and methodologies as scheduled	❑	❑	❑		1, 1.1(2)(b) & (e)
3. Assures personnel understand functions	❑	❑	❑		1, 1.1(2)(d)

	Response			References	
IN–PROCESS PHASE	Yes	No	N/A	21 CFR	OECD
4. Assures deviations communicated, documented	❏	❏	❏		2, 2.2(1)(e)
D. **Study Director**				**58.33**	
E. **Quality Assurance Unit**				**58.35**	
F. **Facilities**				**58.41**	
1. Provide separation, preventing adverse effect	❏	❏	❏		3, 3.1(2)
G. **Animal Care Facilities**				**58.43**	
1. Sufficient rooms, areas, to assure proper:					
1.1. Separation of species or test systems	❏	❏	❏		3, 3.2(1)
1.2. Isolation of projects	❏	❏	❏		3, 3.2(1)
1.3. Quarantine	❏	❏	❏		3, 3.2(2) 5, 5.2(2)
1.4. Routine or specialized housing	❏	❏	❏		3, 3.2(2)
2. Separate rooms ensuring isolation for studies with biohazards, radioactives, infectious agents	❏	❏	❏		3, 3.2(1)
3. Separate areas for diagnosis, treatment of diseases	❏	❏	❏		3, 3.2(2)
4. Proper collection, disposal of waste	❏	❏	❏		3, 3.5
5. Vermin, odors, disease hazards minimized	❏	❏	❏		3, 3.5
6. Designed, located to minimize interference	❏	❏	❏		3, 3.1(1)
H. **Animal Supply Facilities**				**58.45**	

IN–PROCESS PHASE	Response			References	
	Yes	No	N/A	21 CFR	OECD

I. Facilities for Handling Test and Control Articles   58.47

1. To prevent contamination, mix-ups, separate areas for:

	Yes	No	N/A	21 CFR	OECD
1.1. Receipt, storage of test/control articles (TCA)	❑	❑	❑		3, 3.3(1)
1.2. Mixing of TCA with carrier	❑	❑	❑		3, 3.3(1)
1.3. Storage of TCA mixtures	❑	❑	❑		3, 3.3(2)
2. Storage TCA/mixtures separate from test system	❑	❑	❑		3, 3.3(2)
3. Adequate to preserve identity, strength, purity, stability of TCA/mixtures	❑	❑	❑		3, 3.3(2)

J. Laboratory Operation Areas   58.49

K. Specimen and Data Storage Facilities   58.51

L. Equipment Design   58.61

	Yes	No	N/A	21 CFR	OECD
1. Equipment used in generation, measurement, assessment, environmental control, appropriate design, capacity to function according to protocol	❑	❑	❑		4(1) 5, 5.1(1)
2. Suitably located for operation, inspection, cleaning, maintenance	❑	❑	❑		4(1)

IN–PROCESS PHASE

		Response			References	
		Yes	No	N/A	21 CFR	OECD
M.	**Maintenance and Calibration of Equipment**				**58.63**	
1.	Adequately inspected, cleaned, maintained	❑	❑	❑		4(1)
2.	Adequately tested, calibrated/standardized	❑	❑	❑		4(2)
3.	SOPs provide sufficient detail of methods, materials, schedules for routine maintenance, calibration	❑	❑	❑		4(2)
4.	SOPs address when to take appropriate remedial action, if malfunction	❑	❑	❑		
5.	SOPs identify responsible person	❑	❑	❑		
6.	Written records of inspections, maintenance, testing, calibration, and/or standardization	❑	❑	❑		4(2)
7.	Records contain whether routine, followed SOPs	❑	❑	❑		
8.	Records maintained of nonroutine repairs	❑	❑	❑		
9.	Records document defect, discovery, remedy	❑	❑	❑		
N.	**Standard Operating Procedures**				**58.81**	
1.	SOPs written to ensure data quality, integrity	❑	❑	❑		7, 7.1
2.	Deviations authorized by study director	❑	❑	❑		1, 1.2(2)(e)
3.	Deviations documented	❑	❑	❑		7, 7.3
4.	Changes in SOPs authorized	❑	❑	❑		7, 7.1
5.	Relevant SOPs, manuals immediately available	❑	❑	❑		7, 7.2
6.	Literature used as supplement to SOPs, not in lieu of	❑	❑	❑		7, 7.2

IN–PROCESS PHASE	Response			References	
	Yes	No	N/A	21 CFR	OECD

O. Reagents and Solutions | | | | **58.83** |

		Yes	No	N/A		OECD
1.	Labeled for identity, concentration, storage, expiration	❐	❐	❐		4(4) 7, 7.4(2)(c)
2.	Deteriorated, expired reagents not used	❐	❐	❐		4(4)

P. Animal Care | | | | **58.90** |

		Yes	No	N/A		OECD
1.	At initiation, free of disease or condition that might interfere	❐	❐	❐		5, 5.2(2)
2.	During study, diseased animals isolated	❐	❐	❐		5, 5.2(2)
3.	Treatment does not interfere	❐	❐	❐		5, 5.2(2)
4.	Diagnosis, treatment authorization documented	❐	❐	❐		5, 5.2(2)
5.	Proper identification	❐	❐	❐		5, 5.2(5)
6.	Required returns receive appropriate I.D.	❐	❐	❐		5, 5.2(5)
7.	All specific I.D. information on outside of housing unit	❐	❐	❐		5, 5.2(5)
8.	Different species in separate rooms, if necessary	❐	❐	❐		3, 3.2(1) 5, 5.2(1)
9.	Different studies not housed in same room	❐	❐	❐		3, 3.2(1) 5, 5.2(1)
10.	If mixed housing, adequate differentiation	❐	❐	❐		3, 3.2(1) 5, 5.2(1)
11.	Adequate cleaning, sanitizing, of cages, racks, etc.	❐	❐	❐		5, 5.2(6)

IN–PROCESS PHASE	Response			References	
	Yes	No	N/A	21 CFR	OECD

12.	Periodic analyses of feed, water for contaminants	☐	☐	☐		5, 5.2(6)
13.	Bedding not interfering	☐	☐	☐		5, 5.2(6)
14.	Bedding changes adequate	☐	☐	☐		5, 5.2(6)
15.	Pest control materials documented	☐	☐	☐		5, 5.2(6)
16.	Cleaning, pest materials that interfere not used	☐	☐	☐		5, 5.2(6)

Q.	**Test and Control Article Characterization**				58.105	
1.	Stability determined concomitantly per SOPs providing for periodic analysis of each batch if not determined before initiation of study	☐	☐	☐		6, 6.2(4) & (5)
2.	SOPs provided, periodic reanalysis of stabilities	☐	☐	☐		6, 6.2(4)
3.	Storage containers: name, abstract or code number, batch, expiration date, storage conditions	☐	☐	☐		6, 6.1(3)
4.	Storage containers assigned for study duration	☐	☐	☐		

R.	**Test and Control Article Handling**				58.107	
	SOPs for handling TCA to ensure:					
1.	Proper storage	☐	☐	☐		6, 6.1(2)
2.	Distribution that precludes change, damage	☐	☐	☐		6, 6.1(2)
3.	Maintenance of identification during distribution	☐	☐	☐		6, 6.1(2)

GLP Audit Manual ©2000, Interpharm Press

IN–PROCESS PHASE	Response Yes	No	N/A	References 21 CFR	OECD

4. Documentation of receipt, distribution, return, each batch ☐ ☐ ☐ — 6, 6.1(1)

S. Mixtures of Articles with Carriers — 58.113

1. TCA mixed with carrier, analytical methods:

   1.1. Determine uniformity, and determine periodically, concentration TCA in mixture ☐ ☐ ☐ — 6, 6.2(2) & (5)

   1.2. Determine stability of TCA mixture ☐ ☐ ☐ — 6, 6.2(2) & (5)

2. Stability of TCA in mixture determined either before initiation ☐ ☐ ☐ — 6, 6.2(5)

   or

3. Concomitantly per SOPs providing for periodic analysis of TCA in the mixture ☐ ☐ ☐ — 6, 6.2(5)

4. Component expirations clearly shown ☐ ☐ ☐ — 6, 6.1(3)

5. Earliest expiration shown ☐ ☐ ☐ — 6, 6.1(3)

T. Protocol for the Conduct of a Nonclinical Laboratory Study — 58.120

1. Revisions, reasons documented and signed by director ☐ ☐ ☐ — 8, 8.1(2)(a)

U. Conduct of a Nonclinical Laboratory Study — 58.130

1. Conducted according to protocol ☐ ☐ ☐ — 8, 8.3(2)

2. Monitored in conformity with protocol ☐ ☐ ☐ — 1, 1.2(2)(e) 2, 2.2(1)(c)

		ABBREVIATED GLP AUDIT CHECKLIST				
		Response			**References**	
IN–PROCESS PHASE		Yes	No	N/A	21 CFR	OECD
3. Specimens identified by test system, study, nature, date of collection		❐	❐	❐		8, 8.3(1)
4. Information located, precluding record, storage error		❐	❐	❐		8, 8.3(1)
5. Gross findings available to pathologist examining specimen histopathologically		❐	❐	❐		8, 8.3(3) 10. 10.1(a)
6. Data recorded directly, promptly, legibly, in ink		❐	❐	❐		8, 8.3(3)
7. Entry dated on the date of entry, signed by same person		❐	❐	❐		8, 8.3(3)
8. Changes do not obscure, reason given, dated and signed at time of change		❐	❐	❐		8, 8.3(4)
9. Individual responsible for automated data entries identified at time of input		❐	❐	❐		8, 8.3(5)
10. Changes in automated entries do not obscure, reason given, dated, responsible individual identified		❐	❐	❐		8, 8.3(5)
V. Reporting of Nonclinical Laboratory Study Results					**58.185**	
W. Storage and Retrieval of Records and Data					**58.190**	
X. Retention of Records					**58.195**	

GLP Audit Manual ©2000, Interpharm Press

Observations/Comments

_____

_____

_____

_____

_____

_____

_____

_____

_____

_____

_____

_____

_____

_____

_____

_____

_____

_____

_____

_____

_____

_____

_____

_____

_____

Inspected/Audited by: _____ **Date:** _____

COMPLETION PHASE ABBREVIATED GLP AUDIT CHECKLIST

The *Completion Phase Abbreviated GLP Audit Checklist* illustrated here includes the regulatory parts listed below, except for those designated with an "*," which are excluded. Those excluded from this checklist are included in others that are considered more appropriate. In some instances, because of the type of study, or management's preference for increased redundancy, additional regulatory parts may be amended to form a more suitable checklist, unique to a specific inspection or auditing requirement.

A. General Provisions

B. Personnel

C. Testing Facility Management

D. Study Director*

E. Quality Assurance Unit*

F. Facilities

G. Animal Care Facilities*

H. Animal Supply Facilities*

I. Facilities for Handling Test and Control Articles*

J. Laboratory Operation Areas

K. Specimen and Data Storage Facilities*

L. Equipment Design

M. Maintenance and Calibration of Equipment

N. Standard Operating Procedures

O. Reagents and Solutions

P. Animal Care*

Q. Test and Control Article Characterization*

R. Test and Control Article Handling*

S. Mixture of Articles with Carriers*

T. Protocol for and Conduct of a Nonclinical Laboratory Study

U. Conduct of a Nonclinical Laboratory Study

V. Reporting of Nonclinical Laboratory Study Results*

W. Storage and Retrieval of Records and Data*

X. Retention of Records*

This is the third of four critical phases designated by the Quality Assurance Unit (QAU) for inspection during a nonclinical study. All checklists have references to the appropriate part of the Code of Federal Regulation, 21 CFR, Part 58, Good Laboratory Practice, and the OECD Principles of Good Laboratory Practice, for cross-referencing, allowing one to switch from the abbreviated checklists to the specific questions of the *GLP Audit Checklist—Complete,* or to the official regulatory document itself.

The checklists are retained by some organizations; on each list, space has been provided for the inspector/auditor signature and date. The regulations do require completion and retention of the Study Inspection Record and Inspection Report, examples of which are found in Section III. These documents provide the necessary signed records of each periodic inspection [58.35(b)(3)].

COMPLETION PHASE ABBREVIATED GLP AUDIT CHECKLIST

INSPECTION HISTORY: PHASE III

Facility Location

Name: _____

Address: _____

Date(s) of Inspection Auditing: _____

Protocol

Title: _____

Number: _____   Date Issued: _____

Scientific Report

Number: _____   Date Issued: _____

Assigned Inspector/Auditor: _____

Title: _____

Alternate Inspectors/Auditors: _____

Name: _____

Inspection Reports

Number: _____   Date Issued: _____   Date Returned: _____

Number: _____   Date Issued: _____   Date Returned: _____

Number: _____   Date Issued: _____   Date Returned: _____

Phase Study Inspection Record Completed: _____   Date:_____

Quality Assurance Unit Statement

Issued: _____ Date: _____

Discussion Sessions

Attendees: _____ Date: _____

Study Director: _____

Comments

_____

_____

_____

_____

_____

_____

_____

_____

_____

_____

_____

_____

_____

_____

_____

_____

_____

_____

_____

_____

_____

_____

GLP Audit Manual ©2000, Interpharm Press

	Response			References	
COMPLETION PHASE	**Yes**	**No**	**N/A**	**21 CFR**	**OECD**

A. General Provisions

1. Consultant laboratory, contractor, grantee notified of nonclinical study conduct per GLP — ☐ ☐ ☐ — 58.10

2. Laboratory permits inspections of facilities, records, specimens required — ☐ ☐ ☐ — 58.15

B. Personnel — 58.29

1. Education, training, experience for assignments — ☐ ☐ ☐ — — 1, 1.1(2)(b)

2. Summaries, training, experience, job descriptions — ☐ ☐ ☐ — — 1, 1.1(2)(c)

3. Appropriate clothing worn, changes properly made — ☐ ☐ ☐

C. Testing Facility Management — 58.31

1. Assuring resources and methodologies as scheduled — ☐ ☐ ☐ — — 1, 1.1(2)(b) & (e)

2. Assuring personnel understand functions — ☐ ☐ ☐ — — 1, 1.1(2)(d)

3. Assuring deviations communicated, documented — ☐ ☐ ☐ — — 2, 2.2(1)(c)

D. Study Director — 58.33

E. Quality Assurance Unit — 58.35

F. Facilities — 58.41

1. Provide separation preventing adverse effect — ☐ ☐ ☐ — — 3, 3.1(2)

COMPLETION PHASE	Response			References	
	Yes	No	N/A	21 CFR	OECD

G.	Animal Care Facilities				58.43	
H.	Animal Supply Facilities				58.45	
I.	Facilities for Handling Test and Control Articles				58.47	
J.	Laboratory Operation Areas				58.49	
1.	Separate space provided for routine and specialized procedures	☐	☐	☐		
K.	Specimen and Data Storage Facilities				58.51	
L.	Equipment Design				58.61	
1.	Equipment used for generation, measurement, assessment, environmental control, appropriate design, capacity to function according to protocol	☐	☐	☐		4(1) 5, 5.1(1)
2.	Suitably located for operation, inspection, cleaning, maintenance	☐	☐	☐		4(1)
M.	Maintenance and Calibration of Equipment				58.63	
1.	Adequately inspected, cleaned, maintained	☐	☐	☐		4(1)
2.	SOPs provide sufficient detail of methods, materials, schedules, for routine maintenance, calibration	☐	☐	☐		4(2)
3.	SOPs address when to take appropriate remedial action, if malfunction	☐	☐	☐		
4.	SOPs identify responsible person	☐	☐	☐		

GLP Audit Manual ©2000, Interpharm Press

COMPLETION PHASE	Response			References	
	Yes	No	N/A	21 CFR	OECD
5. Records written of inspections, maintenance, etc.	☐	☐	☐		4(2)
6. Records contain whether routine, followed SOPs	☐	☐	☐		4(2)
7. Records maintained of nonroutine repairs	☐	☐	☐		
8. Records document defect, discovery, remedy	☐	☐	☐		
N. Standard Operating Procedures				**58.81**	
1. SOPs written to ensure data quality, integrity	☐	☐	☐		7, 7.1
2. Deviations authorized by study director	☐	☐	☐		1, 1.2(2)(e)
3. Deviations documented	☐	☐	☐		7, 7.3
4. Changes in SOPs authorized	☐	☐	☐		7, 7.1
5. Relevant SOPs, manuals immediately available	☐	☐	☐		7, 7.2
6. Literature used as supplement to SOPs, not in lieu of	☐	☐	☐		7, 7.2
O. Reagents and Solutions				**58.83**	
1. Labeled for identity, concentration, storage, expiration	☐	☐	☐		4(4) 7, 7.4(2)(c)
2. Deteriorated, expired reagents not used	☐	☐	☐		4(4)
P. Animal Care				**58.90**	
Q. Test and Control Article Characterization				**58.105**	

COMPLETION PHASE	Response			References	
	Yes	No	N/A	21 CFR	OECD
R. **Test and Control Article Handling**				58.107	
S. **Mixture of Articles with Carriers**				58.113	
T. **Protocol for and Conduct of a Nonclinical Laboratory Study**				58.120	
1. Revisions, reasons documented and signed by director	❐	❐	❐		8, 8.3(2)
U. **Conduct of a Nonclinical Laboratory Study**				58.130	
1. Conducted according to protocol	❐	❐	❐		8, 8.3(2)
2. Monitored in conformity with protocol	❐	❐	❐		1, 1.2(2)(e) 2, 2.2(1)(c)
3. Specimens identified by test system, study, nature, date of collection	❐	❐	❐		8, 8.3(1)
4. Information located, precluding record, storage error	❐	❐	❐		8, 8.3(1)
5. Gross findings available to pathologist examining specimen histopathologically	❐	❐	❐		8, 8.3(3) 10, 10.1(a)
6. Data recorded directly, promptly, legibly, in ink	❐	❐	❐		8, 8.3(3)
7. Entry dated on date of entry, signed by same person	❐	❐	❐		8, 8.3(3)
8. Changes do not obscure, reason given, dated and signed at time of change	❐	❐	❐		8, 8.3(4)

	COMPLETION PHASE	Response			References	
		Yes	No	N/A	21 CFR	OECD
9.	Individual responsible for automated data entries identified at time of input	❏	❏	❏		8, 8.3(5)
10.	Changes in automated entries do not obscure, reason given, dated, responsible individual identified	❏	❏	❏		8, 8.3(5)
V.	**Reporting of Nonclinical Laboratory Study Results**				**58.185**	
W.	**Storage and Retrieval of Records and Data**				**58.190**	
X.	**Retention of Records**				**58.195**	

Observations/Comments

_____

_____

_____

_____

_____

_____

_____

_____

_____

_____

_____

_____

_____

_____

Inspected/Audited by: _____ Date: _____

REPORTING PHASE ABBREVIATED GLP AUDIT CHECKLIST

The *Reporting Phase Abbreviated GLP Audit Checklist* illustrated here includes the regulatory parts listed below, except for those designated with an "*," which are excluded. Those excluded from this checklist are included in others that are considered more appropriate. In some instances, because of the type of study, or management's preference for increased redundancy, additional regulatory parts may be amended to form a more suitable checklist, unique to a specific inspection or auditing requirement.

A. General Provisions

B. Personnel

C. Testing Facility Management

D. Study Director

E. Quality Assurance Unit

F. Facilities*

G. Animal Care Facilities*

H. Animal Supply Facilities*

I. Facilities for Handling Test and Control Articles

J. Laboratory Operation Areas*

K. Specimen and Data Storage Facilities

L. Equipment Design

M. Maintenance and Calibration of Equipment

N. Standard Operating Procedures

O. Reagents and Solutions

P. Animal Care

Q. Test and Control Article Characterization

R. Test and Control Article Handling

S. Mixtures of Articles with Carriers

T. Protocol for and Conduct of a Nonclinical Laboratory Study

U. Conduct of a Nonclinical Laboratory Study

V. Reporting of Nonclinical Laboratory Study Results

W. Storage and Retrieval of Records and Data

X. Retention of Records

This is the fourth of four critical phases designated by the Quality Assurance Unit (QAU) for inspection during a nonclinical study. All checklists have references to the appropriate part of the Code of Federal Regulations, 21 CFR, Part 58, Good Laboratory Practice, and the OECD Principles of Good Laboratory Practice, for cross-referencing, allowing one to switch from the abbreviated checklists to the specific questions of the *GLP Audit Checklist—Complete,* or to the official regulatory document itself.

The checklists are retained by some organizations; on each list, space has been provided for the inspector/auditor signature and date. The regulations do require completion and retention of the Study Inspection Record and Inspection Report, examples of which are found in Section III. These documents provide the necessary signed records of each periodic inspection [58.35(b)(3)].

REPORTING PHASE ABBREVIATED GLP AUDIT CHECKLIST

AUDITING HISTORY: PHASE IV

Facility Location

Name: _____

Address: _____

Date(s) of Inspection Auditing: _____

Protocol

Title: _____

Number: _____ Date Issued: _____

Scientific Report

Number: _____ Date Issued: _____

Assigned Inspector/Auditor: _____

Title: _____

Alternate Inspectors/Auditors: _____

Name: _____

Inspection Reports

Number: _____ Date Issued: _____ Date Returned: _____

Number: _____ Date Issued: _____ Date Returned: _____

Number: _____ Date Issued: _____ Date Returned: _____

Phase Study Inspection Record Completed: _____ Date:_____

Quality Assurance Unit Statement

Issued: _____   Date: _____

Discussion Sessions

Attendees: _____   Date: _____

Study Director: _____

Comments

_____

_____

_____

_____

_____

_____

_____

_____

_____

_____

_____

_____

_____

_____

_____

_____

_____

_____

_____

REPORTING PHASE

	Response			References	
	Yes	No	N/A	21 CFR	OECD

A. General Provisions

1. Consultant laboratory, contractor, grantee notified of nonclinical study conduct per GLP

| | ❑ | ❑ | ❑ | 58.10 | |

2. Laboratory permits inspections of facilities, records, specimens required

| | ❑ | ❑ | ❑ | 58.15 | |

B. Personnel — 58.29

1. Education, training, experience for assignments

| | ❑ | ❑ | ❑ | | 1, 1.1(2)(b) |

2. Summaries, training, experience, job descriptions

| | ❑ | ❑ | ❑ | | 1, 1.1(2)(c) |

3. Sufficient personnel to conduct protocol properly

| | ❑ | ❑ | ❑ | | 1, 1.1(2)(b) |

C. Testing Facility Management — 58.31

1. If necessary, replaces study director promptly

| | ❑ | ❑ | ❑ | | 1, 1.1(2)(g) |

2. Assure proper tests of TCA or TCA mixtures

| | ❑ | ❑ | ❑ | | 1, 1.1(2)(p) |

3. Assure personnel understand functions

| | ❑ | ❑ | ❑ | | 1, 1.1(2)(d) |

4. Assure deviations communicated, documented

| | ❑ | ❑ | ❑ | | 2, 2.2(1)(e) |

	Response			References	
REPORTING PHASE	Yes	No	N/A	21 CFR	OECD

D. Study Director — 58.33

Assure:

		Yes	No	N/A		OECD
1.	Protocol changes approved per protocol section	☐	☐	☐		1, 1.2(2)(a) & e 8, 8.1(1) & (2)(a)
2.	Experimental data, including unanticipated responses, accurately recorded and verified	☐	☐	☐		1, 1.2(2)(f)
3.	Unforeseen circumstances affecting quality and integrity noted; corrective action taken, documented	☐	☐	☐		1, 1.2(2)(e)
4.	Test systems as specified in protocol	☐	☐	☐		1, 1.2(2)(e)
5.	Applicable GLP regulations followed	☐	☐	☐		1, 1.2(2)(h)
6.	All raw data, documentation, specimens transferred to archives by close of study	☐	☐	☐		1, 1.2(2)(i)

E. Quality Assurance Unit — 58.35

		Yes	No	N/A		OECD
1.	Responsible for monitoring each study	☐	☐	☐		2, 2.1(2)
2.	Assures management facilities, equipment, personnel, methods, records, etc., in conformance	☐	☐	☐		2, 2.1(1)
3.	Maintains copy of master schedule as required	☐	☐	☐		2, 2.2(1)(a)
4.	Maintains copies of relevant protocols	☐	☐	☐		2, 2.2(1)(a)

 GLP Audit Manual ©2000, Interpharm Press

		Response			References	
REPORTING PHASE		Yes	No	N/A	21 CFR	OECD

		Yes	No	N/A	21 CFR	OECD
5.	Inspects at intervals to assure and document integrity of study	☐	☐	☐		2, 2.2(1)(c)
6.	Immediately informs study director of significant problems likely to affect integrity	☐	☐	☐		2, 2.2(1)(e)
7.	Periodically submitted to management and study director reports, including corrective actions	☐	☐	☐		
8.	Determine no deviation from protocols or SOPs without authorization and documentation	☐	☐	☐		2, 2.1(1)(c)
9.	Reviews final report, assuring it is accurate and reflects raw data	☐	☐	☐		2, 2.2(1)(d)
10.	Issues QAU statement for inclusion in final report	☐	☐	☐		2, 2.2(1)(f)
11.	Maintains record with inspection dates, study, phase, inspector	☐	☐	☐		2, 2.2(1)(f)
12.	Assures inspections done according to GLP	☐	☐	☐		2, 2.1(1)
13.	Maintains QAU records in one location of facility	☐	☐	☐		
F.	**Facilities**				**58.41**	
G.	**Animal Care Facilities**				**58.43**	
H.	**Animal Supply Facilities**				**58.45**	

		Response			References	
REPORTING PHASE		Yes	No	N/A	**21 CFR**	**OECD**

I.	**Facilities for Handling Test and Control Articles**				**58.47**	
1.	To prevent contamination, mix-ups, separate areas for:					
	1.1. Receipt, storage of test/control articles (TCA)	❏	❏	❏		3, 3.3(1)
	1.2. Mixing of TCA with carrier	❏	❏	❏		3, 3.3(1)
	1.3. Storage of TCA mixtures	❏	❏	❏		3, 3.3(2)
2.	Storage TCA/mixtures separate from test system	❏	❏	❏		3, 3.3(2)
3.	Adequate to preserve identity, strength, purity, stability of TCA/mixtures	❏	❏	❏		3, 3.3(2)

J.	**Laboratory Operation Areas**				**58.49**	

K.	**Specimen and Data Storage Facilities**				**58.51**	
1.	Archives, limited to those authorized only, for storage, retrieval of raw data and specimens	❏	❏	❏		10, 10.3

L.	**Equipment Design**				**58.61**	
1.	Equipment used for generation, measurement, assessment, environmental control, appropriate design, capacity to function according to protocol	❏	❏	❏		4(1) 5, 5.1(1)
2.	Adequately inspected, cleaned, maintained	❏	❏	❏		4(1)

REPORTING PHASE

		Response			References	
		Yes	No	N/A	21 CFR	OECD
M.	**Maintenance and Calibration of Equipment**				**58.63**	
1.	Adequately tested, calibrated/standardized	❒	❒	❒		4(2)
2.	SOPs address when to take appropriate remedial action, if malfunction	❒	❒	❒		
3.	Written records of inspections, maintenance, testing, calibration, and/or standardization	❒	❒	❒		4(2)
4.	Records contain whether routine, followed SOPs	❒	❒	❒		
5.	Records maintained of nonroutine repairs	❒	❒	❒		
6.	Records document defect, discovery, remedy	❒	❒	❒		
N.	**Standard Operating Procedures**				**58.81**	
1.	SOPs written to ensure data quality, integrity	❒	❒	❒		7, 7.1
2.	Deviations authorized by study director	❒	❒	❒		1, 1.2(2)(e)
3.	Deviations documented	❒	❒	❒		7, 7.3
4.	Changes in SOPs authorized	❒	❒	❒		7, 7.1
5.	SOPs established, but not limited to:					
	5.1. Animal room preparation	❒	❒	❒		7, 7.4(4)(a)
	5.2. Animal care	❒	❒	❒		7, 7.4(4)(b)
	5.3. Receipt, identification, storage, handling, mixing, and method of sampling TCA	❒	❒	❒		7, 7.4(1)
	5.4. Test systems observations	❒	❒	❒		7, 7.4(4)(c)
	5.5. Laboratory tests	❒	❒	❒		7, 7.1

		Yes	No	N/A	21 CFR	OECD
5.6.	Handling of animals found moribund or dead	☐	☐	☐		7, 7.4(4)(d)
5.7.	Necropsy or postmortem examinations	☐	☐	☐		7, 7.4(4)(d)
5.8.	Collection and identification of specimens	☐	☐	☐		7, 7.4(4)(e)
5.9.	Histopathology	☐	☐	☐		7, 7.4(4)(e)
5.10.	Data handling, storage, and retrieval	☐	☐	☐		7, 7.4(3)
5.11.	Maintenance and calibration of equipment	☐	☐	☐		7, 7.4(2)(a)
5.12.	Transfer, placement, and identification of animals	☐	☐	☐		7, 7.4(4)(b)
6.	Literature used as supplement to SOPs, not in lieu of	☐	☐	☐		7, 7.2
7.	Historical file of SOPs and all revisions	☐	☐	☐		1, 1.1(2)(k) 10, 10.1(f)
O.	**Reagents and Solutions**				**58.83**	
P.	**Animal Care**				**58.90**	
1.	Diagnosis, treatment authorization documented	☐	☐	☐		5, 5.2(2)
2.	Different species in separate rooms, if necessary	☐	☐	☐		3, 3.2(1) 5, 5.2(1)
3.	Adequate cleaning, sanitizing of cages, racks, etc.	☐	☐	☐		5, 5.2(6)

				ABBREVIATED GLP AUDIT CHECKLIST	
		Response			References
REPORTING PHASE	Yes	No	N/A	21 CFR	OECD
4. Periodic analyses of feed, water for contaminants	❑	❑	❑		5, 5.2(6)
5. Such analyses maintained as raw data	❑	❑	❑		8, 8.3(3)
6. Pest control materials documented	❑	❑	❑		5, 5.2(6)
7. Cleaning and pest materials that interfere not used	❑	❑	❑		5, 5.2(6)
Q. Test and Control Article Characterization				58.105	
1. Identity, strength, purity, composition, and other characteristics defined TCA prior to initiation	❑	❑	❑		6, 6.2(1) & (2)
2. Methods of derivation documented	❑	❑	❑		
3. Marketed products characterized by labeling	❑	❑	❑		
4. Stabilities before initiation,	❑	❑	❑		6, 6.2(4) & (5)
or					
5. Concomitantly per SOPs providing for periodic analysis of each batch	❑	❑	❑		6, 6.2(4) & (5)
6. Storage containers: name, chemical abstract or code number, batch, expiration date, storage conditions	❑	❑	❑		6, 6.1(3)
7. Reserve samples retained for proper period	❑	❑	❑		6, 6.2(6)

REPORTING PHASE	Response			References	
	Yes	No	N/A	21 CFR	OECD

R. Test and Control Article Handling — 58.107

SOPs for handling TCA to ensure:

		Yes	No	N/A		OECD
1.	Proper storage	☐	☐	☐		6, 6.1(2)
2.	Distribution that precludes change, damage	☐	☐	☐		6, 6.1(2)
3.	Maintenance of identification during distribution	☐	☐	☐		6, 6.1(2)
4.	Documentation of receipt, distribution returned, each batch	☐	☐	☐		6, 6.1(1)

S. Mixtures of Articles with Carriers — 58.113

		Yes	No	N/A		OECD
1.	For TCA mixed with carrier, analytical methods:					
	1.1. Determine uniformity, and determine periodically, concentration of TCA in mixture	☐	☐	☐		6, 6.2(2) & (5)
	1.2 Determine stability of TCA in mixture	☐	☐	☐		6, 6.2(2) & (5)
2.	Stability of TCA in mixture determined either before initiation,	☐	☐	☐		6, 6.2(5)
	or					
3.	Concomitantly per SOPs providing for periodic analysis of TCA in the mixture	☐	☐	☐		6, 6.2(5)

T. Protocol for and Conduct of a Nonclinical Laboratory Study — 58.120

		Yes	No	N/A		OECD
1.	Revisions, reasons documented and signed by director	☐	☐	☐		8, 8.1(2)(a)

REPORTING PHASE

		Response			References	
		Yes	No	N/A	21 CFR	OECD

		Yes	No	N/A	21 CFR	OECD
U.	**Conduct of a Nonclinical Laboratory Study**				**58.130**	
1.	Conducted according to protocol	❏	❏	❏		8, 8.3(2)
2.	Monitored in conformity with protocol	❏	❏	❏		1, 1.2(2)(e) 2, 2.2(1)(c)
3.	Specimens identified by test system, study, nature, date of collection	❏	❏	❏		8, 8.3(1)
4.	Information located, precluding record, storage error	❏	❏	❏		8, 8.3(1)
5.	Gross findings available to pathologist examining specimen histopathologically	❏	❏	❏		8, 8.3(3) 10, 10.1(a)
6.	Data recorded directly, promptly, legibly, in ink	❏	❏	❏		8, 8.3(3)
7.	Entry dated on date of entry, signed by same person	❏	❏	❏		8, 8.3(3)
8.	Changes do not obscure, reason given, dated and signed at time of change	❏	❏	❏		8, 8.3(4)
9.	Individual responsible for automated data entries identified at time of input	❏	❏	❏		8, 8.3(5)
10.	Changes in automated entries do not obscure, reason given, dated, responsible individual identified	❏	❏	❏		8, 8.3(5)

		Response			References	
REPORTING PHASE		Yes	No	N/A	21 CFR	OECD

V. Reporting of Nonclinical Laboratory Study Results — 58.185

1.	Final report prepared for each study	☐	☐	☐		9, 9.1(1)
2.	Final report includes, but not limited to:					
2.1.	Name, address of facility performing study	☐	☐	☐		9, 9.2(2)(a) & (b)
2.2.	Study initiation and completion dates	☐	☐	☐		9, 9.2(3)
2.3.	Objectives stated in protocol, including changes	☐	☐	☐		9, 9.2(5)(a) & (b)
2.4.	Statistical methods for analyzing data	☐	☐	☐		9, 9.2(6)(c)
2.5.	TCA identified by name, chemical abstracts or code number, strength, purity, composition, or appropriate characteristics	☐	☐	☐		9, 9.2(1)(b)
2.6.	Stability of TCA under conditions administered	☐	☐	☐		9, 9.2(1)(d)
2.7.	Description of methods used	☐	☐	☐		9, 9.2(5)(a)
2.8.	Description of test system used	☐	☐	☐		9, 9.2(5)(b)
2.9.	If applicable, number of animals, sex, body weight range, source, species, strain, substrain, age	☐	☐	☐		
2.10.	Procedure used for animal identification	☐	☐	☐		
2.11.	Description of dosage, regimen, route, duration	☐	☐	☐		

REPORTING PHASE	Response			References	
	Yes	No	N/A	21 CFR	OECD
2.12. Circumstances affecting quality or integrity	☐	☐	☐		
2.13. Name of study director, other scientists, and all supervisory personnel	☐	☐	☐		9, 9.2(2)(c), (d), (e)
2.14. Description of transformations, calculations, or operations performed on data	☐	☐	☐		9, 9.2(6)(c)
2.15. Summary and analysis of data, statement of conclusions drawn from data	☐	☐	☐		9, 9.2(6)(a), (c), (d)
2.16. Signed, dated reports of individual scientists involved in study	☐	☐	☐		9, 9.1(2)
2.17. Locations of specimens, raw data, and report are stored	☐	☐	☐		9, 9.2(7)
2.18. QAU statement	☐	☐	☐		9, 9.2(4)
3. Final report signed and dated by study director	☐	☐	☐		9, 9.1(3)
4. Corrections, additions in form of amendment by study director	☐	☐	☐		9, 9.1(4)
5. Amendment identifies the modified part in final report with reasons	☐	☐	☐		
6. Amendment signed by person responsible	☐	☐	☐		9, 9.1(4)

REPORTING PHASE	Response			References	
	Yes	No	N/A	21 CFR	OECD

W. Storage and Retrieval of Records and Data | | | | **58.190** |

1. Raw data, documentation, protocols, final reports, specimens—except mutagenicity specimens and wet specimens of blood, urine, feces, and biological fluid—retained ☐ ☐ ☐ | | | | | 10, 10.1(a)

2. Storage, retrieval of raw data, protocols, specimens, and reports ☐ ☐ ☐ | | | | | 3, 3.4 \
10, 10.2

3. Deterioration minimized by conditions of storage ☐ ☐ ☐ | | | | | 3, 3.4

4. Reference in archive to any contracted storage ☐ ☐ ☐

5. Individual identified as responsible for archive ☐ ☐ ☐ | | | | | 1, 1.1(2)(l)

6. Entry to archive by authorized personnel only ☐ ☐ ☐ | | | | | 10, 10.3

7. Archived material indexed to permit expedient retrieval ☐ ☐ ☐ | | | | | 10, 10.2

X. Retention of Records | | | | **58.195** |

1. Except for wet specimens, samples of TCA, the documentation records, raw data, and specimens retained in archive are retained as required:

   1.1. At least 2 years following NDA approval ☐ ☐ ☐ | | | | | 10, \
10.1(a)–(g)

   1.2. At least 5 years following date on which study submitted in support of an NDA ☐ ☐ ☐ | | | | | 10, \
10.1(a)–(g)

	Yes	No	N/A	21 CFR	OECD
1.3. Other situations at least 2 years following date study completed, terminated, or discontinued	❐	❐	❐		10, 10.1(a)–(g)
2. Wet specimens, samples of TCA retained as long as quality of preparation can be evaluated	❐	❐	❐		10, 10.1("h")
3. Master schedule, protocols, QAU records retained by QAU as easily accessible for time applicable	❐	❐	❐		10, 10.1(b)
4. Summaries of training, experience, job descriptions, testing facility employment records retained for time applicable	❐	❐	❐		10, 10.1(c)
5. Records, reports of maintenance, calibration, and inspection of equipment retained for time applicable	❐	❐	❐		10, 10.1(d)
6. Records this part retained as original or true copies (e.g., photocopies, microfilm, microfiche) or other accurate reproductions of original records	❐	❐	❐		
7. Materials specified above transferred to sponsor should contracted facility discontinue business	❐	❐	❐		10, 10.4
8. FDA notified in writing should No. 7 occur.	❐	❐	❐		

Observations/Comments

_____

_____

_____

_____

_____

_____

_____

_____

_____

_____

_____

_____

_____

_____

_____

_____

_____

_____

_____

_____

_____

_____

_____

_____

_____

_____

Inspected/Audited by: _____ **Date:** _____

SECTION III
DOCUMENT/REPORT EXAMPLES

QUALITY ASSURANCE DOCUMENTS

NORTHERN RESEARCH, INC.
QUALITY ASSURANCE

MASTER SCHEDULE

PROJECTIONS:

QAU Inspections	I
Protocol Issued	P
Final Report	R
Interim Report	r

Test Article	Study Identity	Test System	Study Initiation	Study Director	Nature of Study and Current Status/Final Report	2000										2001	
						M	A	M	J	J	A	S	O	N	D	J	F
AC-123	99-322	Mouse	01/07/00	Smith	Acute PO Tox/In-Preparation	R											
AC-123	00-107	Rat	03/09/00	Robbins	One-Month IV Tox/In-Process	I	I	R									
BD-149	99-257	Dog	09/19/99	Robbins	Three-Month IV Tox/ Issued 3/26/00	R											
BD-149	99-314	Monkey	12/17/99	Williams	Six-Month IV Tox/In-Process	I			I				R				
RS-160	00-059	Rat	03/25/00	Smith	Acute IV Tox/Protocol	P	I	R									
RS-160	00-103	Rat	04/09/00	Smith	One-Month IV/Protocol		P	I		R							
TM-172	99-337	Salmonella	03/05/00	Gentry	Ames Test/In-Preparation	I	R										
VW-178	99-157	Dog	08/20/99	Robbins	Twelve-Month Diet/In-Process			I	I		I						R
ST-205	98-145	Mouse	08/06/98	Williams	2-Year Carcinogenicity Diet In-Process	r		I			I			I			
ST-205	98-147	Rat	09/20/98	Harris	2-Year Carcinogenicity Diet In-Process	I	r		I			I			I		I
WS-310	99-258	Dog	10/02/99	Anderson	Three-Month PO Tox/In-Preparation	R											

NORTHERN RESEARCH, INC.
QUALITY ASSURANCE

STUDY INSPECTION RECORD

Study Title:	*Three-Month Intravenous Toxicity*
Test Article:	*BD-149*
Study Identity:	*99-257*
Study Species:	*Dog*
Route:	*IV*
Study Director:	*G. Robbins, Ph.D.*

Study Phase	Inspection Dates	Signature/ Date	Inspection Report No./Date	Corrective Action Acceptance Date
Initiation	*9/19, 20/99*	*A.D. Moulohin 9/20/99*		
	9/24, 25/99	*A.D. Moulohin 9/25/99*		
In-Process	*10/8, 15–17/99*	*A.D. Moulohin 10/17/99*		
	10/23–25/99	*A.D. Moulohin 10/25/99*	*99-093 10/25/99*	*11/6/99*
	11/12–15/99	*A.D. Moulohin 11/15/99*		
Completion	*12/17–20/99*	*A.D. Moulohin 12/20/99*	*99-112 12/20/99*	*1/8/00*
	1/13, 14/00	*A.D. Moulohin 1/14/00*		
	2/5/00	*A.D. Moulohin 2/5/00*		
Reporting	*2/28/00*	*A.D. Moulohin 2/28/00*		
	3/4–6/00	*A.D. Moulohin 3/6/00*	*00-027 3/6/00*	*3/20/00*

STUDY INSPECTION RECORD

Study Title:	*Three-Month PO Toxicity*
Test Article:	*WS-310*
Study Identity:	*99-258*
Study Species:	*Dog*
Route:	*PO*
Study Director:	*J. B. E. Anderson, Ph.D.*

Study Phase	Procedure	Inspection Dates	Signature/ Date	Inspection Report No./Date	Corrective Action Acceptance Date
Initiation	*PRO*	*10/2/99*	*P. Verke 10/2/99*		
	AWS	*10/2/99*	*P. Verke 10/2/99*		
	FOR	*10/9/99*	*P. Verke 10/9/99*		
In-Process	*AWS*	*10/7/99*	*P. Verke 10/7/99*		
	DOS	*10/9/99*	*P. Verke 10/9/99*	*99-097 10/9/99*	*10/17/99*
	FOR	*11/12/99*	*P. Verke 11/12/99*		
	BCN	*12/2/99*	*P. Verke 12/2/99*		
	OBS	*12/6/99*	*P. Verke 12/6/99*		
	DOS	*12/10/99*	*P. Verke 12/10/99*		
Completion	*NEC*	*1/2, 3/00*	*P. Verke 1/3/00*		
	CCH/HEM	*1/2,7/00*	*P. Verke 1/7/00*	*00-002 1/7/00*	*1/23/00*
	MJC	*2/19/00*	*P. Verke 2/19/00*		
Reporting	*RPJ*	*4/7-9/00*	*P. Verke 4/9/00*		
		4/16/00	*P. Verke 4/16/00*	*00-017 4/16/00*	*4/27/00*

NORTHERN RESEARCH, INC.
QUALITY ASSURANCE

INSPECTION REPORT

To: _____W. Harris, PhD_____ and _____K. Bellows, DVM_____

_____STUDY DIRECTOR_____ _____MONITOR_____

Please respond to the observation below, and return this report to the
Quality Assurance Unit by 9/6/99.

Report Number: 99-092	
Study Number: 98-147	Page: 1 of 1
Test Article: ST-205	Date: 8/6/99

This is a *CONFIDENTIAL* legal document that is to be maintained by Quality Assurance for the period of time as prescribed by the Good Laboratory Practice Regulations.

STUDY PHASE(S) INSPECTED				INSPECTION DATE(S)	
Initiation:					
In-Process:	✓			5/24; 6/12, 20; 7/8; 8/2/99	
Completion:					
Reporting:					
		B. W. Herrick		*B. W. Herrick 8/6/99*	

OBSERVATION: The scale used in measuring body weights had an expired calibration date. Weighings were done 8/2/99; the labeled calibration expiration date of this scale, Asset #434, was 7/25/99.

RESPONSE: All the scales and balances within this facility are in a scheduled calibration program. The Calibration Record for the above scale indicates that it was calibrated on 7/24/99; however, the calibrator had failed to replace the calibration tag with one having the correct information. Metrology had been contacted and will make the correction.

Prepared By:	*H. James*	Date:	*8/8/99*	
Approved By:	*K. Bellows*	Date:	*8/9/99*	
Proposed Completion Date(s):	*8/14/99*			
Study Director:	*W. Harris*	Date:	*8/12/99*	

QAU REVIEW:

Action Plan Acceptance Date:	*August 13, 1999*
By:	*B. W. Herrick*
Completed:	*8/19/99*

GLP Audit Manual ©2000, Interpharm Press

NORTHERN RESEARCH, INC.

QUALITY ASSURANCE UNIT STATEMENT

Periodic Inspections of Study Number 99-257 entitled Three-Month IV Toxicity Study of BD-149 in Dogs were conducted on all phases by the Quality Assurance Unit of Northern Research, Inc.

The study was inspected on the following dates:

September 19, 20, 24, 25, 1999

October 8, 15–17, 23–25, 1999

November 12–15, 1999

January 13, 14, 2000

February 5, 2000

February 28, 2000

March 4, 6, 2000

Results of inspections were reported to Management and the Study Director on 10/25, 12/20/99, and 3/6/00.

To the best of the signator's knowledge there were no significant deviations from the Good Laboratory Practice Regulations that affected the quality of integrity of the study.

James A. Kensington   _3/6/00_

James A. Kensington   Date

Manager, Quality Assurance

INSPECTION REPORT EXAMPLES

GLP Audit Manual ©2000, Interpharm Press

NORTHERN RESEARCH, INC.
QUALITY ASSURANCE

INSPECTION REPORT

To: _____J. C. Smith, PhD_____ and ___W. C. Jones, DVM, PhD___

_____STUDY DIRECTOR_____   _____MONITOR_____

Please respond to the observation below, and return this report to the
Quality Assurance Unit by 4/27/00.

Report Number: 00-019

Study Number: 00-059   Page: 1 of 1

Test Article: RS-160   Date: 3/26/00

This is a *CONFIDENTIAL* legal document that is to be maintained by Quality Assurance for the period of time as prescribed by the Good Laboratory Practice Regulations.

STUDY PHASE(S) INSPECTED			INSPECTION DATE(S)
Initiation:	✓		3/25/00
In-Process:			
Completion:			
Reporting:			
		A. V. Moulchin	A. V. Moulchin 3/26/00

OBSERVATION: The study protocol did not include the following information as required by the GLP (58.120):

The procedure for identification of the test system

The statement of the proposed statistical methods to be used

RESPONSE: Protocol Change #2 has been approved and issued. It includes both the procedure for identification of the rats and the proposed statistical methods to be used.

Prepared By:	R. Derich	Date:	3/26/00
Approved By:	W. C. Jones	Date:	3/27/00
Proposed Completion Date(s):	3/27/00		
Study Director:	J. C. Smith	Date:	3/27/00

QAU REVIEW:

Action Plan Acceptance Date:	March 30, 2000
By:	A. V. Moulchin
Completed:	3/30/00

NORTHERN RESEARCH, INC.
QUALITY ASSURANCE

INSPECTION REPORT

To: _____D. Williams, PhD_____ and _____K. Bellows, DVM_____
_____STUDY DIRECTOR_____ _____MONITOR_____

Please respond to the observation below, and return this report to the
Quality Assurance Unit by 4/16/00.

Report Number: 00-006

Study Number: 99-314   Page: 1 of 1

Test Article: BD-149   Date: 3/6/00

This is a *CONFIDENTIAL* legal document that is to be maintained by Quality Assurance for the period of time as prescribed by the Good Laboratory Practice Regulations.

STUDY PHASE(S) INSPECTED			INSPECTION DATE(S)
Initiation:			12/17, 18, 24/99; 3/2, 6/00
In-Process:	✓		
Completion:			
Reporting:			
		P. Verke	P. Verke 3/6/00

OBSERVATION: SOP 43-74A, Subcutaneous Injections of Primates, requires the use of a 23-gauge needle for administration of the test article mixture or control substances for a 2-minute duration. A 20-gauge needle was being used, and the mixtures were administered for a duration of 1 minute.

RESPONSE: For these larger primates (4.5–5.5 kg) the increased cannula size and administration rates are appropriate. The SOP will be modified, reflecting the increases in cannula size and administration rates correlated with primate weight averages.

Prepared By:	W. Strykowski	Date:	3/12/00
Approved By:	K. Bellows	Date:	3/14/00
Proposed Completion Date(s):	4/15/00		
Study Director:	D. Williams	Date:	3/17/00

QAU REVIEW:

Action Plan Acceptance Date:	3/20/00
By:	P. Verke
Completed:	4/16/00

NORTHERN RESEARCH, INC.
QUALITY ASSURANCE

INSPECTION REPORT

To: _____G. Robbins, PhD_____ and _____L. Kannal, PhD_____

_____STUDY DIRECTOR_____ _____MONITOR_____

Please respond to the observation below, and return this report to the
Quality Assurance Unit by 1/21/00.

Report Number: __99-112__

Study Number: __99-257__ _____ Page: __1 of 1_____

Test Article: __BD-149__ Date: __12/20/99__

This is a *CONFIDENTIAL* legal document that is to be maintained by Quality Assurance for the period of time as prescribed by the Good Laboratory Practice Regulations.

STUDY PHASE(S) INSPECTED			INSPECTION DATE(S)	
Initiation:			12/17–20/99	
In-Process:				
Completion:	✓			
Reporting:				
		A. D. Moulchin	A. D. Moulchin 12/23/99	

OBSERVATION: During necropsy, spinal fluid was collected from each animal. No departmental SOP existed describing this procedure.

RESPONSE: The method describing this procedure has been written and will be documented in a new SOP.

Prepared By:	J. Rodriguez	Date:	12/27/99
Approved By:	L. Kannal	Date:	1/2/00
Proposed Completion Date(s):	1/17/00		
Study Director:	G. Robbins	Date:	1/3/00

QAU REVIEW:

Action Plan Acceptance Date:	1/8/00
By:	A. D. Moulchin
Completed:	1/21/00

NORTHERN RESEARCH, INC.
QUALITY ASSURANCE

INSPECTION REPORT

To: _____G. Robbins, PhD_____ and _____L. McCarley, MS_____

_____STUDY DIRECTOR_____ _____MONITOR_____

Please respond to the observation below, and return this report to the
Quality Assurance Unit by 4/6/00.

Report Number: __00-027__

Study Number: __99-257__   Page: __1 of 1__

Test Article: __BD-149__   Date: __3/6/00__

This is a *CONFIDENTIAL* legal document that is to be maintained by Quality Assurance for the period of time as prescribed by the Good Laboratory Practice Regulations.

STUDY PHASE(S) INSPECTED			INSPECTION DATE(S)	
Initiation:			2/28; 3/4, 6/00	
In-Process:				
Completion:				
Reporting:	✓			
		A. D. Moulchin	A. D. Moulchin 3/9/00	

OBSERVATION: The following were observed in the appended Statistical Report:

In Table 2, Urinanalysis, the mean pH of the T_2 females and T_3 males are respectively recorded as 6.5 and 7.3. Calculations associated with the raw data indicate respective mean pH of 6.7 and 7.0.

In Table 5, Absolute and Relative Animal Organ Weights, the mean kidney weight of the T_4 males is recorded as 44.23 g. Calculations associated with the raw data indicate a mean of 47.57 g.

RESPONSE: The errors noted above in Tables 2 and 5 of the Statistical Report have been corrected and the database changed to reflect the raw data.

Prepared By:	R. Chin, PhD (Statistician)	Date:	3/12/00	
Approved By:	B. Murphy, PhD (Stats. Mgr.)	Date:	3/17/00	
Proposed Completion Date(s):	3/20/00			
Study Director:	G. Robbins	Date:	3/18/00	

QAU REVIEW:

Action Plan Acceptance Date:	3/20/00
By:	A. D. Moulchin
Completed:	3/23/00

NORTHERN RESEARCH, INC.
QUALITY ASSURANCE

AUDIT REPORT

Report Number:	99-105		Page 1
Date:	11/26/99		
Department Audited:	Analytical Chemistry		
Location:	Coeur d'Alene, ID		
Date(s) Audited:	11/18–21/99		
Auditors:	M. Simpson		*M. Simpson 11/26/99*

OBSERVATION:	1—The Test Method validation file for determining the potency of test article ST-211 had been approved by the department manager on 10/16/99. Animals in study 99-125 were dosed with this compound seven days prior to approval of the validated test method on 10/9/99. 2—SOP 796-35.02, MS-345 Determination-HPLC, approved 5/3/97, requires that for each standardization and test, the chromatographic system components be identified on the charts. Examination of the applicable HPLC charts, run 11/06/99, indicate that only the column has been identified.
RESPONSE:	1—The Test Method for the determination of potency for ST-211 includes more than three determinations consistently using the same methodology and performed by the same scientist. This had been reviewed in Dr. Shaw's Laboratory Notebook, accepted and acknowledged by management on 10/3/99. The formal Test Methodology Report was signed, as indicated, on 10/16/99. An explanation will be written, filed both in the validation file and with the documentation for study 99-125. 2—These identity requirements had inadvertently been excluded; however, they had been recorded in the scientist's Laboratory Notebook, and have now been transcribed to the proper charts. This individual and other members of the department have been cautioned to be more meticulous.

Prepared By:	*R. Shaw, PhD, Methods Development*	Date:	*11/28/99*
Approved By:	*W. Kimura, PhD, Laboratory Manager*	Date:	*12/4/99*

QAU REVIEW:

Report Return Date:	*12/5/99*		
Accepted By:	*J. A. Kensington*	Plan Accepted Date:	*12/6/99*
Title:	*Manager QA*		
Auditor:	*M. Simpson*	Plan Completed Date:	*12/18/99*

NORTHERN RESEARCH, INC.
QUALITY ASSURANCE

AUDIT REPORT
Contract Laboratory

Report Number:	00-030		Date:	6/18/00
Facility Audited:	Southwestern Laboratories (SWL)			
Location:	Albuquerque, NM			
Date(s) Audited:	6/11/00			
Auditors:	B. E. Anderson		*B. E. Anderson 6/18/00*	
AUDIT SCOPE:	At the request of Drug Safety Evaluation (DSE), the QAU evaluated the organization, facilities, operations, and SOPs of SWL with intensive surveillance of their QAU.			

AUDIT COVERAGE:

Facilities:	There are four buildings at this site, three used for nonclinical studies. Primates were in Bldg. B; observed building and room maintenance, cage washing, and facility security. Bldg. A contained the QAU, the archives, and computer operations. Evaluated QAU facilities and files. Observed raw data, report retention, and histology tissue storage.
Operations:	Observed technicians IM dosing, measuring body weights and food consumption, recorded by clinical signs; general facility animal husbandry and computer operations. Concurrently reviewed the QAU auditor's documentation process.
Documentation:	Reviewed the SOPs of the study function groups and the QAU. Also reviewed the Master Schedule, CVs of personnel, pest applicator's log, cage wash equipment log, instrument calibrations, and primate TB treatment test records.
Recommendation:	Our evaluation of SWL indicates that it has an excellent staff of scientific and technical personnel. Their management personnel stress total compliance and administer the most stringent international standards to all of their contracted regulated studies. We would recommend contracting with SWL, contingent on their acquiring a remote secondary backup tape storage facility.
Observation:	SWL stores its computer backup tapes within the computer operation's facilities of Bldg. A. It does not have off-site location for the secure storage of secondary backup tapes.
Response:	In my phone conversation of 6/22/00 with Dr. Simmons, President of SWL, he informed me they have a project commencing on 7/1/00, providing for secondary computer tape storage in a local bank vault.

Prepared By:	*K. Patel, DVM, Toxicologist*		Date:	*6/23/00*
Approved By:	*R. Schubert, PhD, Director DSE*		Date:	*6/25/00*

QAU REVIEW:

Report Return Date:	*6/23/00*		
Accepted By:	*J. A. Kensington*	Plan Accepted Date:	*6/24/00*
Title:	*Manager QA*		
Auditor:	*B. E. Anderson*	Plan Completed Date:	*7/6/00*

SECTION IV
REFERENCES/REGULATORY TEXTS

FRIDAY, DECEMBER 22, 1978
PART II

DEPARTMENT OF HEALTH, EDUCATION, AND WELFARE

Food and Drug Administration

■

NONCLINICAL LABORATORY STUDIES

Good Laboratory Practice Regulations

[110-03-M]

Title 21—Food and Drugs

CHAPTER I—FOOD AND DRUG' ADMINISTRATION, DEPARTMENT OF HEALTH, EDUCATION, AND WELFARE

[Docket No. 76N-0400]

NONCLINICAL LABORATORY STUDIES

Good Laboratory Practice Regulations

AGENCY: Food and Drug Administration.

ACTION: Final Rule.

SUMMARY: The agency is issuing final regulations regarding good laboratory practice in the conduct of nonclinical laboratory studies. The action is based on investigatory findings by the agency that some studies submitted in support of the safety of regulated products have not been conducted in accord with acceptable practice, and that accordingly data from such studies have not always been of a quality and integrity to assure product safety in accord with the Federal Food, Drug, and Cosmetic Act and other applicable laws. Conformity with these rules is intended to assure the high quality of nonclinical laboratory testing required to evaluate the safety of regulated products.

EFFECTIVE DATE: June 20, 1979.

FOR FURTHER INFORMATION CONTACT:

Paul D. Lepore, Bureau of Veterinary Medicine (HFV-102), Food and Drug Administration, Department of Health, Education, and Welfare, 5600 Fishers Lane, Rockville, MD 20857, (301-443-4313).

SUPPLEMENTARY INFORMATION: The Food and Drug Administration (FDA) is establishing regulations in a new Part 58 (proposed as Part 3e) in Title 21 (21 CFR Part 58) regarding good laboratory practice. These constitute the first of a series of regulations concerning investigational requirements which are being developed as a result of the FDA Bioresearch Monitoring Program. Proposed regulations, providing interested persons 120 days to submit comments, were published in the FEDERAL REGISTER of November 9, 1976 (41 FR 51206). In addition, public hearings were held on February 15 and 16, 1977 for the presentation of oral testimony on the proposal. Twenty-two oral presentations were given (transcripts are on file with the Hearing Clerk, Food and Drug Administration), and 174 written comments were received. The comments have been categorized and include the fol-

lowing: manufacturers of regulated products (64), associations (40), medical centers (20), private testing or consulting laboratories (18), educational institutions (15), government agencies (8), individuals (8), and an airport director (1).

In the proposal, regulations were designated as a new Part 3e. This final rule incorporates them into a new Part 58 (21 CFR Part 58). The following redesignation table correlates the new sections with those proposed, and, in most instances, reference to the new sections will be used hereinafter.

New Section	Old Section
Subpart A	
58.1	3e.1
58.3	3e.3
58.10	3e.10
58.15	3e.15
Subpart B	
58.29	3e.29
58.31	....
58.33	3e.31
58.35	3e.33
Subpart C	
58.41	3e.41
58.43	3e.43
58.45	3e.45
58.47	3e.47
58.49	3e.49
58.51	3e.51
58.53	3e.53
Subpart D	
58.61	3e.61
58.63	3e.63
Subpart E	
58.81	3e.81
58.83	3e.83
58.90	3e.90
Subpart F	
58.105	3e.105
58.107	3e.107
58.113	3e.113
Deleted	3e.115
Subpart G	
58.120	3e.120
58.130	3e.130
Subpart J	
58.185	3e.185
58.190	3e.190
58.195	3e.195
Subpart K	
58.200	3e.200
58.202	3e.202
58.204	3e.204
58.206	3e.206
58.210	3e.210
58.213	3e.213
58.215	3e.215
58.217	3e.217
58.219	3e.219

As a part of the overall bioresearch monitoring program that was described in the proposal, a pilot inspection program was carried out to assess the current status of laboratory practice of nonclinical testing facilities to aid in evaluating the relevance of the proposed regulations, and to identify any unanticipated difficulties in implementing an agency-wide monitoring and compliance program for the testing facilities.

The pilot inspection program began in December of 1976 and covered a representative sample of testing facilities. The results of these inspections have been evaluated, and the result of the analysis have been made available to the public as OPE Study 4 "Results of the Nonclinical Toxicology Laboratory Good Laboratory Practice Pilot Compliance Program." Notice of availability of this report was published in the FEDERAL REGISTER of October 28, 1977 (42 FR 56799).

TABLE OF CONTENTS FOR PREAMBLE

GENERAL ISSUES (PARAGRAPHS 1 THROUGH

General Provisions

Scope (paragraphs 10 through 16).
Definitions (paragraphs 17 through 36).
Applicability to studies performed under grants and contracts (paragraphs 37 through 38).
Inspection of testing facility (paragraphs 39 through 48).

Organization and Personnel

Personnel (paragraphs 49 through 57).
Testing facility management (paragraph 58).
Study director (paragraphs 5 through 74).
Quality assurance unit (paragraphs 75 through 92).
Access to professional assistance (paragraph 93).

Facilities

General (paragraphs 94 through 95
Animal care facilities (paragraphs 9 through 101).
Animal supply facilities (paragraph 102 through 104).
Facilities for handling test and control articles (paragraphs 105 through 106).
Laboratory operation areas (paragraphs 107 through 110).
Specimen and data storage facilities (paragraph 111).
Administrative and personnel facilities (paragraph 112).

Equipment

Equipment design (paragraphs 1 though 115).
Maintenance and calibration of equipment (paragraphs 116 through 119).

Testing Facilities Operation

Standard operating procedures (paragraphs 130 through 145).
Reagents and solutions (paragraphs 146 through 149).
Animal care (paragraphs 1 through 167).

GENERAL ISSUES

1. Many of the written responses to
he proposal were in two parts: a dis-
ussion of broad issues and a critique
the regulations by section and para-
aph. Over a thousand individual
ems have been considered.
2. Thirty-two comments requested
publication of the proposed regula-
ons as guidelines.
The Commissioner of Food and
rugs advises that publishing guide-
nes rather than regulations was con-
dered and rejected before publica-
on of the proposal. The question was
onsidered again in preparation of this
rder, and again rejected. The serious-
ess of problems encountered in test-
g facilities demands the use of an ap-
roach that will achieve compliance
irectly and promptly. Only by speci-

fying the requirements for compliance
in detailed, enforceable regulations
can the Commissioner be assured of
the quality and integrity of the data
submitted to the agency in support of
an application for a research or marketing permit.

3. Some comments objected to the
incorporation by reference of other
laws, recommendations, and guidelines
as being either redundant or without
the authority conferred by rulemaking
procedures as required by the Administrative Procedure Act. It was also asserted that such incorporation could
lead to confusion.

The Commissioner agrees that these
regulations should not duplicate regulations and requirements subject to
the purview of other agencies. Therefore, reference to animal care provisions of the Animal Welfare Act of
1970 (Pub. L. 91-570) and recommendations contained in Department of
Health, Education, and Welfare
(HEW) Publication No. (NIH) 74-23
have been deleted from §§ 58.43(a) and
58.90(a) (21 CFR 58.43(a) and
58.90(a)). Also, all provisions that referred to regulations of the Occupa-
tional Safety and Health Administration or were concerned with the
health and safety of employees have
been revised or deleted, i.e., 21 CFR
58.33(a) (by deletion of proposed 21
CFR 3e.31(a)(11)), 21 CFR 58.53(b), 21
CFR 58.81 (by deletion of proposed 21
CFR 3e.81(b)(10)), and 21 CFR
58.120(a) (by deletion of proposed 21
CFR 3e.120(a)(17)). Reference to the
regulations of the Nuclear Regulatory
Commission has been removed from
§ 58.49; and proposed § 3e.115, dealing
with the handling of carcinogenic substances, has been deleted. In addition,
the Commissioner has deleted reference to the various animal care guide-
line cited in the proposal.

4. Some comments said the regulations should not be retroactive to previous studies or those ongoing and
should include reasonable transitional
provisions for their implementation.

To give nonclinical laboratory facilities adequate time to implement required changes in their organization
and physical plant, a period of 180
days after publication in the FEDERAL
REGISTER is provided for these regulations to become fully effective. The
regulations are not retroactive. All
studies initiated after the effective
date shall be subject to the regulations. The remaining portions of studies in progress on the effective date of
the regulations shall be conducted in
accordance with these regulations.

5. A number of comments challenged
the general legal authority of FDA to
issue good laboratory practice regulations. Other comments challenged the
legal authority to require record retention or quality assurance units, or to

specify the content of required record
or location of storage.

The Commissioner finds that the a
thority cited in the preamble to th
proposal (41 FR 51219; Nov. 19, 1976
provides a sound legal basis for th
regulations. Although many matte
covered in these regulations are no
explicitly mentioned in any of th
laws administered by the Commission
er, the Supreme Court has recognize
in *Weinberger v. Bentex Pharmaceut
cals, Inc.*, 412 U.S. 645, 653 (1973), th
FDA has authority that "is implicit i
the regulatory scheme, not spelled ou
in haec verba" in the statute. A
stated in *Morrow v. Clayton*, 326.F.2
36, 44 (10th Cir. 1963):

However, it is a fundamental principle
administrative law that the powers of an a
ministrative agency are not limited to tho
expressly granted by the statutes, but i
clude, also, all of the powers that may I
fairly implied therefrom.

See *Mourning v. Family Publicatio
Service, Inc.*, 411 U.S. 356 (1973); s
also *National Petroleum Refiners Ass
ciation v. F.T.C.*, 482.F.2d 672 (D.(
Cir. 1973). The Commissioner co
cludes that there is ample authorit
for the promulgation of good labor
tory practice regulations. No commer
presented any explanation or inform
tion to the contrary, let alone a coger
argument that FDA lacks legal at
thority under existing statutes. Th
standards prescribed represent ampl
fication of the legal requirements r
garding evidence of safety necessary
approve an application for a researc
or marketing permit and parallel, to
great extent, steps that FDA ha
found have been taken by members
the regulated industry to improve nor
clinical laboratory operations.

6. One comment argued that th
opinion of the Court of Appeals
*American Pharmaceutical Associatic
v. Weinberger*, 530 F.2d 1054 (D.C. Ci
1976), should be read to limit FDA
authority to issue regulations und
section 701(a) of the act (21 U.S.(
371(a)).

The Commissioner disagrees wit
the argument advanced in the cor
ment. As discussed in the preamble
the proposed regulation, the agency
authority to issue regulations und
section 701(a) of the act has bee
upheld by the courts. (See *Weinberg
v. Hynson, Westcott & Dunning, In
412 U.S. 609 (1973); see also *Nation
Confectioners Association v. Califan
No. 76-1617 (D.C. Cir. Jan. 20, 1976
Upjohn Co. v. Finch, 422 F.2d 944 (6
Cir. 1970); *Pharmaceutical Manufa
turers Association v. Richardson*, 3
F. Supp. 301 (D. Del. 1970).) The que
tion is not FDA's authority to issu
regulations under section 701(a) of tl
act per se, but whether regulatio
issued under section 701(a) of the a
appropriately implement other se

tions of the act. As articulated in the original proposal, and as discussed in the previous two paragraphs, the Commissioner has determined that these regulations are essential to enforcement of the agency's responsibilities under sections 406, 408, 409, 502, 503, 505, 506, 507, 510, 512, 513, 514, 515, 516, 518, 519, 520, 706, and 801 of the Federal Food, Drug, and Cosmetic Act, as well as the responsibilities of FDA under sections 351 and 354-360F of the Public Health Service Act.

7. A number of comments said various sections of the act did not specify the submission of safety data or did not deal with "applications for research or marketing permits."

The Commissioner has reviewed the comments and finds that the comments are based on a misunderstanding of the phrase, "applications for research or marketing permits." This concept is discussed in relation to § 58.3(e) below. Each cited provision contains authority for FDA either to require submission of, or to use, nonclinical safety data to justify a decision to approve the distribution of a regulated product.

8. A number of comments said the cost of implementing the proposed regulations would be prohibitive to smaller testing laboratories and would, at the least, result in a substantial increase in the cost of product testing.

The Commissioner agrees that implementation of these regulations will increase the cost of nonclinical laboratory testing. The Commissioner finds, however, that such costs are justified on the basis of the resultant increase in the assurance of the quality and integrity of the safety data submitted to the agency. The agency has previously concluded (see the FEDERAL REGISTER of November 19, 1976 (41 FR 51220)) that this document does not contain regulations requiring preparation of an inflation impact statement under Executive Orders 11821 and 11929, Office of Management and Budget Circular A-107 and the guidelines issued by the Department of Health, Education, and Welfare. For a notice on the availability of the agency's economic impact assessment regarding rules for good laboratory practice for nonclinical laboratory studies, see the FEDERAL REGISTER of February 7, 1978 (43 FR 5071). The revisions in this final rule, along with the findings of the pilot program, which showed that many of the inspected facilities were already substantially in compliance with the proposed regulations, should allay some of the concerns of small facilities regarding cost or feasibility of compliance.

9. Many comments suggested changes in language, grammar, terminology, punctuation, sentence structure, and other editorial changes to clarify or improve upon the requirements as stated in the regulations or to eliminate redundancies or inconsistencies. Comments that raised significant policy questions, suggested changes in the substance of the regulation, or otherwise required, in the Commissioner's opinion, a specific response, are discussed individually below. Many of the suggested changes, however, were editorial and stylistic and do not warrant a detailed discussion.

The Commissioner has reviewed each of these numerous editorial and language changes to determine whether it offered an improvement in clarity or definition, eliminated an obvious error or redundancy, promoted consistency with other portions of the regulations, or otherwise identified textual problems that had not been previously noted by FDA. Where the proposed alternative language or other changes suggested by the comments were superior to the proposal, they were adopted in substance or verbatim. Where they did not offer any improvement, the Commissioner declined to accept them.

GENERAL PROVISIONS

SCOPE

10. Numerous comments addressed the stated scope of the proposed regulations (§ 58.1). Six comments said the proposed scope was vague. Ten comments said the scope should be limited to long-term animal toxicity studies. Twenty-two comments indicated that the scope should be limited to animal safety studies to be submitted to FDA. Individual comments recommended limiting the scope to studies performed on marketed products, studies performed on animals and other biological test systems, or studies submitted in support of a color additive petition, food additive petition, investigational new drug application, new drug application, or new animal drug application.

In the preamble to the proposed regulations, the Commissioner set forth the reasons for the broad terminology employed in the statement of scope, stating "these regulations are intended to ensure, as far as possible, the quality and integrity of test data that are submitted to FDA and become the basis for regulatory decisions made by the Agency." In the proposed rule (41 FR 51210), the Commissioner specifically invited comments on which laboratories and/or studies should be subject to the regulations, and further, on whether the scope of the regulations should be defined in terms of the type of testing facility rather than the type of study performed. Based on the review of the comments, the Commissioner has chosen to describe the scope of the regulations in language

that is only slightly changed from t proposal. Further clarification scope is achieved by the specific de nition of the key terms, "nonclini laboratory study" and "application research or marketing permit" § 58.3. Taken together, these pro sions eliminate any vagueness in t scope of these regulations.

The Commissioner has rejected t request to narrow the scope by listi in the regulation specific types studies covered. Any such list, if it cluded all types of studies used by t agency to assess the safety of all t products it regulates, would be cu bersome and might exclude speci types of studies that could become i portant to future safety decisions. T Commissioner emphasizes that t decision does not mean, however, t the scope of the regulations is unlim ed. The scope of the GLP regulatic is limited in several ways.

First, they apply only to nonclini laboratory studies that are submitt or are conducted for submission to t agency in support of a research marketing permit for a regulated pr uct. Language has been added th provides that the scope includes st ies "intended" to support applicatic for research or marketing permi This language was included in the p amble to the proposed regulation (FR 51209), and the Commissioner h added the language to the regulati because it helps to make clear in vance when a study should comp with the regulation and when a stu should be listed on a testing facilit master schedule sheet as a nonclini laboratory study subject to these re; lations (§ 58.35(b)(1)). Tests never tended to be submitted to the ager in support of (i.e., as the basis for) t approval of a research or marketi permit, such as exploratory saf studies and range-finding experimen are not included even though th may be required to be submitted part of an application or petition.

Second, the definition of "noncli cal laboratory study" (§ 58.3(d)) mal it very clear that studies utilizi human subjects, clinical studies, field trials in animals are not includ

Third, the scope of coverage is n limited to safety studies, i.e., th which can be used to predict adve effects of, and to establish safe characteristics for, a regulated pr uct. "Functionality studies" have b excluded in the final rule.

Fourth, the definition of "t system" (§ 58.3(i)) taken together w the definition of "nonclinical labo tory study" makes it clear that scope of coverage is confined to st ies performed on animals, plants, croorganisms or subparts thereof.

Products regulated by the agen for which safety data may be requir over a wide range of diverse items that pose quite different types of risk. Examples include implantable medical devices; indirect food additives which may occur in food in very small quantities; direct food additives which may be consumed on a daily basis in larger quantities; human drugs intended for prescription or over-the-counter use; animal drugs intended for use in pets and other companion animals of social importance, drugs used in food-producing animals (drug residues can become a part of food); radiation products used in the diagnosis and/or treatment of a disease or condition; radiation products (e.g., microwave ovens and television sets) widely used by the public; vaccines; and blood components and derivatives.

The guarantee of the safety of each of these product classes requires conducting a broad spectrum of safety tests, all of which should be subject to the same standards. Therefore, the Commissioner rejects the proposal to limit the scope of these regulations to long-term animal toxicity studies. Median lethal dose (LD₅₀) and other short-term tests are covered by the regulations because they may serve as part of the basis for approval of, for example, use of an indirect food additive or an investigational new drug in man.

In vitro biological tests are included insofar as such tests have a bearing on product safety, even though they are not now used in agency decisions, because they may in the future become important indicators of safety. Examples of such tests include short-term mutagenicity tests as well as various other tissue culture and organ tests.

Also included in the scope of these regulations are studies of safety of regulated products on target animals, acute toxicity studies on a final product formulation, studies of test articles that are completed in 14 days or less, studies conducted on test articles used in "minor food producing species of animals," and studies on test articles which are not widely used.

11. Several comments closely related to the concerns expressed in paragraph 10 of this preamble requested that further language be added to the regulation exempting certain specific types of studies from coverage.

The Commissioner has reviewed the requests and has chosen not to change the language of the regulation itself to exclude specific study types other than those already mentioned (e.g., studies utilizing human subjects). The regulations apply to any study conducted to provide safety data in support of an application for a research or marketing permit for an FDA-regulated product, and a specific type of study which may be important in the overall safety evaluation of one type

of regulated product may not be important in evaluating another. The Commissioner believes it useful to identify in this preamble further examples of studies that are—or are not—within the scope of the GLP regulations.

Examples of studies that are not within the scope of these GLP regulations include:

a. Clinical tests performed solely in conjunction with product efficacy.

b. Chemical assays for quality control.

c. Stability tests on finished dosage forms and products.

d. Tests for conformance to pharmacopeial standards.

e. Pharmacological and effectiveness studies.

f. Studies to develop new methodologies for toxicology experimentation.

g. Exploratory studies on viruses and cell biology.

h. Studies to develop methods of synthesis, analysis, mode of action, and formulation of test articles.

i. Studies relating to stability, identity, strength, quality, and purity of test articles and/or control articles that are covered by good manufacturing practice regulations.

Further examples of types of tests not covered include:

a. Food additives: Tests of functionality and/or appropriateness of the product for its intended use; tests of extractability of polymeric materials that contact food; and all chemical tests used to derive the specifications of the marketed product.

b. Human and animal drugs: Basic research; preliminary exploratory studies; pharmacology experiments; studies done to determine the physical and chemical characteristics of the test article independent of any test system; and clinical investigations.

c. Medical devices: All studies done on products that do not come in contact with or are not implanted in man.

d. Diagnostic products: Essentially all are excluded.

e. Radiation products: Chemical and physical tests.

f. Biological products: All tests conducted for the release of licensed biologicals described in Part 601 (21 CFR Part 601) of this chapter.

These examples do not represent all the exclusions from the regulations, but provide guidance in applying the agency's safety considerations to specific situations. The defined scope of the regulations is necessarily broad to encompass the wide range of types of safety tests, types of testing facilities and regulated products for which proper safety decisions are important.

12. More than 20 comments sought the addition of specific language exempting various classes of FDA-regulated products, such as medical de-

vices, from coverage by the regulations.

The Commissioner has general elected not to permit exempti based on broad categories of regula products because no compelling sons have been presented that wo support the contention that assura of safety is less desirable for one c of regulated products than for other. Proper safety decisions are portant for all these products; acco ingly, the processes by which s safety data are collected should subjected to identical standards quality and integrity.

13. Several comments said that animal care provisions should ap only to these nonclinical studies us laboratory animals and should apply to nonclinical studies which volve large animals.

It is clear that the animal care pr sions are directed toward the use laboratory animals, and therefore tain of these provisions may not ap to studies not involving laboratory mals, such as tissue residue and tabolism studies conducted in cat Although these studies do fall wit the definition of a nonclinical labo tory study, the animals used in suc study are not generally kept in a la ratory setting. Because the husban requirements for laboratory anin differ greatly from those for large mals, the agency does not require t large animals be reared and m tained under the same conditions laboratory animals. The regulati are revised to include terms such "when applicable" and "as requir in those provisions for which a v latitude of acceptable husbandry p tice exists.

14. Three comments said the reg tions should apply to all stu whether submitted in support of o a challenge to an "application for a search or marketing permit."

The Commissioner agrees, in pri ple, that all nonclinical studies sho be performed in a manner designe ensure the quality and integrity of data. FDA is requiring that, at time a study is submitted, there be cluded with the study either a st ment that the study was conducte compliance with Part 58 requireme or, if the study was not conducte compliance with those requiremen statement that describes in detail deviations. This requirement me that, at the time a study not cond ed in compliance with the requ ments is submitted, the agency evaluate the effects of the noncon ance and take one of the following tions: (1) Determine that the nonc pliance did not affect the validit the study and accept it, or (2) de mine that the noncompliance have affected the validity of the st and require that the study be validated by the person submitting it, or (3) reject the study completely. The standard of review applied to studies that contain data adverse to a product is no different. That is, a study that failed to comply with these regulations might, nonetheless, contain valid and significant data demonstrating a safety hazard. Thus, FDA is not proposing a double standard, but is, rather, seeking to address those studies that present the most serious regulatory problems.

The preamble to the proposed regulation (41 FR 51215) discussed this issue as follows:

Valid data and information in an otherwise unacceptable study which are adverse to the product, however, may serve as the basis for regulatory action.

This disparity in treatment merely reflects the fact that a technically bad study can never establish the absence of a safety risk but may establish the presence of a previously unsuspected hazard. It reflects current agency policy, even in situations where the scientific quality of an investigational drug study is not in question. FDA may receive data but not use it in support of a decision to approve testing or commercial distribution because of ethical improprieties in the conduct of the study. (See 21 CFR 312.20).

A positive finding of toxicity in the test system in a study not conducted in compliance with the good laboratory practice regulations, may provide a reasonable lower bound on the true toxicity of the substance. The agency must be free to conclude that scientifically valid results from such a study, while admittedly imprecise as to incidence or severity of the untoward effect, cannot be overlooked in arriving at a decision concerning the toxic potential of the product. The treatment of studies conducted by a disqualified testing facility is discussed in paragraph 231a, below.

15. Exemptions from coverage by these regulations were requested for various types of facilities. Requests were received that they not apply to academic, medical, clinical, and not-for-profit institutions.

The public health purpose of these regulations applies to all laboratory studies on which FDA relies in evaluating the safety of regulated products, regardless of the nature of the facilities in which the studies are conducted. The Commissioner finds that granting an exemption based on type of facility would frustrate the intent of the good laboratory practice regulations. Many other comments urged that such exemptions not be considered because the standards applied to nonclinical testing should be uniform. Many of the requests for exemption were based on the idea that academic or not-for-profit institutions conduct primarily basic research and ought,

therefore, to be specifically excluded. Insofar as academic institutions are concerned, the Commissioner notes that such institutions conduct significant amounts of commercial testing pursuant to contracts. He also notes that significant levels of noncompliance with GLP requirements have been found in such institutions. Moreover, as noted in paragraph 11, basic research on drugs is outside the scope of these regulations. In short, no justification has been presented to warrant granting an exemption to such a facility, and any such exemption from the regulations by the type of facility collecting safety data would not provide equal application of the principles of good laboratory practice. Product safety decisions are equally important whether data are collected by the largest commercial nonclinical laboratory facility or by the smallest nonprofit facility. Therefore, the data collected in all types of facilities should be subjected to the same standards of quality and integrity. The results of the pilot program show that the proposed regulations represent achievable standards.

16. Exemption of or different standards for studies conducted outside the United States were requested.

These regulations are designed to protect the public health of the American people by assuring the scientific integrity and validity of laboratory studies that the agency relies on in evaluating the safety of regulated products. The same assurance is needed, whether the studies relied on are foreign or domestic in origin. The Commissioner notes that FDA clearly may refuse to accept studies from any nonclinical testing facility, foreign or domestic, that does not follow the requirements set forth in these regulations. To exempt from the requirements imposed on studies conducted in domestic testing facilities a nonclinical study conducted in a testing facility outside the United States that is submitted to FDA in support of an application for a research or marketing permit or to impose different standards for such studies, would only have the effect of discriminating against U.S. firms. Although inspection of a foreign facility may not be made without the consent of that facility, FDA will refuse to accept any studies submitted by any facility that does not consent to inspection. These same conditions apply to other FDA regulations, e.g., the current good manufacturing practice regulations (21 CFR Part 210), a program of inspection of foreign facilities for compliance with those regulations has been conducted by FDA for several years. A similar inspection program of foreign laboratory facilities conducting studies within the scope of this regulation will

be conducted; several foreign labo... ries were inspected during the ... program, and mechanisms for su... spections are being worked out ... representatives of the responsible ... ulatory authorities in foreign ... tries.

DEFINITIONS

The Commissioner received ... dreds of comments regarding d... tions (§ 58.3). General comment... listed immediately below; comm... regarding specific definitions foll... numerical order.

17. Several comments asked ... commonly used terms such as "ba... "area," "laboratory," "patholo... "quality data," "data integrity," ... pervisor," and "management" b... fined or clarified.

The Commissioner finds that, ... the exception of "batch," the ... set out above do not require indiv... definitions. The term "patholog... used in its ordinary sense, as ar... terms "supervisor" and "managen... and the phrases "quality data" ... "data integrity." As a general rul... regulation defines separately ... those words which will be used ... sense which differs from that giv... currently accepted dictionaries ... words whose meaning will be li... by the regulation. A new defir... has been added for the term "b... because it is used in these regula... in a context different from ... agency regulations, e.g., the good ... ufacturing practice regula... "Batch" in these regulations me... specific quantity of a test or co... article that has been characterize... cording to § 58.105(a).

18. Several comments on § 5... questioned the applicability of ... term "test substance" to medica... vices, radiation products, in vitro ... nostic products, and botanical m... als.

The Commissioner has reviewe... comments carefully and finds ... many of the comments submitte... garding the term "test subst... argued that the term, as defined ... not accurately reflect the scop... tended to be covered. Because ... term "substance," in common u... refers to chemical compounds an... logical derivatives of more or les... fined composition, and because ... term is not commonly understo... include devices or electronic pro... the Commissioner has changed ... term "test substance" to "test art... The term "article" is intended t... clude all regulated products ... may be the subject of an applic... for a research or marketing perm... defined in § 58.3(e).

The Commissioner has delete... reference to botanical material ... cause all botanical materials subj...

A jurisdiction are adequately encompassed by the other articles specifically mentioned in the definition.

9. Clarification of the term "control substance" (§ 58.3(c)) was requested. Several comments asked whether the term was to include carrier substances and solvents and vehicles. Other comments suggested this term could be confused with the same term used by the Drug Enforcement Administration.

The term is changed to "control article" to parallel the revised definition of test article. This change avoids any potential conflict with definitions used by other agencies. The term is intended to define those materials given to control groups of test systems for establishing a basis of comparison. The Commissioner recognizes that for certain nonclinical laboratory studies, no control groups are used, and therefore this definition would not apply. For example, testing the safety of implantable pacemakers in animals would require either no control animals or animals that have only been "sham-operated." The definition includes carrier materials when such carrier materials are given to control groups within test system and likewise for administered vehicles and solvents. The term also applies to articles used as positive controls.

20. Many comments on § 58.3(d) addressed the definition of the term "nonclinical laboratory study." A great many, if not the majority, of the comments sought to change the definition by adding language excluding certain specific tests, products, or types of laboratories.

The Commissioner notes that many of these comments overlap with or are identical to comments submitted in response to § 58.1 (Scope). To the extent that the comments and issues are the same, they have been dealt with in the discussion of § 58.1, above. Other comments are dealt with specifically below.

21. Many comments stated that the proposed language which included studies intended to assess the functionality and/or effectiveness of a test article should be deleted. One comment stated that efficacy testing in nonclinical tests is, by definition, preliminary and should be excluded to be consistent with the scope defined in § 58.1. Other comments stated that the language was too broad and too ambiguous and could be interpreted to include many studies which were not safety studies at all.

The Commissioner has considered these comments and agrees that the language related to functionality and/ effectiveness is too broad. He has, therefore, deleted the sentence.

22. Several comments requested that the last sentence of § 58.3(d) be modi-

fied by deleting the proposed examples of tests.

The Commissioner finds that the examples included in the proposal tended to confuse rather than clarify. The examples, therefore, have been deleted.

23. Section 58.3(e), which defines the various types of submissions to FDA, was criticized for use of the term "application for research or marketing permit." Several comments said the term was misleading because not all products are regulated through the use of "permits."

The Commissioner believes the term is appropriate for the purpose of these regulations. As stated in the proposal, this definition includes all the various requirements for submission of scientific data and information to the agency under its regulatory jurisdiction, even though in certain cases no permission is technically required from FDA for the conduct of a proposed activity with a particular product, i.e., carrying out research or continuing marketing of a product. The term is intended solely as a shorthand way of referring to the separate categories of data (identified in the proposal) that are now, or in the near future will become, subject to requirements for submission to the agency.

24. One comment stated that proposed § 3e.3(e)(14) should be deleted because the language was overly broad and because it contradicted the intent expressed in the preamble to limit GLP regulations to safety studies.

The Commissioner notes that the preamble to the proposal (41 FR 51209) stated that studies conducted to determine whether a drug product conforms to applicable compendial and license standards were excluded from the regulation. Safety data submitted to obtain the initial licensing of a biological product are covered by these regulations in § 58.3(e)(13). Once a biological is licensed, however, it becomes subject to testing procedures similar to compendial testing procedures. The Commissioner finds that postlicensing testing of biologicals is conducted more for quality control purposes than for establishing the basic safety of the biologic product and has, accordingly, deleted postlicensing testing from the definition of research and marketing permit.

25. Several comments stated that in vitro diagnostic tests (proposed § 3e.3(e)(15)) should not be included because in vitro diagnostic products do not come in contact with patients and do not, therefore, require preliminary animal safety testing.

Because in vitro diagnostic products do not require any nonclinical laboratory tests for agency approval, the Commissioner agrees that in vitro diagnostic products need not be included

in the definition "application for a research or marketing permit." Proposed § 3e.3(e)(15) has, therefore, been deleted from the final regulation.

26. Several comments objected to the inclusion of medical devices in § 58.3(e) (16), (17), and (18), stating that medical devices were not "test substances," that medical devices should not be included because the rules for data submission for such devices were as yet undefined, and that inclusion of medical devices would be unduly restrictive. These comments suggested either total or partial exclusion from coverage under the good laboratory practice regulations.

For reasons stated previously, the Commissioner does not agree that medical devices, as a category, should be excluded. Implantable devices may be composed of polymeric material that contain components capable of leaching from the device into the body of the recipient or may themselves be adversely affected by body constituents. In either case, safety studies would be necessary to demonstrate that components of the device did not cause harm or that the body constituents did not promote breakdown or malfunction of the device.

27. Comments also requested deletion of all terms relating to radiation products in § 58.3(e) (20), (21), and (22), stating that to include such products would restrict experimentation unduly, and arguing that radiation products were not "test substances."

The Commissioner rejects these comments. The quality and integrity of the safety data are no less important for radiation products than they are for other agency-regulated products. He does not agree that including radiation products will unduly restrict experimentation. The remaining argument is covered in the discussion of "test article" above. A new paragraph § 58.3(e)(19) is added to cover data and information regarding an electronic product submitted as part of the procedure for obtaining an exemption from notification of a radiation safety defect or failure of compliance with radiation performance standard, described in Subpart D of Part 1003 (21 CFR Part 1003).

28. Many comments stated that the term "sponsor" in § 58.3(f) was too broadly defined. For example, two comments stated that the definition as written, would cover a company which provides a grant to a university a fact which, if true, would inhibit giving grants. One comment said that the definition is so broad that it could be interpreted to apply to stockholders.

The Commissioner advises that a person providing a grant may be a sponsor. In the area of nonclinical laboratory studies, most grantors will mately submit the data to the agency. The Commissioner does not agree that because the definition of "sponsor" includes grantors it will inhibit the giving of grants. No data were submitted to support this argument. The Commissioner further advises that the definition does not include stockholders.

29. Other comments on § 58.3(f) asked whether the regulation allowed for multiple sponsors and whether government agencies could be sponsors.

"Person," as defined in § 58.3(h), includes government agencies, partnerships, and other establishments such as associations. Therefore, a government agency can clearly be a sponsor. In addition, the Commissioner advises that the definition does not preclude joint sponsorship of a study.

30. Several comments asked that the definition of "testing facility" in § 58.3(g) be revised to indicate clearly that a facility conducting a study subject to the regulations should be subject only to the extent that the facility is involved with and responsible for the study.

The Commissioner concludes that no revision to the definition is necessary. The definition clearly does indicate that a facility is covered by the regulations only to the extent that the facility is conducting or has conducted nonclinical laboratory studies.

31. Numerous comments addressed the definition of "test system" in § 58.3(l). Eighteen comments stated that the definition, as written, could be interpreted to require testing of beakers and test tubes. Two comments pointed out that the "test system" is not the container being tested for extractables, but rather it is the animal, microorganism, or cellular components used to test the extractables for safety.

The Commissioner has carefully reviewed the proposed definition in light of the comments and has made a number of changes. The terms "cellular and subcellular" have been replaced for clarity with "subparts thereof" which refers to animals, plants, and microorganisms. The revised definition now reads: "'Test system' means any animal, plant, microorganism, or subparts thereof, to which the test or control article is administered or added for study. 'Test system' also includes appropriate groups or components of the system not treated with the test or control articles." The revisions should make the definition clearly consistent with § 58.3(d) ("nonclinical laboratory study"), which states that studies to determine physical or chemical characteristics of a test article or to determine potential utility are not included. Therefore, testing of beakers and test

tubes, which fall into the category of physical and chemical tests, is excluded.

32. Section 58.3(j), which defines "specimen," drew several comments. These included requests for precise definition of the terms "material" and "tissue" and requests for a clearer definition of the term "specimen."

The Commissioner is modifying the term "specimen" to include any material derived from a test system for examination or analysis. Under these circumstances, blood, serum, plasma, urine, tissues, and tissue fractions are all included if they are intended for further examination or analysis. The definition includes all materials that yield data related to the safety decision on a regulated product.

33. Many comments were received on the definition of "raw data" in § 58.3(k). Included were requests to clarify the term "certified" and to state whether carbons, photocopies, and written reports of dictated material could be classified as "raw data". Other issues concerned whether financial information and first drafts of reports were "raw data."

The Commissioner concludes that the proposed definition should be clarified. The word "exact" is substituted for the word "certified." "Certified" connotes a legal document that requires notarization; "exact" has no such connotation and more precisely reflects the Commissioner's intention. The definition is further clarified by inserting, after the first sentence, a new sentence which reads: "In the event that exact transcripts of raw data have been prepared (e.g., tapes which have been transcribed verbatim, dated, and verified accurate by signature), the exact copy or exact transcript may be substituted for the original source as raw data." This clarification will permit data collection by tape recorders without requiring the retention of the original tapes. Carbons and photocopies satisfy the regulations, provided they are exact and legible copies of the original information. Neither financial information nor first drafts of reports are raw data within the meaning of the term.

34. Several comments said only recorded data contributing substantially to the study should be retained and, similarly, only computer printouts contributing substantially should be retained. Several comments requested clarification of the method for storing machine-generated data and definition of "on line data recording system."

Because the parenthetical example ("derived from on-line data recording systems") served more to confuse than to clarify, it has been deleted. However, an "on line data recording system" pertains to an instrument that can feed data directly into a computer

that analyzes and stores the infor tion. The product of this activity ally consists of a memory unit ph computer program for extracting information from the unit. Hard-c computer printouts are unnecess provided the computer memory program are accompanied by a pr dure that precludes tampering w the stored information.

The Commissioner cannot ag that only those portions of the d that contribute substantially to study need to be retained. Such an proach would require a judgment t made which, if in error, could lea improper or incorrect study re struction. The purpose of retair the raw data is to permit the qua assurance unit and agency invest tors to reconstruct each phase c nonclinical laboratory study. Disc ing essential records would frust this purpose. Raw data may be st in separate areas provided the arch indexes give the data location.

35. Many comments addressed "q ity assurance unit" in § 58.3(l).

The Commissioner has revie these comments and concludes t they are more concerned with the cept of the quality assurance than with the definition. The c ments are therefore dealt with detail in that section of the prean concerned with § 58.35 of the reg tions. (See paragraphs 75 throug below.)

36. Several comments addre "study director" in § 58.3(m). T comments requested clarification, mission to have more than one st director per study, and that the t "implementation" be changed to " duct."

The Commissioner has revised definition to read: "'Study Dire means the individual responsible the overall conduct of a nonclir laboratory study." The revision i tended to emphasize that the stud rector is responsible for the er study, as well as being responsible the interpretation, analysis docum tation, and reporting of results.

The Commissioner concludes the other comments received on definition of "study director" dressed the concept rather than definition, and these comments dealt with under the discussion § 58.33 (see paragraphs 59 throug below).

APPLICABILITY TO STUDIES PERFORM
UNDER GRANTS AND CONTRACTS

37. Two comments requested sion of § 58.10 to specify clearly the sponsor is ultimately respon for data validity, even if the data obtained by a sponsor from a gra or contractor.

The Commissioner concludes that no vision of § 58.10 is necessary. All pers involved in a nonclinical labora- ry study are responsible for part or l of the study, depending upon the :tent of their participation. Athough sponsor who submits studies to FDA ars the responsibility for the work rformed by a subcontractor or antee, that fact in no way relieves a antee or subcontractor from individ- il responsibility for the portion of .e study performed for the sponsor. .deed, the purpose of the require- ent that the sponsor notify a grant- or subcontractor that the work ing performed is a part of a nonclin- il laboratory study which must be nducted in compliance with the od laboratory practice regulations is assure that all parties submitting .ta are aware of their responsibilities der the regulation.

38. Several comments requested ex- aption for certain specialized serv- es which are not commonly availa- e, e.g., ototoxicity studies with diure- :s. The comments stated that these ecialized services would probably t be available to them if the strin- nt requirements of the regulations id to be met by the service organiza- n.

The Commissioner concludes that rtain specialized services cannot be :empted from these regulations. The ecialized services may contribute in rge measure to the agency decision approve a research or marketing rmit. If the studies are intended to ovide safety data in support of an plication for a research or market- g permit, their conduct falls within ie scope of these regulations.

INSPECTION OF A TESTING FACILITY

39. Comments on the inspection pro- sions (§ 58.15) expressed concern re- irding the competence and scientific ialifications of FDA investigators.

The agency has endeavored, through specialized training program, to sure that FDA investigators are mpetent to perform good laboratory actice inspections. The EILP pro- am is new, and training and evalua- on will continue to improve it. The sults of the pilot inspection program id the manner in which it was co- icted should provide added assur- ice to testing facility management garding the competence of FDA in- stigators. The quality of the pro- am is not, however, dependent on ie competence or training of any ngle individual. Inspection of find- igs are always subject to supervisory view within the agency, and no offi- al action may be taken without con- irrence of a number of qualified per- ins.

40. Several comments stated that gency inspection should be limited to those facilities under current FDA legal authority.

The scope of the regulations and the definition of a "nonclinical laboratory study" define those studies covered by the regulations. The agency intends to inspect all facilities which are conduct- ing such studies. Many of these facili- ties are subject to inspection under ex- press statutory authority vested in FDA. As noted in the preamble to the proposal (41 FR 51220):

Inspections of many, perhaps most, test- ing facilities will not be conditioned upon consent. Under section 704(a) of the act, FDA may inspect establishments including consulting laboratories, in which certain drugs and devices are processed or held, and may examine research data that would be subject to reporting and inspection pursu- ant to section 505 (i) or (j) or 507 (d) or (g) of the act. In addition, any establishment registered under section 510(h) of the Act is subject to inspection under section 704 of the act. Thus, most manufacturing firms that conduct in-house non-clinical labora- tory studies on drugs and devices, and those contract laboratories working for such firms, would be subject to FDA inspection whether or not they consented.

Facilities that are not subject to statu- tory inspection provisions will be asked to consent to FDA inspection. The absence of any statutory authori- zation does not bar FDA from asking permission to conduct an inspection, and the agency should not bar itself from seeking permission. Thus, the proposal in the comment is not accept- ed.

41. Several comments requested that FDA make its enforcement strategy known as promised in the preamble to the proposal.

The enforcement strategy was dis- cussed in the preamble to the proposal (41 FR 51216) and is amplified in the compliance program which imple- ments this regulation. The compliance program is publicly available and may be obtained by sending a written re- quest to the agency official whose name and address appear at the begin- ning of this preamble as the contact for further information.

42. Two comments on § 58.15 as pro- posed requested that the requirement that the testing facility permit inspec- tion by the sponsor be deleted. The comments argued that the rights and obligations of a sponsor and its labora- tory are a matter of contract between them alone, and not a proper subject for government regulation.

The Commissioner has considered this issue, is persuaded that the com- ments are correct, and has deleted the phrase "the sponsor of a nonclinical laboratory study." At the same time, however, the Commissioner reempha- sizes that, because a sponsor is respon- sible for the data he or she submits to the agency, the sponsor may well wish to assure that the right to inspect a testing facility is included in any co: tract.

43. Other comments suggested th the sponsor should accompany tl FDA investigator during an inspectic of a contract testing facility and th FDA access to data should require tl sponsor's consent.

The Commissioner disagrees wit these comments. An agency investig: tor may be inspecting the results studies from several sponsors durir an inspection. The logistics required notify and arrange for several spor sors to accompany an investigator, to obtain sponsor consent to inform tion release, would be unworkabl FDA's practice of unannounced i spections has proved to be an effectiv and efficient use of scarce resource Because of resource limitations, FD cannot inspect each facility as often it would like to, and the Commission finds that the possibility of una nounced FDA inspections at any tin motivates compliance.

44. Many comments were concern that trade secret information obtain during the inspection would be r leased by FDA.

The Commissioner notes that tra secrets obtained as a result of an i spection are fully protected under tl provisions of section 301(j) of the a (21 U.S.C. 331(j)), as well as 18 U.S. 1905 and the Freedom of Informatic Act (5 U.S.C. 552(b)(4)) and the FDA implementing regulations (21 CF 20.61). Interested parties may refer the agency's public information reg lations (21 CFR Part 20), whic govern agency release of documents.

45. One comment requested that tl results of government laboratory i spections be made public.

The Commissioner notes that no d tinctions will be made between gover ment or nongovernment laboratorie The results of an inspection of testii facilities will be available after all quired followup regulatory action h been completed.

46. The phrase "and specimens" h been added to § 58.15(a). The Comm: sioner finds that examination of spe mens may be required to enable tl agency, where necessary, to reco struct a study from the study recorc

47. Many comments stated that tl inspection of records should n extend to certain records compiled the quality assurance unit.

The Commissioner agrees and h exempted from routine inspectic those records of the quality assuran unit which state findings, note prc lems, make recommendations, evaluate actions taken following r ommendations. These exemptio from inspection are discussed in gre er detail under the discussion § 58.35.

48. A new paragraph (b) has been added to § 58.15. This paragraph is similar to proposed § 58.200 and reiterates that a determination that a nonclinical laboratory study will not be considered in support of an application for a research or marketing permit does not relieve an applicant from any obligation under any applicable statute or regulation (e.g., 21 CFR Parts 312, 314, 514, etc.) to submit the results to FDA. If a testing facility refuses inspection of a study, FDA will refuse to consider the study in support of an application for a research or marketing permit. This refusal, however, does not relieve the sponsor from any other applicable regulatory requirement that the study be submitted.

ORGANIZATION AND PERSONNEL

PERSONNEL

49. A number of comments addressed the definition of training, education, and experience in § 58.29. Several comments considered such references too vague; several others suggested that appropriate qualifications be established by professional peer groups.

It would be inappropriate, if not impossible, for FDA to specify exactly what scientific disciplines, education, training, or expertise best suit a specific nonclinical laboratory study. These factors, which vary from study to study, are left to the discretion of responsible management and study directors. They are responsible for personnel selection and for the quality and integrity of the data these personnel will collect, analyze, document, and report. The Commissioner urges, however, that management and study directors carefully consider personnel qualifications as they relate to a particular study. The agency has uncovered instances, discussed in the preamble of the proposal (41 FR 51207), in which the conduct of a study by inadequately trained personnel resulted in invalid data. Although the Commissioner recognizes the value of certification by professional peer groups, he does not agree that the concept is appropriate for regulatory purposes.

50. Several comments said the study director should be given responsibility for assurance of qualifications of personnel.

The Commissioner agrees that, generally, the study director will be responsible for ensuring that personnel selected to conduct a nonclinical laboratory study meet necessary educational, training, and experience requirements. The Commissioner notes, however, that management also has selection and hiring responsibilities and privileges.

51. One comment stated that the requirement of § 58.29 that each individual engaged in the conduct of a study have sufficient training or experience to enable the individual to perform the assigned function should be limited to those personnel engaged in supervision and collection and analysis of data.

The Commissioner disagrees. These factors are important and should be considered for personnel other than supervisors or those engaged in collection and analysis of data. The approach suggested by the comment would ignore the fact that specific expertise is required, for example, by animal caretakers, physical science technicians, and by persons using pesticides near animal-holding areas. While the degree of education, training, and experience necessary for these positions will be quite different from the qualifications necessary for supervisors or scientific staff, the need for sufficient training or experience is no less important.

52. One comment pointed out the appropriateness of changing the term "person" to "individual" in § 58.29(a).

Because the term "person" as defined in § 58.3(h) includes partnerships, corporations, etc., the Commissioner agrees that "individual" is the proper term and has so amended § 58.29(a).

53. Seventeen comments questioned the use of, or objected to reference to, the term "curriculum vitae" for nontechnical personnel such as animal caretakers, as required in proposed § 58.29(b).

Another comment asserted that the requirement infringed on management's prerogatives without specifying how any such infringement occurred. One comment stated that the requirement that such records be retained after termination of employment was unnecessarily cumbersome.

The Commissioner does not agree that the requirement infringes on management's prerogatives. However, the Commissioner agrees with the remaining comments and has revised the section. "Curriculum vitae" has been changed to "summaries of training and experience plus job descriptions." Reference to the maintenance of records of terminated employees is deleted from this section because the requirement is redundant to the record retention requirements set forth in § 58.195(e).

54. Ten comments said the wording of § 58.29(c), relating to "sufficient numbers of personnel" and to "timely" conduct of the study, was vague.

The Commissioner purposely left the paragraph broad in context and coverage because differences in types of studies preclude any specific approach to defining numbers of personnel. The precise number of personnel required for a specific study, as [...] for all ongoing studies, is a m[...] ment decision. FDA experience[...] ever, indicates that a shorta[...] qualified personnel can lead to [...] equate or incomplete monitorin[...] study and to delayed preparatio[...] analysis of results, and the numl[...] personnel conducting a study [...] be sufficient to avoid these probl[...]

55. Ten comments requested de[...] of § 58.29(d) or clarification of tl[...] guage regarding employee [...] habits, stating that the section v[...] vague and that an employer v[...] sponsible for health habits o[...] work. One comment submitted [...] nate language.

The Commissioner adopts [...] modifications the alternate lan[...] The paragraph now requires onl[...] personnel take necessary pe[...] sanitation and health precauti[...] avoid contamination of test an[...] trol articles and test systems.

56. Several comments asked th[...] term "laboratory" in § 58.29(e), [...] plied to protective clothing, be c[...] because it is too restrictive. [...] comments suggested that the r[...] ment that clothing be chang[...] often as necessary to prevent co[...] nation be eased by changing "pr[...] to "help prevent." Four relate[...] ments requested modification [...] fect only "contamination affect[...] lidity of studies."

The Commissioner agrees [...] elimination of "laboratory" as a[...] to clothing. The provision of s[...] ized clothing is, however, an esta[...] and well-known procedure for pr[...] ing contamination in a variety [...] ations. The Commissioner di[...] with any suggested modificat[...] this section which weakens the [...] of the regulation. The objectiv[...] prevent contamination of th[...] system.

57. A number of commen[...] dressed several aspects of § 58.2[...] garding personal illnesses, pe[...] health records, types of illness[...] records of illnesses. Comment[...] disclosure of medical records [...] invasion of privacy and of litt[...] evance to the proper conduct of [...] clinical laboratory study.

The Commissioner agrees tha[...] mentation of personal illnesse[...] constitute an unwarranted inva[...] privacy, and this requirement [...] leted. The Commissioner di[...] with the requests for deletion [...] entire paragraph, noting the re[...] ship between personnel healt[...] possible contamination of te[...] tems. Revised § 58.29(f) require[...] viduals with illnesses that m[...] versely affect the quality and in[...] of nonclinical laboratory studie[...] excluded from direct contact wi[...] and control articles and test sy[...]

ersonnel should be instructed to
rt such medical conditions to their
ediate supervisor, who should pro-
test systems from personnel re-
ng as ill.

TESTING FACILITY MANAGEMENT

Many comments on the responsi-
es of the study director objected
some of the responsibilities as-
d to the study director were more
erly assigned to management.
e Commissioner agrees that sever-
the responsibilities previously as-
d to the study director should be
ned to the testing facility man-
ient. For clarification, a new
1 is added to the regulations. It is
agement's responsibility to assure
for each study there is a study di-
r and an independent quality as-
ice unit, as required by the regu-
ns. It is also management's re-
sibility to ensure that any devi-
is from the regulations which are
rted by the quality assurance unit
in turn, reported to the study di-
r and that corrective actions are
taken and documented. Designa-
management responsibilities in
manner merely clarifies the fact
the study director should be
ed as the chief scientist in charge
study. Duties which are more ad-
strative than scientific are the re-
sibility of management; however,
agement may delegate appropri-
administrative duties to the study
tor.

STUDY DIRECTOR

More than 50 comments ad-
sed the scope of responsibilities
osed for the study director. Many
ments stated that these responsi-
ies were much too broad for one
on.
the proposal, the Commissioner
nced the concept of a single fixed
t of responsibility for overall con-
of each nonclinical laboratory
y. Experience has demonstrated
if responsibility for proper study
luct is not assigned to one person,
e is a potential for the issuance of
licting instructions and improper
ocol implementation. The study
tor is charged with the technical
ction of a study, including inter-
ation, analysis, documentation,
reporting of results. As discussed
aragraph 58, several of the respon-
ities proposed for the study direc-
have been transferred to testing
lity management. This transfer
ild allay concerns regarding the
nitude of the responsibilities as-
ed to the study director.
. Nine comments object to the
a "ultimate" as applied to the
ly director's responsibility.
ne Commissioner agrees that "ulti-
e" responsibility for the study

rests with facility management and/or
the sponsor. Therefore, the word "ultimate" has been replaced by "overall"
in § 58.33.

61. Several comments argued that
more than one study director should
be allowed for each study.

The Commissioner rejects these
comments. As noted above, there must
be a single point of responsibility for
overall technical conduct of the study.
The potential for conflicting instructions and confusion in study implementation is too great to diffuse the
responsibility by, for example, study
direction by a committee. The regulation does not, however, preclude the
study director from directing more
than one study.

62. Many comments stated that the
requirements would interfere with
management's prerogatives to organize
and conduct studies as it so chooses.

The requirement that the study director be the single point of responsibility for technical conduct of the
study need not interfere with normal
delegation of authority by management.

63. Five comments argued that the
proposed requirements for study director and quality assurance unit were
duplicative.

The Commissioner has carefully reviewed the proposal and comments
and has clearly separated the responsibilities in the final regulation to avoid
duplication. The first sentence in
§ 58.33 has been revised to specify
clearly that each study shall have a
study director. The second sentence
has been revised to amplify the concept: "The study director has overall
responsibility for the technical conduct of the study, as well as for the interpretation, analysis, documentation
and reporting of results and represents the single point of study control."

64. One comment suggested revising
§ 58.33(a) to specify that the sponsor
must approve the protocol and the
study director must approve any
change.

The Commissioner advises that
§ 58.120(a)(15) requires that the sponsor approve the protocol, and
§ 58.120(c) requires that the study director approve any changes or revisions to the protocol. The language in
§ 58.33(a) has been revised to reference
§ 58.120.

65. Five comments objected to the
proposed requirement that the study
director assure that test and control
articles or mixtures be appropriately
tested. The comments argued that this
was not a proper function of the study
director.

The Commissioner agrees that this
responsibility is more properly assigned to testing facility management.

Therefore, the requirement has been
transferred to § 58.31(d).

66. Three comments suggested that,
rather than the study director assuring that test systems are appropriate,
the study director should assure that
the test systems are as specified by the
protocol.

The Commissioner agrees that the
determination of the appropriateness
of the test system is a scientific decision beyond the scope of these regulations. Section 58.33(d) has been revised to state: "Test systems are as
specified in the protocol."

67. Four comments argued that the
scheduling of personnel, resources,
facilities, and methodologies was not a
proper requirement for the study director.

The Commissioner agrees that this
scheduling is beyond the scope of the
study director's responsibilities and
has, therefore, transferred it to the responsibilities of testing facility management in § 58.31(e).

68. Two comments object to the requirement that personnel clearly understand the functions they are to perform.

The Commissioner finds that it is essential that personnel be adequately
trained to assure the integrity and validity of the data. However, the Commissioner concludes that training is a
proper responsibility of testing facility
management and has transferred the
requirement to § 58.31(f).

69. Three comments suggested deletion of the phrase "and verified" from
the proposed requirement that the
study director assure that all data are
accurately recorded and verified. Four
comments requested definition of the
term "verified."

The Commissioner disagrees with
the requested deletion. Recording and
verifying data are key operations in
the successful completion of a study.
The Commissioner intends that the
study director assure that data are
technically correct and accurately recorded. "Verified" is used in its ordinary sense of "confirmed" or "substantiated." The process by which verification is achieved may be determined by
the study director.

70. One comment stated that proposed § 3e.31(a)(8) merely repeated
proposed § 3e.31(a)(7).

The Commissioner finds that the
two sections can be combined for clarity. Accordingly, § 58.33(c) now reads
"unforeseen circumstances that may
affect the quality and integrity of the
nonclinical laboratory study are noted
when they occur, and corrective action
is taken and documented."

71. One comment stated that the requirement that the study director
assure that responses of the test
system are documented is unreasonable.

The Commissioner disagrees. Assuring that all experimental data including unforeseen responses to the test system are accurately observed and documented is a critical part of study conduct and is a responsibility properly assigned to the study director.

72. Two comments stated that the requirement that the study director assure that good laboratory practice regulations are followed either should be modified to make it more flexible or should be deleted. One comment suggested that the study director should be allowed to delegate the responsibility.

The Commissioner rejects these comments. The regulations constitute an effective means to aid study directors in achieving better control of complex studies. Responsibility for assuring compliance properly rests with the study director. While delegation of authority is always the prerogative of a manager, responsibility cannot be delegated.

73. Several comments stated that the requirement that the study director assure that study documentation is transferred to the archives is redundant to § 58.190.

The Commissioner does not agree that the sections are redundant. Section 58.190 requires that the study records be retained, and § 58.33(f) requires that the study director assure that the records are transferred for retention. The phrase "and other information to be retained" has been deleted from § 58.33(f) because the phrase is subsumed by raw data, documentation, protocols, specimens and final reports.

74. Thirteen comments questioned the proposed approach to study director replacement, specifically objecting to the requirement that justification of such replacement be documented and retained as raw data. The comments argued that justification carries a negative connotation and that replacement of a study director is a management prerogative.

The Commissioner is persuaded that replacement of the study director should remain within the discretion of management and that the requirement that justification for such replacement be documented and retained is an inappropriate subject for these regulations. Consequently, the requirement for justification for such replacement has been deleted. The requirement that the study director be replaced promptly when necessary has been transferred to § 58.31(b).

QUALITY ASSURANCE UNIT

75. More than 100 comments objected to part or all of § 58.35 as proposed. Many comments questioned the need for a quality assurance unit as proposed. Some comments stated that the

establishment of such a unit would increase the administrative burden and costs of performing nonclinical studies to the point of forcing small facilities out of business. Others stated that the provisions would interfere with management's prerogatives to organize the facility or with the informed scientific judgment of principal investigators or study directors.

The Commissioner has retained the requirement that each testing facility have a quality assurance unit (QAU) to monitor the conduct and reporting of nonclinical laboratory studies. In view of the potential gain to management, to sponsors, and to FDA, through the added assurance of well-conducted studies, increased costs, if any, are justified. The quality assurance unit need not be a separate organizational entity composed of personnel permanently assigned to that unit. All nonclincial studies falling within the scope of this regulation must be monitored by a quality assurance unit composed of at least one person. Within this framework, management retains its organizational prerogatives. Because different individuals may be responsible for quality assurance functions at different times, it is important that quality assurance unit records be centrally located, and § 58.35(e) has been modified to so require. The regulations permit a study director for a particular study to serve as a part of the quality assurance unit or as the quality assurance unit for a different study. However, for any given study a separation must exist between individuals actually engaged in the conduct of a study and those who inspect and monitor its progress. In those situations in which several different individuals are performing the quality assurance functions for different studies, each such individual must maintain that portion of the master schedule sheet which relates to the study he or she is monitoring. This means that several people may be responsible for maintaining the master schedule sheet. Because the function of the quality assurance unit is administrative rather than scientific, the Commissioner does not agree that the functions of a QAU will interfere with the study director's control of the overall technical conduct of the study. In order to emphasize this point, the following language has been added to § 58.35(a): "For any given study the quality assurance unit shall be entirely separate from and independent of the personnel engaged in the direction and conduct of that study."

76. Sixteen comments objected to the word "unit" in the term "quality assurance unit" and suggested alternate words such as "function" or "program."

The Commissioner has elect[ed] preserve the word "unit" to cor[respond] to similar wording in other regula[tions] such as the current good manuf[actur]ing practice regulations. The Com[mis]sioner agrees, however, with the [ratio]nale of the comments that the h[impor]tant objective of this section is [that] there be a quality assurance fun[ction] operating for each nonclincial s[tudy]. As indicated in paragraph 75, exact organizational means by [which] this function is achieved is the pr[erog]ative of facility management and [will] vary from facility to facility.

77. Numerous comments add[ressed] the composition of the quality [assur]ance unit. Four comments soug[ht in]clusion of criteria for education[, train]ing, and experience of QAU pers[onnel]. Seven comments indicated that [com]pliance with this section was imp[racti]cal because of a shortage of p[eople] qualified to staff such a unit.

The Commissioner has not att[empt]ed to specify the qualificatio[ns of] quality assurance personnel be[cause] qualifications should be determin[ed by] management and will vary acco[rding] to the type of facility and the ty[pe of] studies conducted by each facilit[y. Be]cause the function of the quali[ty as]surance unit is to assure comp[liance] with procedural and administrati[ve re]quirements rather than to overse[e the] technical aspects of study co[nduct,] QAU personnel need not be limi[ted to] professional personnel and/or [scien]tists. The Commissioner does [not] agree, therefore, that there exist[s a se]rious shortage of qualified peop[le to] fulfill this function.

78. Two comments indicated [that] the quality assurance unit shou[ld be] composed of outside consultan[ts in] order to assure the independen[ce of] the function. One comment requ[ested] that quality assurance unit me[mber]ship be restricted to employees [of the] facility.

The Commissioner notes tha[t the] quality assurance functions m[ay be] performed by outside consul[tants.] This fact should enable small fa[cilities] or facilities conducting nonclinci[al lab]oratory studies for submission t[o the] FDA on an irregular basis to me[et the] quality assurance requirements [in a] cost-effective manner. At the [same] time, the Commissioner does not [feel] that the QAU function must b[e per]formed by an outside body. The [orga]nizational separation of the QA[U from] the study team should be suffici[ent to] assure objective monitoring b[y the] QAU.

79. Four comments question[ed the] last sentence in § 58.35(a) as pro[posed,] stating that it seemed to require [moni]toring of some studies by two QA[Us—] that of the sponsor and that [of the] contract facility.

FEDERAL REGISTER, VOL. 43, NO. 247—FRIDAY, DECEMBER 22, 1978

Commissioner has deleted from 5(a) the sentence in question. QAU of the testing facility is responsible for fulfilling the quality assurance functions for studies conducted within that facility. In cases where portions of a study, feed analysis, are performed by a contract facility which, because it is itself a nonclincial facility, does have a QAU, the person letting contract, and not the contract facility is responsible for the performance of the quality assurance functions.

Commissioner believes that the mechanism by which a sponsor is assured of the quality of nonclinical studies performed for it under contract is a matter that can be left to the contracting parties and need not be discussed in these regulations.

Three comments suggested that testing facilities be licensed or certified in lieu of having an ongoing quality assurance unit.

Commissioner considered such approach and rejected it before establishing the proposed regulations. (41 FR 51208–51209.) No persuasive arguments for changing this decision were presented in the comments. Diversity in the size and nature of clinical testing facilities subject to provisions of these regulations makes licensing or certification procedures impractical. The regulation is intended to assure the quality and validity of the data obtained by each nonclinical laboratory study, and the QAU provides a mechanism to monitor each ongoing study. Licensing a testing facility could not achieve the same.

Many comments objected to the provisions of § 58.35(b)(1) which require that the quality assurance unit maintain a master schedule sheet of nonclinical laboratory studies. Some comments believed the requirement was excessive, while others questioned the proposed format and content of the list. One comment pointed out that not every study includes all items listed.

Commissioner is convinced that maintenance of a master schedule is essential to the proper function of the Quality Assurance Unit. Only through such a mechanism can management be assured that the facilities are adequate and that there are sufficient numbers of qualified personnel available to accomplish the protocols of all nonclinical studies being conducted at a facility at any given time.

On careful review of the items required to be listed, the Commissioner agrees that the requirement that animal species be identified may be deleted because the requirement that "test system" be listed adequately

covers this point. He has, in addition, deleted the examples of study types because he agrees that including the information is not necessary to achieve objectives of this section. The Commissioner has further reworded this section to eliminate reference to whether the final report has been approved for submission to the sponsor because the language was strictly applicable only to studies done under contract. The revised language simply requires that the status of the final report be listed.

82. Nine comments objected that § 58.35(b)(2) required too much duplicative paper.

The Commissioner has concluded that the QAU must maintain copies of study protocols to assure that they are followed and amended in accordance with the further provisions of these regulations. The Commissioner agrees that the requirement that the QAU maintain copies of all standard operating procedures would substantially increase the volume of records needed to be retained by this unit. Because there should be many copies of standard operating procedures present throughout the facility which should be freely available to QAU members, the Commissioner has deleted the requirement that these be maintained by the QAU.

83. Fifteen comments suggested that § 58.35(b)(3) be deleted on the basis that FDA should not dictate how the QAU achieves its objectives. One comment suggested that "inspect" be changed to "audit."

The Commissioner remains convinced of the need for a formal mechanism through which the QAU maintains oversight of the conduct of a study. Such a mechanism must be based on direct observation in order that the independence of the QAU be preserved. The Commissioner has retained the word "inspect" in preference to "audit." "Inspect" more accurately conveys the intent that the QAU actually examine and observe the facilities and operations for a given study while the study is in progress, whereas "audit" could be interpreted to mean simply a detailed review of the records of a study. Because the QAU function is to observe and report the state of compliance with the regulations and to determine whether the protocol is being followed rather than to verify the results of a study, "inspect" more properly conveys the agency's intent.

84. Fourteen comments addressed the need to inspect "each phase of a study * * * periodically," seeking clarification or different language. Nine of these comments called for the use of random sampling procedures in choosing studies or phases of studies to inspect in order to decrease the work-

load and resource requirements of the QAU.

The Commissioner does not agree that random sampling would be an adequate method of evaluation in the nonclinical laboratory setting. In situations which involve the repetition of similar or identical procedures, random sampling can provide an adequate means of quality control. Here, however, the differences among study operations and among the personnel conducting them invalidate any assumption that the conduct of one phase of one study is representative of the conduct of that phase of another or of other phases of a single study. The term "each phase" is intended to emphasize the need for repeated surveillance at different times during the conduct of a study so that each critical operation is observed at least once in the course of the study. The term "periodically" is retained to indicate the need for more than one inspection of certain repetitive continuing operations that are part of the conduct of longer term studies such as animal observations and diet preparation.

85. Many comments objected to the proposed requirement that any problems found by the QAU be brought to the attention of management and appropriate responsible scientists. Some felt that this would require that excessive resources be spent on minor problems. Others felt that notification of appropriate supervisory personnel rather than management was sufficient.

The Commissioner agrees that only those problems likely to affect the outcome of the study need to be brought to the immediate attention of personnel who are in a position to resolve those problems, and the language of § 58.35(b)(3) has been changed accordingly. The term "management" in its ordinary usage means appropriate supervisory personnel and has not, therefore, been changed.

86. More than 40 responses to proposed § 3e.33(b)(4) objected to the specific time frames required for evaluation. Several comments suggested that the paragraph be deleted. Others objected to the specific requirements, and still others stated that appropriate times for evaluations should be selected by management.

The Commissioner advises that periodic inspection is necessary and that the time periods specified are the minimum required to assure that a study is being conducted in compliance with the regulation. Should deviations be found during the periodic inspections, there may still be time to take corrective action. The Commissioner has, however, determined that inspection of studies lasting less than 6 months need only be conducted at intervals adequate to assure the integrity of the study and that specific time intervals for such studies need not be set out in this regulation. The requirement that studies lasting more than 6 months be inspected every 3 months remains unchanged. The section has been added to § 58.35(b)(3).

87. Several comments requested that the phrase "complete evaluation" in proposed § 3e.33(b)(4) be clarified.

The Commissioner has changed the term "complete evaluation" to "inspect." The function of the QAU is to inspect studies at specified intervals to maintain records required by this regulation, and to report to management and the study director deviations from the protocol and from acceptable laboratory practice. Evaluation of any reported deviations is left to the study director and to management.

88. Fifteen comments sought deletion of § 58.35(b)(4), which requires the periodic submission of status reports to management and the study director. Three comments questioned the need to note problems and corrective action taken.

The Commissioner has retained this provision as proposed. Only through the submission of such status reports can management be assured of the continuing conformity of study conduct to the provisions of these regulations. Because § 58.35(b)(3) has been revised to require that only significant problems be reported immediately to management, the periodic status report becomes even more important as a means of informing management of minor problems and normal study progress. The status reports are needed to document problems and corrective actions taken so that management can be certain that quality is being maintained and that management intervention is not required. The timing of such reports may be determined by management.

89. Six comments objected that the term "prior" preceding "authorization" in § 58.35(b)(5) was too restrictive. The comments pointed out that unforeseen circumstances may prevent prior authorization for deviations from standard procedure and that the QAU should be concerned with the documentation of the deviation, not with whether prior authorization existed. Two comments stated that the QAU cannot assure that deviations do not occur but can determine, by inspection, whether deviations were documented.

The Commissioner is persuaded that prior authorization cannot always be obtained. For example, a fire in the facility would necessitate immediate action. The Commissioner agrees that documentation of the deviation rather than prior authorization is the important point and has deleted "prior" and added "documentation." In addition,

"assure" has been changed to "determine" to respond to the comments and to reflect more accurately the Commissioner's intent. Section 58.35(b)(5) now reads: "Determine that no deviations from approved protocols or standard operating procedures were made without proper authorization and documentation."

90. Several comments objected to the wording of § 58.35(b)(6), which states that the QAU shall review the final study report. The comments stated that such review requires a scientific judgment and is not an appropriate function for the QAU to perform. One comment suggested that the requirement should be modified to allow for random sampling rather than a complete review of all studies.

The Commissioner agrees that the QAU should not attempt to evaluate the scientific merits of the final report. Therefore, he has modified the paragraph. The QAU must however ensure that the final report was derived from data obtained in accordance with the protocol. Data in the final report significantly contributing to the quality and integrity of a nonclinical laboratory study shall be reviewed. A random sampling approach is not acceptable.

90a. The Commissioner has added to § 58.35 new paragraph (b)(7) which requires that the QAU prepare and sign a statement to be included with the final report which specifies that dates inspections of the study were made and findings reported to management and the study director. This requirement clarifies the fact that QAU review should extend through the completion of the final report and provides a mechanism for documenting that the review has been completed. A conforming section has been added to the final report requirements of § 58.185 as new paragraph (a)(14).

91. Many comments argued that requiring all portions of a quality assurance inspection to be available for FDA inspection might serve to negate their value as an effective management tool for ensuring the quality of the studies during the time in which the studies are being conducted.

The Commissioner shares the concerns of the comments that general FDA access to QAU inspection reports would tend to weaken the inspection system. He believes that FDA's review of quality assurance programs is important, and he recognizes the need to maintain a degree of confidentiality if QAU inspections are to be complete and candid. Therefore, the Commissioner has decided that, as a matter of administrative policy, FDA will not request inspections and copying of either records of findings and problems or records of corrective actions recommended and taken; and §§ 58.15

and 58.35(c) have been revised to rate those records subject to re inspection by FDA from those re not subject to such inspection. Ex from routine FDA inspection are ords of findings and problems as as records of corrective actions re mended and taken. All other re are available. Although the Con sioner is deleting the requiremen new § 58.35(d) that testing fa management shall, upon request authorized employee, certify in ing that the inspections are being formed and that recommended a is being or has been taken. Upo ceiving such a request, manageme required to submit the certificatio compliance. A person who subm false certification is liable to pro tion under 18 U.S.C. 1001.

The one exception to FDA's p of not seeking access to recor findings and problems or of corre actions recommended and take that FDA may seek productio these reports in litigation under cable procedural rules, as for o wise confidential documents.

92. Many comments objected requiring internal quality assu audits to be available to the a might violate the constitutional lege against compelled self-incri tion.

The Commissioner disagrees the comments. It is settled tha privilege against compelled se crimination is an individual priv relating to personal matters; the lege is not available to a colle entity, such as a business enter or to an individual acting in a r sentative capacity on behalf of lective entity. *California Ba Ass'n v. Schultz*, 416 U.S. 21, 55 (1 *Bellis v. United States*, 417 U. (1974); *United States v. Kordel* U.S. 1, 8 (1970); *Curcio v. U States*, 354 U.S. 118, 122 (1957); *U States v. White*, 322 U.S. 694, (1944); *Wilson v. United States* U.S. 361, 382–384 (1911); *Ha Henkel*, 201 U.S. 43, 74–75 (1906). for individuals, the privilege ag compelled self-incrimination is in cable where a reporting requirem applied to an "essentially noncri and regulatory area of inquiry." self-reporting is the only fe means of securing the required mation, and where the requireme not applied to a "highly sele group inherently suspect of cri activities" in an "area permeated criminal statutes." *California v. United States*, 390 U.S. 39 (1968 *bertson v. SACB*, 382 U.S. 70, 79 (*Shapiro v. United States*, 335 U (1948).

FEDERAL REGISTER, VOL. 43, NO. 247—FRIDAY, DECEMBER 22, 1978

ESS TO PROFESSIONAL ASSISTANCE

Comments on proposed § 3e.35 ested rephrasing the statement to fy that professional assistance be orized by the study director, that either in person or by telephone, it be available within a reason- period, and that reference to ability of a veterinary clinical pa- gist be included. Other comments ested that the concept was dupli- e of the function of the study di- r and should be deleted.

e Commissioner proposed this re- ment to assure that a scientist or professional would be available spond to requests for assistance nsultation from less experienced nnel. However, because manage- is responsible for assuring that nnel are available and that per- el clearly understand the func- they are to perform, and because tudy director has overall respon- ty for the technical conduct of tudy, access to professional assist- is a matter best left to manage- 's discretion. Therefore, the sec- is deleted from the final regula-

FACILITIES

GENERAL

Many comments requested defi- 1 or clarification of the terms de- g separation (i.e., separate area, ed area, separate space, and spe- ed area), which are used in 41, 58.43, 58.47, 58.49, and 58.90.

e Commissioner's intent in pro- g that there be defined (and, e required, separate or special- areas in a testing facility was to e the adequacy of the facility for ucting nonclinical laboratory es. This intent is more clearly d in the revised second sentence 58.41, which now reads: "It shall signed so that there is a degree of ation that will prevent any func- or activity from having an adverse t on the study." The important is that the facility be designed so the quality and integrity of the data is assured. The manner in h the separation is accomplished be determined by testing facility igement.

equate separation may be, in var- situations, a function of such fac- as intended use of the specific of the facility, space, time, and rolled air. The broad variety of systems, test and control articles, the size and complexity of testing ities preclude the establishment ecific criteria for each situation. these reasons the Commissioner nes to include in the regulation r a definition or specific examples nethods for achieving adequate ration.

95. One comment suggested that a number of additional animal care and facility requirements be added to the regulations. The suggestions included, e.g., ambience to assure nonstressful conditions; ventilation and room access arranged to prevent cross con- tamination; and surveillance of animal health before and during a test or ex- periment.

The Commissioner concludes that no additional requirements need to be added because the regulation, as it stands, adequately covers the addi- tions proposed by the comments. For example, ventilation and room access arranged to prevent cross contamina- tion are addressed by the degree of separation requirement in § 58.41.

ANIMAL CARE FACILITIES

96. Many comments suggested that accreditation of animal care facilities by a recognized organization should provide adequate evidence that a test- ing facility is in compliance with § 58.43(a). One comment suggested ac- creditation by recognized organiza- tions for analytical laboratories.

Although the Commissioner is aware of the value of accreditation programs, he cannot delegate FDA's responsibili- ty for determining compliance with these regulations to an organization over which FDA has no authority. Few, if any, accreditation programs cover the same areas covered by this regulation. Furthermore, the Commis- sioner is unaware of any facility ac- creditation program which is manda- tory. The agency's obligation to in- spect a testing facility for overall com- pliance would not be altered by the fact that a facility was otherwise ac- credited.

97. Numerous comments objected to the requirements concerning separa- tion of species, isolation of projects, and quarantine of animals as impracti- cal and not necessary in all instances, e.g., separation of species in large animal studies and quarantine of all newly acquired animals. Some of the comments stated that the require- ments of this section allow no latitude for judgment concerning their applica- bility.

The Commissioner reiterates that all requirements may not be applicable or necessary in all nonclinical laboratory studies and that the degree to which each requirement should apply in each case can be determined by informed judgment. Because of the variability of nonclinical laboratory studies, a degree of flexibility in applying the re- quirements of § 58.43(a) is necessary, and the language of § 58.43(a) is amended to read: "A testing facility shall have a sufficient number of animal rooms or areas, as needed, to assure proper: (1) separation of species or test systems, (2) isolation of individ-

ual projects, (3) quarantine of animals, and (4) routine or specialized housing of animals." As noted in the general discussion at the beginning of this pre- amble, all references to other stand- ards ("The Animal Welfare Act") have been deleted.

98. Several comments suggested that § 58.43(b) be amended to include isola- ton of test systems with infectious dis- eases as well as isolating studies con- ducted with infectious or otherwise harmful test articles.

The Commissioner agrees that test systems with infectious diseases should be isolated. Proposed § 3e.49(b) provided for specialized areas for han- dling volatile agents and hazardous aerosols. Section 3e.49(b) also provided for special procedures for handling other biohazardous materials. Pro- posed § 3e.49(c) provided for special facilities or areas for handling radioac- tive materials.

To clarify all these requirements, the Commissioner has amended § 58.43(b) to read: "A testing facility shall have a number of animal rooms or areas separate from those described in paragraph (a) of this section to ensure isolation of studies being done with test systems or test and control articles known to be biohazardous, in- cluding volatile substances, aerosols, radioactive materials, and infectious agents." The provisions in proposed § 3e.49(b) and (c) regarding specialized areas for handling volatile agents, haz- ardous materials and radioactive mate- rials are deleted from § 58.49.

99. One comment on § 58.43(c) sug- gested that, in addition to the area designated for the care and treatment of diseased animals, a separate area should be provided for animals with contagious diseases.

The Commissioner agrees, and the paragraph is amended to allow for an area for treatment of animals with contagious diseases, and it is to be sep- arate from the area designated for the care and treatment of diseased ani- mals.

100. Several comments questioned the requirement for separate areas for diseased animals, indicating that often such animals are sacrificed rather than treated.

The Commissioner does not agree that a separate area is not always needed for diseased animals. Although diseased animals may be sacrificed, this is not always the case, and it may not always be possible immediately to sacrifice diseased animals. Thus, a sep- arate area should be available for such animals until sacrifice can be accom- plished.

101. One comment requested that § 58.43(e), which deals with facility design, construction, and location to minimize disturbances that interfere with the study, should also define the

acoustic and sound-insulating requirements necessary to satisfy this requirement.

The Commissioner concludes that it is ·impractical to attempt to define acoustic and sound insulation requirements. It would be equally impractical to attempt to define all other types of possible disturbances that might interfere with a study.

ANIMAL SUPPLY FACILITIES

102. One comment asked that § 58.45 be clarified by specifically excluding "carriers" from the storage requirements.

The term "carrier," as used in § 58.113, is the material with which the test article is mixed, e.g., feed. The Commissioner concludes that it is necessary to provide facilities for proper storage of carriers and declines, therefore, to exclude them from the storage requirements.

103. One comment requested deletion of the section, stating that it discusses items not appropriate for FDA concern.

Improper storage of feed, carriers, bedding, supplies, and equipment can adversely affect the results of a study. Therefore, the Commissioner finds these matters to be of legitimate concern to FDA and declines to delete the section.

104. Two comments stated that separate storage space need not be required as long as material is properly stored and does not interfere with the conduct of the study.

The Commissioner agrees with these comments, in principle, but is convinced that storage areas for feed and bedding should be separate from the areas housing the test system to preclude mixups and contamination of the test systems. The section has been modified by adding the words "as needed."

FACILITIES FOR HANDLING TEST AND CONTROL ARTICLES

105. One comment stated that § 58.47, as worded, represented an impossible standard and suggested that use of the "designed to prevent" concept would be more realistic.

The Commissioner rejects this comment. The inherent purpose or "design" of all regulations is to prevent or require some action, and the use of the phrase "designed to prevent" would be an awkward and ambiguous modification of § 58.47.

106. Numerous comments objected to creating the number of separate or defined areas proposed by § 58.47, stating that the volume of testing would make it infeasible to create all the separate areas. One comment asked whether eight separate areas were required.

The Commissioner reiterates that the purpose of this section is to assure that there exists a degree of separation that will prevent any one function or activity from having an adverse effect on the study as a whole. Because of the wide variety of studies covered by these regulations, a degree of flexibility is appropriate in applying these requirements, and the degree to which each requirement should apply in each case may vary. To make this clear, the term "defined" has been deleted from § 58.47. Section 58.47(a) now reads: "As necessary to prevent contamination or mixups, there shall be separate areas for * * *." There is no specific requirement for eight separate areas.

LABORATORY OPERATION AREAS

107. A number of comments stated that § 58.49 required clarification, that in some instances more than one activity could be permitted in the same room, and that certain of the requirements would not be appropriate in every case.

The Commissioner agrees that the section as proposed was subject to misinterpretation. Because of the nature and scope of the types of studies subject to these regulations, it would be inappropriate to set specific uniform requirements for all studies. Therefore, the provisions are revised to make it clear that reasonable judgments regarding area and space requirements may be made on the basis that a particular function or activity will not adversely affect other studies in progress. Proposed § 58.49(b) has been revised, and the references to biohazardous materials has been added to the list of activities in § 58.49(a). (See the discussion at paragaph 98 above.)

108. Two comments suggested that the wording of § 58.49(a) be changed to refer to "adequate" rather than "separate" laboratory facilities, stating that animal studies require that laboratory facilities be available on the immediate premises. One comment requested that provisions be made for the use of outside laboratory facilities.

The Commissioner concludes that the term "separate" is proper in the context of § 58.49(a). He does not agree that laboratory facilities must be available on the immediate premises of the testing facility, and finds that many laboratory functions can be conducted properly in separate buildings or by independent laboratories located outside the testing facility.

109. Two comments on § 58.49(b) stated that the requirement that space and facilities separate from the housing areas for the test systems be provided for cleaning, sterilizing, and maintaining equipment and that sup-

plies should apply only to ↑ equipment.

The Commissioner does not ↑ The objective of the requirement prevent the occurrence of thos verse effects which might result study from the activities of clea sterilizing, and maintaining. No ↑ ingful distinctions based on "majc "not major" equipment can be ma

110. One comment on § 58. stated that the proposed wordin not have useful application in al systems or studies and that the se should be rewritten to focus on tl tended principle and not on the w achieve it.

The section has been revised. I reads, "separate space shall be vided for cleaning, sterilizing, maintaining equipment and suj used during the course of the st The revised wording grants flexi in application as long as study r are not affected.

SPECIMEN AND DATA STORAGE FACIL

111. Several comments asked w er § 58.51 applied to completed going studies. Concern was als pressed that limiting access to st areas to authorized personnel wa feasible.

This section is amended to apj archive storage of all raw data specimens from completed st The commissioner cannot agree, ever, that limiting access of th chives to authorized personnel o not feasible. Prudence would d such limited access even in th sence of a requirement. The pot for misplaced data and specime too great to allow unlimited acc the archives.

ADMINISTRATIVE AND PERSONNE FACILITIES

112. One comment on § 58 stated that the section was uni sary because adminsitrative fun had been previously define §§ 58.29, 58.33, and 58.35.

The Commissioner notes that section specifies facilities rather duties. References to OSHA r tions have been deleted.

EQUIPMENT

EQUIPMENT DESIGN

113. Five comments on § 58.61 ↑ that the section was fragmente redundant.

The Commissioner agrees with comments and has consolidate section into one paragraph, ↑ reads: "Automatic, mechanical o tronic equipment used in the ge tion, measurement or assessme data and equipment used for fi environmental control shall be (propriate design and adequate (

y to function according to the proto-
l and shall be suitably located for
eration, inspection, cleaning and
aintenance." This consolidation
iminates the fragmentation and re-
ndancy of the proposal and specifies
early that the requirements are lim-
ed to that equipment which, if im-
roperly designed, or inadequately
eaned and/or maintained, could ad-
rsely affect study results.

114. Two comments objected to the
ndefined general terms "adequate"
d "appropriate" in § 58.61.

The Commissioner points out that
road terms are necessary because of
e wide range of equipment used in
e studies covered. Exact design and
pacity requirements for each piece
 equipment are clearly beyond the
ope of these regulations.

115. Four comments on § 58.61 stated
at how cleaning is accomplished is
relevant and that the regulation
ould emphasize accomplishment
ther than ease of accomplishment.

The Commissioner agrees that the
rimary concern is that adequate
eaning be accomplished. However,
st experience has demonstrated that
hen equipment is not designed and
cated to facilitate cleaning and
aintenance, it is much less likely to
 adequately cleaned and maintained.

MAINTENANCE AND CALIBRATION OF
EQUIPMENT

116. Five comments suggested that
58.63(a) should allow the required
nctions to be performed at the time
e equipment is used rather than
pecifying that the functions be per-
rmed regularly.

The Commissioner agrees that per-
rming these functions at the time of
se is satisfactory and is amending
58.63(a) to provide flexibility. The
cond sentence of this section now
ads: "Equipment used for the gen-
ation of data shall be adequately
sted, calibrated and/or standard-
ed."

117. Two comments suggested that
calibrated" should be changed to
standardized" because the word "cali-
rated" normally means a perform-
nce check against known standards,
hereas "standardized" normally
eans to make uniform.

The Commissioner finds that for
me equipment the term "calibrated"
 more appropriate and for other
quipment the term "standardized" is
ore appropriate. Revised § 58.63(a)
llows application of either term.

118. Two comments suggested that
he reference to the use of cleaning
nd pest control materials is misplaced
 § 58.63(a).

The Commissioner agrees that this
se is more appropriately addressed
nder "Testing Facility Operations",

and the requirements have been transferred to § 58.90(l).

119. Comments requested a precise
definition of the equipment for which
§ 58.63(b) requires written standard
operating procedures.

The Commissioner advises that because of the range of study and product types covered, such a list is impractical. The language of this section is
retained as proposed to encompass the
total range of equipment used in conducting nonclinical studies.

120. Eleven comments questioned
the appropriateness of designating a
responsible individual in § 58.63(b).

The Commissioner has changed "individual" to "person" as defined in
§ 58.3(h) to allow for designation of an
organizational unit.

121. One comment indicated the
need for a clear FDA policy regarding
primary calibration standards.

The Commissioner concludes that
proper standards are the responsibility
of management, and these are to be
set forth in the standard operating
procedures.

122. One comment agreed with the
standard operating procedure requirements of § 58.63(b), but suggested a
several year phase-in period.

The Commissioner concludes that
180 days is a sufficient time period for
developing standard operating procedures. Furthermore, the Commissioner's intent to require such procedures
has been known since November 1976,
when the proposed regulation was
published.

123. Seven comments suggested that
the manufacturer's recommendations
should be sufficient for standard operating procedures. Additionally, one
comment pointed out that maintenance could be subcontracted and a
certificate should be allowed.

The Commissioner advises that the
regulation does not preclude the use
of manufacturer's recommendations as
part of the standard operating procedures, nor does it preclude subcon-
tracting maintenance. The Commissioner advises, however, that if a facility decides to subcontract maintenance,
that fact does not relieve the facility
of the responsibility for maintenance.

124. One comment argued that the
requirement that all equipment records specify remedial action to be
taken is excessive, and two comments
said there are too many variables to
specify in advance the remedial action
to be taken.

The Commissioner notes that trou-
ble-shooting charts are available for
most equipment. The remedial action
taken may influence the results of the
study and therefore must be documented.

125. Several comments suggested
that the equipment for which standard operating procedures are required

be limited by rewording in one of t
following ways: "major" equipmer
"equipment used in data collectior
or "delicate, complex equipment."

The Commissioner has consider
the comments and has modified tl
language of § 58.63(b) to require th
standard operating procedures d
scribe in "sufficient" detail the proc
dures to be used in cleaning, testin
and standardizing equipment. Tl
Commissioner points out th
§ 58.81(a) (standard operating proc
dures) states that the written standa
operating procedures are to be tho
which management is satisfied a
adequate to ensure the quality and i
tegrity of study data. While the Cor
missioner does not find it feasible
confine the requirement for standa
operating procedures to "majo
equipment, he does find that the reg
lation clearly contemplates that tl
required procedures need be only
detailed as deemed necessary to assu
the integrity of the study data. Simp
equipment, therefore, should requi
only brief standard operating proc
dures.

126. Five comments suggested th
written records for nonroutine repa
should only be required where t
nature of the malfunction could affe
the validity and integrity of the da

The Commissioner rejects this su
gestion because it is not always pos
ble to make this judgment ahead
time.

127. Many comments argued th
the recordkeeping requirements
§ 58.63(c) are excessive.

The Commissioner has conclud
that the cost of maintaining records
cleaning exceeds the benefits, and tl
requirement is deleted. However, t
requirement for maintaining recor
of all inspections, maintenance, te
ing, calibrating and/or standardizi
operations is retained because the
records may be necessary to reco
struct a study and to assure the val
ity and integrity of the data.

128. One comment proposed that
new sentence, reading as follows,
added to § 58.63(c): "Where approp
ate, the written record noted abc
may consist of a notation temporar
fastened to the piece of equipme
stating when the last specified acti
with respect to the equipment w
taken."

The Commissioner finds that t
suggested approach is not preclud
by the language of the section as wr
ten, but cautions that where such
approach is used, the notations con:
tute records which must be retained
required by § 58.195(f).

129. One comment asked wheth
each client of a contract facility m
receive a copy of the equipment ma
tenance and calibration records.

The Commissioner concludes that the regulation does not so require.

TESTING FACILITIES OPERATION

STANDARD OPERATING PROCEDURES

130. Two comments suggested deleting § 58.81 in whole or in part. Several others said the requirements for standard operating procedures were unnecessary and burdensome.

The Commissioner does not agree. The use of standard operating procedures is necessary to ensure that all personnel associated with a nonclinical laboratory study will be familiar with and use the same procedures. These requirements will prevent the introduction of systematic error in the generation, collection, and reporting of data, and they will ensure the quality and integrity of test data that are submitted to FDA to become the basis for decisions made by the agency. The Commissioner recognizes that the requirements for standard operating procedures may place an additional burden on testing facilities, but finds that the resulting benefits should outweigh the burden. The requirements will benefit the public by producing better quality data and will benefit the testing facility by reducing the need to repeat nonclinical laboratory studies because of errors in the data.

131. A few comments suggested that responsibility for the standard operating procedures should be specified.

The Commissioner has concluded that this function should reside with the management of a facility, and the first sentence of § 58.81(a) is revised accordingly.

132. Several comments suggested that the responsibility for authorizing significant changes in established procedures be vested in someone other than management.

The Commissioner disagrees. Because standard operating procedure will often apply to more than one study in a testing facility, the Commissioner believes that significant changes to a standard operating procedure, which could affect several different studies, should be authorized by management.

133. Several comments stated that standard operating procedures should not apply to certain types of test systems, that the requirement would introduce difficulties in open-ended exploratory experimentation and electromedical equipment testing, that the approach would not lend itself to rapidly changing methodology such as mutagenicity testing, and that requiring chemical standard operating procedures for each test and procedure was not realistic.

The Commissioner agrees that routine standard operating procedures should not apply to exploratory stud-

ies involving basic research. He does not agree, however, that electromedical equipment testing should be exempt unless such testing does not fall under the definition of "nonclinical laboratory study." Standard operating procedures are feasible for studies using methods which change rapidly and for studies using any test system. In the case of chemical procedures, the Commissioner finds that it is realistic to require written standard operating procedures for each test.

134. One comment recommended that the phrase "written standard operating procedures" in § 58.81(a) be changed to "documented appropriate operating procedures." The same comment suggested that the term "ensure" in the first sentence of § 58.81(a) be changed to "maintain."

The Commissioner disagrees with both suggestions. The term "standard operating procedures" refers to routine and repetitive laboratory operations. "Appropriate operating procedures," as a phrase, implies that such procedures could be changed at will. The Commissioner also rejects the suggestion that "ensure" be changed to "maintain." The purpose of written standard operating procedures is to ensure the quality and integrity of the data generated in the course of nonclinical laboratory study. The term "maintain" assumes the procedures already in existence are sufficient to ensure the quality and integrity of the data when, in fact, they may not be sufficient.

135. One comment said that the term "adequate" in the first sentence of § 58.81(a) is a nonprecise term.

The Commissioner agrees, but finds that a testing facility may have a broad range of divergent standard operating procedures for many different studies and that it is impractical to define the adequacy of such procedures for all types of tests. A determination of the adequacy of each standard operating procedure is the responsibility of the management of the testing facility.

136. Numerous comments asked what changes or deviations from standard operating procedures should be documented in the raw data, as required in § 58.81(a). One comment said any deviation should be documented, whether authorized or not.

Every deviation or change in a standard operating procedure should be documented in the raw data. The second sentence of § 58.81(a) has been revised for clarity. It now reads: "All deviations in a study from standard operating procedures shall be authorized by the study director and shall be documented in the raw data."

137. Seven comments indicated that it is inappropriate to require that

every minor deviation be docum and reported in writing to the Q.

The Commissioner agrees th cause the QAU is no longer requi maintain copies of standard ope procedures, it is inappropriate quire that every deviation be re in writing to the QAU. It is suf that all deviations from standard ating procedures be authorized study director and documented raw data. No exceptions can be for "minor" deviations. Becaus deviation or change may affe outcome of a study, it is not po to judge in advance whether or deviation is, in fact, "minor."

138. Several comments ind that the requirement for standa erating procedures should be g in nature.

The Commissioner disagrees. proposal, the Commissioner cite dence from agency investigatic certain testing facilities that failed to maintain written sta operating procedures of the kin lined in § 58.81(b). As a result, c technical personnel were unawa the proper procedures required for care and housing of anima ministration of test and contro cles, laboratory tests, necrops; histopathology, and handling of The Commissioner has conclude a specific delineation of standard ating procedures will allow fo form performance of testing dures by personnel and conseque provement in the quality of the

139. Two comments indicated the requirements for standard o ing procedures set out in § 58.81 through (12) largely concern a studies and that this should be s cated in this section.

The Commissioner agrees that of the provisions listed in § 58 are applicable only to studies inv animals. Such is true, howev many provisions throughout the lations, and no special mention fact is required here. The Comm er emphasizes that operations i ing standard operating procedur not limited to those listed in § 58

140. One comment suggested the phrase "and control" be d from the first sentence of § 58.81 which requires standard ope procedures for test and contro cles, because a control article often be a competitor's product.

The Commissioner does not Where a control article is a co cially available product, its spe tions and characterization may cumented by its labeling.

141. Several comments sug that the last sentence of pro § 58.81(b)(3), which reads: "The program shall be designed to es the identity, strength, and pu

test and control substances, to ss stability characteristics, where ible, and to establish storage cons and expiration dates, where appriate" be deleted or suggested : the sentence be transferred to anr section.

ne Commissioner agrees. The sence is deleted from § 58.81(b)(3), and ropriate portions of the sentence transferred to § 58.105(a). The cons expressed in this sentence prop belong in the section of the regus relating to "Test and Control cle Characterization." The phrase ting and administration" has been ted from the first sentence of 81(b)(3) for the same reason. To ify clearly the Commissioner's nt, "method of" has been added to 81(b)(3) to modify "sampling." Red § 58.81(b)(3) now reads: "Receipt, tification, storage, handling, ing and method of sampling of the and control articles."

2. One comment stated that 81(b)(9), "Histopathology," and 81(b)(8), "Preparation of specis," were duplicative.

ne Commissioner has revised 81(b)(8) to read: "Collection and tification of specimens" to distin h the requirement from 81(b)(9), "Histopathology." The a "histopathology" covers the exnation of specimens, not their collon and identification.

3. Eight comments recommended wording of the requirement in proed § 3e.81(b)(12) that standard oping procedures be established for preparation and validation of the l study report.

ne Commissioner concludes that requirement should be deleted bese the reporting provisions of .185 adequately describe the reements for final reports. A new graph, § 58.81(b)(11), covering intenance and calibration of ipment," has been added to reflect requirements of § 58.63(b).

4. Seven comments suggested that § 58.81(c) the requirement that idard operating procedures be liable at all times to personnel in immediate bench area be broadd to be within "easy access." Aner comment said the location of a materials should be left to the fay's discretion.

he Commissioner has concluded t unless standard operating procees are immediately available within laboratory area they are not in "easy access" and may not be suited by personnel when routine rations are being performed. The t sentence in § 58.81(c) has been ed for clarity, but the requirement ains.

5. Several comments were received arding § 58.81(c) and the use of

textbooks as standard operating procedures. One comment suggested that textbooks be considered appropriate as part of a standard operating procedure. Two comments assumed that standard operating procedures would permit the incorporation of textbooks by reference. One comment suggested that supplementary material should be written to augment textbooks. An additional comment suggested that textbooks be used in the absence of standard operation procedures.

Standard operating procedures should be set forth in writing, and textbooks may be used as supplements to written standard operating procedures. Reference to applicable procedures in scientific or manufacturer's literature may be used as a supplement to written standard operating procedures. For example, a standard operating procedure could refer to the pertinent pages of any portion(s) of a textbook or other published literature that might be pertinent to a laboratory procedure performed; these supplementary materials need not be incorporated verbatim in the standard operating procedure, but would be required to be immediately available in the laboratory area for the use of personnel. The last sentence of § 58.81(c) is revised to make this point clear. Additionally, § 58.81(d) regarding a historical file of standard operating procedures has been clarified to read: "A historical file of standard operating procedures, and all revisions thereof, including the dates of such revisions, shall be maintained."

REAGENTS AND SOLUTIONS

146. Numerous comments on § 58.83 said that to require that the labeling of reagents and solutions in laboratory areas include the method of preparation was neither feasible nor necessary.

The Commissioner agrees and is deleting the phrase "method of preparation" from § 58.83 because the method of preparation could be too lengthy to fit readily on the label. The method of preparation of reagents and solutions should, however, be addressed by the standard operating procedures.

147. Several comments stated that the provision for the handling and use of deteriorated materials and materials of substandard quality should specify only that they not be used and should not specify or require their removal from the laboratory because their removal should be left to the discretion of the laboratory.

The Commissioner agrees, and § 58.83 has been revised accordingly.

148. One comment suggested that the phrase "used in nonclinical studies" be substituted for the phrase "in the laboratory areas" in the first sentence of § 58.83.

The Commissioner disagrees with this comment. All reagents and solutions used in a laboratory conducting a nonclinical study should be properly labeled as provided in the regulation to preclude inadvertent mixups of reagents and solutions that are used in such studies with those that are not intended for such use.

149. Two comments suggested that the phrase "Deteriorated materials and materials of substandard quality" in the second sentence of the section be changed to incorporate the terms "reagents" and "solutions."

The Commissioner agrees and is revising the second sentence of § 58.83 accordingly. Revised § 58.83 now reads: "All reagents and solutions in laboratory areas shall be labeled to indicate identity, titer or concentration, storage requirements, and expiration date. Deteriorated or outdated reagents and solutions shall not be used."

ANIMAL CARE

150. Several comments raised the issues of unnecessary animal experimentation and the humane care of animals.

The issue of using animals in laboratory experiments designed to establish the safety of regulated products has been raised many times in the course of agency rulemaking. The position of FDA has been consistent on this issue. The use of animal tests to establish the safety of FDA-regulated products is necessary to minimize the risks from use of such products by humans. The humane care of test animals is a recognized and accepted scientific and ethical responsibility and is encouraged both by various agency guidelines and the Animal Welfare Act. The good laboratory practice regulations should, in fact, encourage the humane treatment of animals used in nonclinical laboratory studies by establishing minimum requirements for the husbandry of animals during the conduct of such studies. In addition, there should occur a reduction in the amount of animal testing that has to be repeated or supplemented because the original studies were inadequate or inappropriate to establish the safety of FDA-regulated products.

151. Numerous comments objected to the incorporation by reference of guidelines and standards proposed in § 58.90(a).

As noted early in the preamble, all references to other standards such as the Animal Welfare Act of 1970 and HEW Publication No. (NIH) 74-23 have been deleted. Section 58.90(a) is revised to read: "There shall be standard operating procedures for the housing, feeding, handling and care of animals."

152. Several comments stated that the quarantine of animals required in

§ 58.90(b) was impossible in some cases, unnecessary under certain conditions, and would prevent the use of certain animals, such as "timed-pregnant" mice. Other comments said the paragraph could be interpreted to require a separate quarantine area or an extensive quarantine time period.

The purpose of this paragraph is to require that the health status of newly received animals be known before they are used. This requires a separate quarantine area where necessary to determine animal health status. The concept of "separate areas" has been previously discussed. In some cases, depending on such factors as the species or type (e.g., time-pregnant) of animal, or the source and the nature of the expected use of the animal, a health evaluation can be made immediately, or soon after arrival, resulting in a very short quarantine period. The regulation does not preclude this type of health evaluation if it is done in accordance with acceptable veterinary medical practice.

153. Several comments stated that quarantine is unnecessary when animals are obtained from reputable or specific pathogen-free sources.

A health evaluation is required of all newly received animals regardless of the supply source, although the source can be a factor in determining the degree or depth of health evaluation required. Seldom can the conditions under which animals are transported from their source be considered certain to preclude the possibility of exposure of the animals to disease.

154. Some comments requested deletion of § 58.90(b) because it duplicates the animal care requirements regulations.

The Commissioner rejects these comments. The agency is responsible for animal care procedures as they pertain to testing facilities conducting nonclinical laboratory studies, and the provisions are appropriately included in § 58.90(b).

155. Several comments said that the requirements of § 58.90(c) and (d) concerning the isolation of known or suspected diseased animals and keeping animals free of disease or conditions that would interfere with the conduct of the study were impractical.

For clarity, these paragraphs are revised and combined in § 58.90(c). This paragraph deals only with those diseases and conditions that might interfere with the study. This excludes a wide range of diseases and conditions and allows the consideration of such factors as etiology and whether the disease is communicable. The section does not require isolation of all animals in a shipment from a study when only one or some of the animals are diseased, and it covers only those animals that are known or suspected to be diseased.

156. Some comments suggested that specific requirements be provided for the management of diseased animals, and one comment said the veterinary staff should be able to treat diseased animals as they deem proper.

The Commissioner concludes that it is beyond the scope and purpose of these regulations to describe detailed requirements concerning the management of diseased animals and that § 58.90(c) is sufficiently explicit to exclude the use of diseased animals that would interfere with the purpose or conduct of a nonclinical laboratory study. The regulation does not prohibit the treatment of diseased animals if such treatment does not interfere with the study. If treatment will interfere with the study, the diseased animals shall be removed from the study.

157. More than 60 comments objected to or requested revision of proposed § 3e.90(e), which called for the unique identification of all animals used in nonclinical laboratory studies. Fifty-four of the comments addressed specific issues related to this concept, e.g., unique identification of mice, costs of such systems, application to suckling rodents, injury to animals from identification systems, effects of dyes or tattoos, a lack of need in single-dose or short-term experiments, and cage identification instead of animal identification with precautions being taken to prevent animal mixups.

In the absence of a proven and acceptable method of unique identification for small rodents, the Commissioner is revising § 58.90(d) to require appropriate identification for warm-blooded animals, excluding suckling rodents, which require manipulations and observations over extended periods of time. Suckling rodents have been excluded from the requirements because of potential cannibalization by the mother. The same information needed to specifically identify each animal is required on the outside of housing containers or cages. Such identification should substantially reduce the possibility for animal mixup. Because of the varied nature of the tests conducted and the test systems used, the manner of identification is left to the discretion of the testing facility.

The Commissioner advises that whenever a study requires that animals be removed from and returned to their home cages, there is a potential for mixup. Thus, if a single-dose or short-term study requires such manipulations, the animals shall receive appropriate identification.

Because the requirement for unique identification has been deleted, the concerns expressed regarding cost, injury to the animals from various identification systems, and the [use] of dyes or tattoos are no long[er]... mane.

158. Two comments que[stion] whether the study director co[uld] practice assure unique identif[ication] as proposed in § 3e.90(e), w[ith] direct observation.

The requirement has been d[eleted] along with the requirement for identification.

159. Two comments requeste[d dele]tion of the last sentence of pr[oposed] § 3e.90(e) regarding the identif[ication] of specimens.

The Commissioner conclude[s that] proper specimen identification [is an] integral part of proper study co[nduct] but that the requirement more [proper]ly belongs under standard op[erating] procedures. Consequently, § 58.8[1] now incorporates this provision.

160. One comment inquired wh[ether] in the event animals of the sa[me spe]cies in different tests were in th[e same] room, FDA would require ide[ntifica]tion of all compounds. This, [it was] felt, would raise confidentiality [ques]tions for a contract testing facili[ty].

The Commissioner advises th[at the] use of coding to identify test [or con]trol articles is not preclud[ed by] § 58.90(e). The concluding phra[se "to] avoid any intermixing of te[st ani]mals," was deleted as redundant.

161. Proposed § 3e.90(g) r[equired] comparison of cage and animal [identi]fication for each transfer, proc[edures] for verification, and written [permis]sion of the study director for [each] transfer. Seventeen comments [object]ed to part or all of these requir[ements] as vague, burdensome, unnec[essary] and redundant.

The Commissioner agrees, a[nd the] paragraph is deleted. Procedu[res for] the transfer and proper placem[ent of] animals are required as standar[d oper]ating procedures in § 58.81(b)(12).

162. Several comments claim[ed that] the requirements of p[roposed] § 3e.90(h), redesignated § 58.90([i)? are] redundant in view of the requi[rement] for standard operating proced[ures in] § 58.81. Other comments state[d that] the incorporation of guidelines [by ref]erence was inappropriate.

The Commissioner conclude[s that] the requirement that animal [cages,] racks, and accessory equipm[ent be] cleaned is appropriately inclu[ded in] this section even though there [may be] overlap with the language of [the] standard operating procedure[s. The] reference to other agency gu[idelines] has been deleted.

163. Three comments assert[ed that] sanitization should not alw[ays be] done, because it could in certai[n cases] interfere with the conduct [of the] study.

The Commissioner agrees an[d points] out that the language in redes[ignated]

.90(f) permits cleaning and sanition at appropriate intervals. The ion now reads: "Animal cages, s and accessory equipment shall cleaned and sanitized at appropriintervals."

4. Many comments objected to posed § 3e.90(i), redesignated .90(g), which requires periodic lysis of feed and drinking water for own interfering contaminants." tain of these comments requested ification or deletion, or expressed cern about the costs involved. ers argued that the use of positive negative controls would accomh the intent of the requirement, or : certificates of analysis from local er supply authorities and feed nufacturers should be permissible. ally, a few comments said analysis eed and water should only be reed when there is reason to believe : a particular contaminant may e an effect on the study, and comits said the analysis requirements uld be specified in the protocol.

ost of the objections raised against analytical requirements of the sec were based on misinterpretation such requirements. The intent of Commissioner was to require analfor contaminants known to be cale of interfering with the nonclinilaboratory study and reasonably ected to be present in the feed or er, and not to require analysis of l and water for all contaminants wn to exist. Certain contaminants ld affect study outcome by maskthe effects of the test article, as observed in recent toxicological lies of pentachlorophenol and hylstilbestrol, in which the feeds d as carriers for the test articles e found to contain varying quantiof pentachlorophenol and estroic activity, respectively, that invalied these studies by producing erc results. The use of positive and ative controls in these examples insufficient to compensate for the ability in contaminant content. refore, the Commissioner agrees h the comments that suggested t analysis of feed and water only be e when there is reason to believe t a particular contaminant may e an effect on the study, and may present in the feed or water, and language of both redesignated .90(g) and § 58.120(a)(9) have been ised to make this clear. This clarifion of the regulations should allay concerns of those comments relatto certificates of analysis, costs, l precise definition of impurities. eptable contaminant limits must specified by the protocol 8.120(a)(9)), and should be deterned at the time the protocol is deoped, taking into account the scienc literature, the availability of suit-

able analytical methodology, and the practicability of controlling the level of the contaminant.

165. One comment suggested additional requirements for, e.g., analysis of nutrients and reserve samples of feed at the testing facility.

Nutrient analysis should be addressed by the facility's standard operating procedures. Requirements for reserve samples of test or control articles/carrier mixture (e.g., feed) are set forth in § 58.113(b). The Commissioner concludes that minimum requirements for those items are set forth in the regulation. The regulation does not preclude the setting of additional requirements by the sponsor and/or the testing facility.

166. Proposed § 3e.90(j) would have required feed to bear an expiration date. Twenty-three comments argued that this requirement is of dubious value, is beyond the current state of the art because of varied storage conditions, and that commercially available feed is not expiration dated, making the requirement impractical or impossible.

The Commissioner agrees with these comments, and this requirement is deleted.

167. Several comments argued that the requirement for weekly changes of bedding should be deleted. The comments stated that, in certain cases, weekly bedding changes are contraindicated.

The Commissioner agrees, and the phrase "at least once per week" is removed from § 58.90(h), which now reads, "Bedding * * * shall be changed as often as necessary to keep the animals dry and clean."

TEST AND CONTROL ARTICLES

TEST AND CONTROL ARTICLE CHARACTERIZATION

168. One comment suggested that § 58.105 be deleted; another suggested that the entire subpart be condensed; and three comments suggested that the section is not generally applicable to nonclinical device studies, particularly with reference to such terms as "identity, strength, quality, and purity."

The Commissioner does not agree that the section should be deleted. Its purpose is to assure that the article being tested has been thoroughly characterized or defined and that either the sponsor or the testing facility has a thorough understanding of what is being tested. The Commissioner agrees that the subpart should be condensed and has shortened it. Section 58.105(a) is modified by the inclusion of the sentence "the identity, strength, purity, and composition or other characteristics which will appropriately define the test or control article." This addition provides for charac-

terization of various products, including devices in terms suited to their identity or uniqueness.

169. One comment argued that the requirement that "other substances contained in the test and control substances" be accounted for, as proposed in § 58.105(a), was vague.

By this provision the Commissioner intended to indicate the need to identify and characterize solvents, excipients, inert ingredients and/or impurities that might be part of the test substance. Because these materials are included by definition in the term "test article," the Commissioner has determined that the original language was unnecessary and has deleted it.

170. Three comments sought definition of the word "batch" as used in § 58.105(a).

The term "batch" is now defined in § 58.3(n).

171. Seventeen comments on § 58.105(a) stated that because some control or reference articles might be a competitor's or a supplier's product, the assay and method of synthesis might not be available or might be confidential.

The Commissioner concludes that, in those cases where a competitor's or supplier's product is used as a control article, such products will be characterized by the labeling and no further characterization is necessary.

172. One comment stated that the testing facility should not be responsible for identity, strength, quality and purity and that this responsibility should rest with the sponsor. This comment also suggested that the requirement, as written, would inhibit the conduct of blind studies.

The Commissioner concludes that it is the responsibility of testing facility management to assure that the requisite tests have been done, either by the sponsor or by the test facility (see § 58.31(d)). In those cases where a testing facility is unable to perform the characterization test or is performing blind studies, the sponsor should perform the required testing and notify testing facility management that the characterization of the test or control article has been performed. The section, as revised, does not inhibit the conduct of blind studies; it does not require that the sponsor give the characterizing information to the testing facility, only that the sponsor notify the testing facility that the required characterization has been done.

173. One comment suggested that the requirements of § 58.105 should only apply if the integrity of the study is threatened, and another suggested that any contaminants in a test or control article should be evaluated only with respect to their impact on study validity.

The Commissioner does not agree that the requirement should be so limited. Thorough characterization of the article under test is essential because the results of the test may be compromised by possible contamination. Only by knowing the identity and quantity of the components can one predict their effect on the study. The evaluation of the impact of test and control article contaminants on the validity of the study is an important part of the thorough characterization of the test and control articles.

174. Thirteen comments suggested that characterization of the test article be permitted during the study, after its completion, or left to such time as specified in the protocol.

The Commissioner concludes that characterization of the test or control article should be determined before the initiation of the study in order to provide a means of controlling variations from batch to batch as well as to make certain that the test article meets the specifications of the protocol. As previously stated, a thorough understanding of the nature of the test article is a basic requirement for assuring the absence of contaminants that may interfere with the outcome of the study. When the stability of the test and control articles has not been determined before initiation of the study, the regulation requires periodic reanalysis of each batch of test and control articles as often as necessary while the study is in progress.

175. One comment stated that the phrase "verifying documentation" in § 58.105(a) was not clear.

The Commissioner has determined that the phrase is not needed, and § 58.105(a) is revised to delete it.

176. Seven comments suggested that stability studies required by § 58.105(b) may not always be necessary; three comments suggested that common vehicles and placebo controls, such as water, should be omitted from stability studies.

Some degree of instability may be associated with every test article that might be the subject of nonclinical laboratory study. The Commissioner concludes, therefore, that stability information must be included as part of the information upon which the agency bases a decision regarding the safety of the article. If the stability of common vehicles is generally recognized and can be documented, stability testing is not required.

177. Twelve comments suggested that the term "production" in proposed § 3e.105(c) should be deleted or changed by substitution of other terms such as "approved" or "released," stating that the use of the word was confusing. Several other comments stated that the requirement that test and control substances be de-

rived from the smallest number of production batches consistent with their stability was not always possible or necessary.

The Commissioner agrees that the section was confusing and finds that the requirement is adequately covered by § 58.105(a). The word "batch" has been defined in § 58.3(n), and proposed § 3e.105(c) has been deleted.

178. One comment suggested that the test and control articles should be derived from a large number of batches to increase the probability that test and control articles are representative.

The Commissioner agrees that, in some cases, combining representative samples of test or control articles from various production sources or lots to form a batch may be desirable. Where this is done, however, the resulting batch, rather than the individual samples, must be characterized in accordance with § 58.105(a).

179. Eight comments on § 58.105(d) suggested that the requirement for reserve sample retention be restricted to those substances whose stability had not been previously determined. Another comment suggested that the section seems to require that a reserve sample of water be retained if water is used as the control article, and another comment suggested that the retention of a reserve sample should be left to the discretion of the sponsor.

The Commissioner does not agree that the decision to retain a reserve sample should be at the discretion of the sponsor. Maintaining a reserve sample is necessary to provide independent assurance that the test system was exposed to the test article as specified in the protocol. Reserve samples need not be reanalyzed routinely if the stability of the test or control article is well established. If, however, the results of a study raise questions as to the composition of the test or control article, retention of reserve samples allows resolution of the question. Retention of a reserve sample of water is required when it serves as the control article in a nonclinical laboratory study.

180. Eight comments on § 58.105(d) suggested that containers should be comparable rather than identical to maintain approximate ratio of mass of article to container volume.

Reserve samples should be stored in containers and under conditions that maximize their useful life. The specifications for containers are deleted from § 58.105(d), however, and are now left to the discretion of the study director.

181. Six comments said § 58.105(d) duplicated §§ 58.105(b) and 58.113(a)(2); three said that the requirement that the reserve sample be analyzed at the time the batch is depleted, at the termination of the

study, or at the expiration date result in unnecessary testing. comment suggested that a portio the remaining article should be te rather than testing the res sample.

The Commissioner agrees that requirement for routine reanalys all test or control articles is unn sary where stability characteri have been well established, and requirement has been deleted. Commissioner does not agree tha cited sections duplicate one ano Section § 58.105(b) concerns the s ity of test and control articles in a rier mixture. But § 58.105(d) con reserve samples of test and contro ticles.

182. A number of comments on posed § 3e.105(f) sought clarifica of the requirements, definition of term "quarantine," and deletion o requirement to reanalyze batche turned from distribution.

The Commissioner has examine provision as proposed and has f that the intent is achieved by the visions of § 58.107 (test and contro ticle handling). Proposed § 3e.1 has, therefore, been deleted.

TEST AND CONTROL ARTICLE HANDLI

183. One comment asserted § 58.105 covered the specifics for dling test and control substances that § 58.107 should be deleted.

The Commissioner disagrees the assertion that § 58.107 re § 58.105. The provisions of § 5 apply to the characterization of and control articles and their sto prior to use. Section 58.107 sets f provisions for the handling and d bution of test and control art during the course of a nonclinical oratory study. The purpose of this tion is to provide further mechar to assure that test and control art meet protocol specifications thro out the course of the study, and test article accountability is m tained.

184. Other comments argued the language of § 58.107 shoul modified and that, as written, the tion was impractical.

The Commissioner does not that the requirements are imprac The section has, however, been e for clarity. Section 58.107(a) reads, "There is proper storage." cause contamination is only one o consequences that may result fron proper handling during distribu the Commissioner has re § 58.107(b) to read: "Distributic made in a manner designed to clude the possibility of contamina deterioration, or damage."

TURES OF ARTICLES WITH CARRIERS

. Many comments stated that the
irements of § 58.113 should only
y to certain types of studies, such
ng term feeding studies, or should
y only in cases where problems of
bility might result from mixing
est article with a carrier.

e Commissioner does not agree.
need to know that the test system
ing exposed to the amounts and
s of test and control articles that
specified in the protocol is
non to all types of studies. The
t of mixing on the concentration
stability of the test or control arti-
1 the mixture cannot be predicted
rehand.

. Six comments stated that the
irement that each batch of a .test
ntrol article that is mixed with a
er be tested for uniformity of
stability, and release, as proposed
8.113, was excessive.

e Commissioner has reviewed the
ons advanced by the comments
has deleted the "for each batch"
irement. Once the uniformity of
nixture has been established for a
1 set of mixing conditions, it is not
ssary to establish the uniformity
ach subsequent batch that is
d according to the same specifica-
. Similar considerations apply to
lity testing. Section 58.113(a)(1)
ductory text and (a) now read:
each test or control article that is
d with a carrier, tests by appro-
e analytical methods shall be con-
ed: (1) to determine the uniform-
f the mixture and to determine,
dically, the concentration of the
or control article in the mixture."
sentence. "[I]f the nonclinical
y is to be performed as a blind
y, enough individual samples of
mixture shall be returned to the
sor for analysis," has been de-
l. The requirement for analysis of
or control article mixtures is ade-
ely addressed by the revised lan-
e of § 58.113(a)(1). The mecha-
1 of satisfying the requirement is
to the testing facility. Blind stud-
are discussed in paragraph 172
e.

7. One comment stated that the
ibility of administration by other
1 the oral route should be consid-

1e Commissioner agrees, and refer-
to the route of administration is
oved.

8. Several comments said the
e and subacute toxicity studies are
n conducted before there is exten-
knowlege about a drug's stability
that in such cases the drug might
repared daily. In addition, it was
rested that § 58.113(a)(2) allow for
urrent stability studies.

1e Commissioner agrees with the
ment and has revised the regula-

tion to allow concurrent studies of stability to proceed with the ongoing
nonclinical laboratory study.

189. Three comments on § 58.113
suggested that establishing expiration
dates for a substance used up in a
week seemed too stringent. Many comments suggested that the expiration
dating requirement be eliminated entirely because batch sizes are established so that they will be used up
prior to deterioration of the test article.

The Commissioner has considered
the comments and has revised, as
noted above, the requirement for labeling each batch of test or control article carrier mixture to permit concurrent stability testing. The Commissioner declines to eliminate entirely
the requirement for listing of expiration dates. Expiration dates should be
used, when known, to minimize the
possibility that subpotent, unstable, or
decomposed test or control article carrier mixtures will be used. New
§ 58.113(c) requires that, where any of
the components of the test or control
article carrier mixture has an expiration date, that date shall be clearly
shown on the container. If more than
one component has an expiration date,
the earliest date shall be shown.

190. Many comments on proposed
§ 3e.113(a)(3) stated that the requirement for tests to determine the release
of the test or control substance from
the carrier needed to be clarified,
might be impossible to do, and were
not always necessary.

The Commissioner has reviewed the
comments and the section and finds
that such testing should be adequately
addressed by the protocol. He has,
therefore, deleted the section.

191. Eleven comments suggested
that the requirement that reserve
samples of each batch of test or control article-carrier mixture be retained
was excessive and impractical.

The Commissioner does not agree.
Maintenance of reserve samples of
these mixtures is necessary for the
same reasons that reserve samples of
test and control articles themselves
are necessary. These reasons are
stated in paragraph 179 above.

192. Proposed § 3e.115 incorporated
principles set forth in other regulations and has, accordingly, been deleted. (See the discussion in paragraph
3.)

PROTOCOL FOR AND CONDUCT OF A NONCLINICAL LABORATORY STUDY

PROTOCOL

193. Several comments said the protocol requirements of § 58.102(a) were
not relevant to specific test articles,
e.g., electronic diagnostic instrumenta-
tion. Other comments objected to requiring a protocol for short-term studies or for routine tests described else-

where in 21 CFR Chapter I. Additional
comments proposed that specific requirements be imposed only where applicable, and one comment said the
protocol should focus on what is intended rather than on how the intended result is to be achieved.

The Commissioner has previously
discussed the types of tests and the
conditions within the scope of Part 58.
Because of the broad range of studies
covered, specific sections may not
apply to all studies. However, the
Commissioner declines to exempt
short-term studies or routine tests
from these requirements. Any study
which qualifies as a nonclinical laboratory study is subject to the requirements. The good laboratory practice
regulations are both process-oriented
and product-oriented, and are designed to ensure, insofar as possible,
the quality and integrity of nonclinical
laboratory data submitted to FDA in
support of regulated products. The
Commissioner recognizes that some of
the requirements of this section have
often not been traditionally included
in a protocol. He has nonetheless concluded that the requirements are essential to ensure that all operations
needed to fulfill the objectives of a
study are performed and that the complete list of information required by
this section is necessary to ensure that
deviations, should they occur, are
readily appparent.

194. One comment asked what was
meant by "all methods" in § 58.120;
one suggested deletion of the word
"approved" to describe the protocol;
and another suggested that reference
to statistical methods in § 58.120(a) be
deleted and that a new paragraph on
statistical methods be added to the list
of information required.

"All methods" refers to all operations necessary to achieve the objectives of the study, e.g., analytical
methods, randomization procedures,
etc. If such methods are from published sources, citation of the source
would fulfill this requirement. If the
methods are not from published
sources, full descriptions would need
to be included in the protocol. The
word "approved" is retained to emphasize that a sponsor or testing facility
should have a mechanism for evaluation and approval of initial protocols
and all amendments. A new paragraph
(a)(16) is provided to emphasize the
need to consider statistical methodology in preparing a protocol.

195. Ten comments objected to the
inclusion, in proposed § 3e.120(a)(3), of
stability methodology as a protocol requirement because such methodology
may not have been developed before
the study was begun. Another comment suggested deletion of this requirement as not relevant to a proto-

col. while three comments suggested revision.

The Commissioner recognizes that stability data may not be available when a study is initiated, and this requirement is deleted from the section. The Commissioner emphasizes, however, that determination of the stability of the test and control articles is a responsibility of the study director, that determination of the stability of the articles per se is required under § 58.105(b), and that determination of the stability of the article/carrier mixes is required under § 58.113.

196. Numerous comments on proposed § 3e.120(a)(4) objected to the listing of the names of laboratory assistants and animal care personnel in the protocol because these jobs are subject to constant turnover or periodic rotation.

The Commissioner agrees that laboratory assistants and animal care personnel need not be identified in the protocol. The list of personnel required to be named is transferred to § 58.185(a)(12).

197. One comment proposed that listing the name of the sponsor and name and address of the testing facility required by § 58.120(a)(3) be restricted to studies done under contract.

The Commissioner does not agree with restricting this requirement to studies done under contract because a testing facility, though a division of the sponsor, may have a specific designation and a location different from the sponsor's, and this information is necessary to determine the exact location of the study.

198. Numerous comments on § 58.120(a)(4) objected to specifying starting and completion dates in the protocol because changing priorities may make such specification impractical. Another comment proposed deletion of the requirement for dates as not relevant to a protocol.

Changing priorities may cause changes in starting dates. For this reason the requirement calls for the proposed dates. If the actual dates differ from the proposed dates, the change should be reflected in a protocol amendment. The dates may be needed in the reconstruction of the study.

199. Ten comments on proposed § 3e.120(a)(7) objected that the proposed date for submission of the final study report to management or to the sponsor was not relevant to a protocol, and one requested a definition of the term "completion date."

The Commissioner agrees that the proposed submission date is not relevant, and the provision is deleted.

200. Numerous comments on § 58.120(a)(6) suggested requiring age of the test system only where applicable or substituting age range for age. Several objected to the requirement for justification for selection of the test system as not relevant to protocol requirements. Additional comments proposed that the requirement for justification be limited to nonroutine systems.

The Commission agrees that age of the test system may not always be critical, and § 58.120(a)(6) now requires number, body weight range, sex, source of supply, species, strain and substrain, and age of the test system only "where applicable." The Commissioner does not agree that justification for selection of the test system is not relevant to a protocol or should be limited to nonroutine systems. Such justification is an integral and essential part of every protocol and to emphasize its importance, the Commissioner is establishing a separate paragraph for this requirement, § 58.120-(a)(5).

201. Several comments on § 58.120(a)(8) (proposed § 3e.120(a)-(10)) objected that the method of randomization was not relevant to the protocol and suggested requiring justification for the selected method only when nonroutine methods are selected; four comments said justification of the method of randomization is unnecessary; and one comment proposed revised language regarding method of randomization.

The Commissioner finds that the method of randomization or other methods of controlling bias are relevant and are essential parts of a protocol, whether the methods used may be described as routine or nonroutine. The suggested revision is adopted in part, and § 58.120(a)(8) now reads: "A description of the experimental design, including the methods for the control of bias."

202. One comment said a description of the diet used in the study (proposed § 3e.120(a)(11), now § 58.120(a)(9)) was unnecessary unless the diet was unusual. The comment further said that the necessity for including solvents and emulsifiers was questionable because these might not be known at the time the protocol is written.

The Commissioner advises that the phrase "and/or identification" in § 58.120(a)(9) permits a commercial animal diet to be identified by its name. The need for using solvents or emulsifiers may not be known when the protocol is written; however, when this information is available and the solvents, etc., are selected, this fact should be reflected in a protocol amendment.

203. Nine comments pointed out that the degree of absorption (proposed § 3e.120(a)(14), now § 58.120(a)(12)) is usually unknown at the time of the preparation of the protocol.

The Commissioner recognizes absorption studies may be conduc concurrently with or as part of nonclinical laboratory study points out that the requirement § 58.120(a)(12) can be fulfilled amending the protocol.

204. Nine comments suggested d tion of the requirement that the tocol include the records to be m tained (proposed § 3e.120(a)(16), § 58.120(a)(14)) because this duplic the requirements under another pr sion of the regulation.

The Commissioner concludes the protocol should include a identifying the records to be m tained and, therefore, does not a that § 58.120(a)(14) should be delet

205. Several comments objecte the § 58.130(c) requirement that s mens be identified. Three comm proposed revisions to eliminate the of specific items (test system, st nature, date of collection) include identification of specimens. Nume comments objected to the identi tion system as overly restrictive, ing that a coding system should permitted.

The Commissioner rejects the gested modifications because the quirements are designed to prec error. The specific items require identify a specimen are the minir necessary to prevent mixup of s mens and permit orderly storage. Commissioner does not agree that system is overly restrictive becau does not preclude a coding system.

206. Numerous comments obje to the requirement, in § 58.130(e) recording data in bound books prenumbered pages as costly, t consuming, overly restrictive, and ficult for long-term studies. Six concerned that much informatio too voluminous to be recorded dir and that reference to other docum should be permitted to ju changes, and two comments obje to recording "dictated observation ink.

The Commissioner agrees that requirement for bound books is to strictive in view of both the varie data recording procedures that ca used in nonclinical laboratory st covered by this part and the n ways in which data are generated collected for these studies. He therefore, revising the section. A vised, § 58.130(e) does not preclude erence to other documents if the ments are clearly identified and av ble. The requirements of the se can be met by maintaining the c tion media or an exact transcripti

207. Three comments proposed § 58.103(e) be revised to reflect

e types of computer entries, i.e., ct on-line recording, input from puter readable forms, and input scribed from recorded raw data. additional comment suggested re-d language to achieve this purpose:

two comments stated that com-r printouts of interim display data i not be maintained when the data wholly contained in subsequent it-ions.

ie revised wording of § 58.130(e) is ully applicable to the various forms omputer data entries. The Com-ioner advises that where the data computer input are in machine-able form, such as marketed-sense s, or are transcribed from record-raw data, the machine-readable is or the recorded raw data would titute raw data within the defini-of this part. Where input is via :t on-line recording, the magnetic ia and the program would consti-raw data within the meaning of part.

3. Three comments objected that a ' signature and date for each y would be burdensome in studies lving daily measurements on each ial.

ction 58.130(e) does not require ng and dating of every individual recorded. An entry can consist of ral observations of several animals e by the same person.

). Three comments suggested dele-of proposed § 3e.130(f), which re-2d the review of all recorded data, use this duplicated the function ie study director.

ie Commissioner agrees that these irements are adequately ad-3ed by § 58.33(b), and the para-h is deleted.

Records and Reports

RTING OF NONCLINICAL LABORATORY STUDY RESULTS

). Seven comments said the re-?ment that the final report in-? all raw data and calculations osed in § 3e.185(a)(3) is not practi-nd that a recapitulation should be uate.

ie Commissioner agrees, and the irement that all raw data be in-2d in the final report is deleted.

l. Two comments on § 58.185(a)(3) d that the scope of the term hod" was not clear.

e Commissioner .advises that hod" does not mean that either actual calculations or a step-by-reiteration of the process be in-2d. The name of the method, the iption of the method, or a refer-to an article or test describing nethod will be sufficient.

!. Several comments on l85(a)(4) stated that the final rt should provide only a reference ie information on "strength, qual-

ity, and purity" rather than the actual values for those characteristics.

The Commissioner does not agree. The final report should include actual values for all characteristics required for proper identification. Because the actual values for strength, quality, and purity are not, in every case, sufficient for adequate identification, the word "quality" has been stricken and the words "and composition or other appropriate characteristics" have been added. The additional language will permit the use of any characteristic which facilitates identification of the test and control article.

213a. Several comments on § 58.185(a)(5) stated that the requirement that stability of the test and control articles be described should be narrowed.

The Commissioner finds that stability information must be submitted as part of the final report. The extent of stability testing required by these regulations is discussed at paragraphs 176, 185, 186, and 189 above.

b. Comments on proposed § 3e.185(a)(8) (now § 58.185(a)(7)) requested that the words "appropriate and necessary" be inserted following the words "procedure used", for identifying the test system.

The Commissioner is modifying § 58.185(a)(7) to require reporting such details where applicable.

214. Seven comments on § 58.185(a)(12) protested the requirement that the final report include reports of each of the individual scientists or other professionals involved in the study.

The Commissioner concludes that the individual reports are required to assure that the final results reported accurately reflect the findings of the individual scientists.

215. A number of comments on § 58.185(a)(3) objected to reporting the location of the raw data in the final report.

For the purpose of information retrieval, the Commissioner is of the opinion that the location of the raw data should be specified.

216. The Commissioner advises that the list of personnel required to be named in the final report as specified in § 58.185(a)(12) has been broadened to include all professionals. (See paragraph 196 above.)

STORAGE AND RETRIEVAL OF RECORDS AND DATA

217. Several comments requested revision and clarification of "other information" in § 58.190(a).

The phrase "and other information" is deleted because it is subsumed by the specific requirements for documentation.

218. Five comments requested clarification of the term "specimen" as used in § 58.190(b).

The term "specimen" is defined in § 58.3(j) and means any material derived from a test system for examination or analysis. This includes wet specimens, histological blocks, and slides that yield information pertinent to the outcome of the study. Such specimens are required to bear sufficient labeling to permit identification and expedient retrieval.

219. Several comments stated that the prohibition against "intermingling" of specimens was unnecessary if specimens are properly labeled and indexed.

The Commissioner agrees and finds that the storage requirements are adequate to achieve their purpose without any further prohibitions. The reference to intermingling of samples is, therefore, deleted.

220. Seven comments said proposed § 3e.190(c) was unclear or redundant and required the maintenance of unnecessary duplicative files by both the testing facility and the sponsor.

The Commissioner agrees with the comments, and the paragraph is deleted.

221. A number of comments requested that § 58.190(c) provide that more than one person be permitted to be responsible for the archives.

The Commissioner reaffirms the need for one individual to be accountable for the maintenance and security of the archives to prevent access by unauthorized personnel. Such access could lead to the loss of, or damage to, records and specimens required to be maintained by these regulations. This provision does not preclude delegation of duties to other individuals who may help maintain the archives.

222. Comments on § 58.190(e) suggested that coding of archival contents should be allowed and objected that the section would require four-way indexing.

The paragraph is revised for clarity. As revised, the use of a coding system is permitted; however, the cross-reference indexing system is retained as a requirement.

223. Section 58.190(g) is deleted because the inspection requirements are adequately addressed by § 58.15.

RETENTION OF RECORDS

224. Several comments stated that the proposed record retention requirements were inconsistent with those previously established.

A new paragraph (a) is added to § 58.195 to make it clear that the record retention requirements of this section do not supersede those of any other regulations in this chapter.

225. Several comments pointed out that IND's are not "approved" and

asked that the record retention requirements for IND's be clarified.

The Commissioner agrees that the record retention requirements, as they apply to both IND's and IDE's, need clarification. In addition to the fact that IND's are not, in a technical sense, "approved," the Commissioner has considered the fact that when either an IND or an IDE is submitted to the agency, the application may contain voluminous data collected over a number of years. It was not the intent of these regulations that such supporting IND or IDE data be destroyed after 2 years because not all studies submitted at the time of filing may be of interest to the agency until several years after submission. Therefore, a new sentence is added to § 58.195(b)(1), which states that the 2-year retention requirement does not apply to studies supporting notices of claimed investigational exemptions for new drugs (IND's) or applications for investigational device exemptions (IDE's). These records are governed by § 58.195(b)(2) and shall be retained for at least 5 years. This additional language clarifies both agency policy and current scientific practice which is, in most cases, to maintain such study records far longer than 5 years.

226. One comment said the variable record retention periods are unworkable, and another said records should be maintained as long as the public is exposed to a chemical.

The record retention period represents the minimum deemed appropriate. For uniformity, all records may be retained for 5 years. Longer retention periods are unnecessary because each nonclinical testing facility will be inspected every 2 years. Studies conducted at facilities that are in substantial compliance with these regulations will be presumed to be valid. When significant deviations are discovered, steps will be taken to validate individual studies before the record retention period expires.

227. Twenty-three comments on § 58.195(b)(3) objected to the record retention requirement as it applies to terminated or discontinued studies, stating that the requirement goes beyond the intent expressed in the definitions or that FDA lacks the authority to require that such studies be retained.

The Commissioner finds that such studies are frequently capable of yielding information applicable to evaluations of related compounds. In the interest of the public health, all such data derived from studies originally intended to be submitted to the agency should be available to the agency. This is particularly important when studies are terminated because of preliminary findings that the test article causes adverse effects at such low levels that

any safe use of the article is precluded. The general question of FDA's authority is discussed in paragraph 5 above.

228. With respect to retention of appropriate samples, including wet specimens, several comments on § 58.195(c) requested that the regulations specifically set forth conditions of storage. Others felt that this requirement would be of doubtful value, and several were concerned that the retention period not exceed that which could adversely affect sample integrity.

The Commissioner states that it would be impractical to attempt to specify the specific storage conditions for sample retention. This should be left to the judgment of the testing facility. It is essential as a check on recorded observations that, wherever possible, samples be retained for confirmation of findings. Such samples should be retained for the minimum period specified in the regulations. The regulation clearly states that fragile samples shall be retained only so long as the quality of the preparation affords evaluation.

229. Three comments on § 58.195(e) objected to archive retention of curricula vitae and job descriptions of all personnel involved in the study.

Section 58.195(e) is revised to permit this information to be retained as part of the testing facility employment records.

230. One comment on § 58.195(f) stated that equipment records should be maintained in an independent log rather than maintained as part of each study.

The Commissioner advises that the language of the section does not preclude such an approach. Records of maintenance and calibration of equipment may be kept in a repair manual or on a tag affixed to the instrument. The reference to cleaning records is deleted.

DISQUALIFICATION OF TESTING FACILITIES

PURPOSE

231. Many comments were received concerning the general concept and purpose of disqualification.

The Commissioner believes that many of these comments were based, at least in part, on misunderstanding of the frequency with which disqualification might be used. The Commissioner believes disqualification is an important alternative to rejection of specific studies and legal prosecution because it can reduce by consolidation the number of FDA investigations and administrative proceedings that might be required if FDA acted only on a study-by-study basis. To clarify the agency's intent regarding the disqualification mechanism and to allay fears that this sanction might be abused,

the Commissioner is revising Sub K of the regulations to define clearly the grounds for disqual tion.

231. Section 58.200(a) has bee vised to clarify the purposes of qualification. The first purpose st in the section is to permit FDA t clude from consideration any com ed studies conducted by a testin cility which has failed to comply good laboratory practice requirem until it can adequately be demons ed that the noncompliance did occur during, or did not affect th lidity of data generated by, a par lar study. Thus, for studies comp before disqualification, the orde disqualification creates a rebut presumption that all studies prev ly conducted by the facility are r ceptable. Such a study may be ac ed, however, upon presentation o dence demonstrating that the non pliance which resulted in the disc fication did not affect the parti study. The second purpose set for the revision of § 58.200(a) is to exc studies completed after the dat disqualification from consider until the facility can satisfy the C missioner that it will conduct st in compliance with the regulat (See also the discussion in parag 241.)

GROUNDS FOR DISQUALIFICATION

232. Many comments argued the disqualification provisions peared to be overly harsh, arbit and ambiguous.

To clarify the agency's intent Commissioner is revising the sec The primary function of the age regulation of nonclinical labor: testing is to assure the quality ar tegrity of data used in making ments about the safety of pro regulated by the agency. The gro for disqualification are based on types of noncompliance that si cantly impair achievement of objectives. Proposed § 3e.2 through (p) is deleted, and § 58.202(a) through (c) clarifies policy that a testing facility ma disqualified only if the Commiss finds all three of the following That the testing facility faile comply with one or more of the s ards set forth in Part 58 or in other FDA regulations rega standards for nonclinical testing: ties (e.g., any supplemental re ments in the IND or IDE regulat (2) that the noncompliance adve affected the validity of the data duced by the study; and (3) that lesser regulatory actions, suc warnings or rejection of data fro dividual nonclinical laboratory st have not been or probably will r adequate to achieve compliance.

ach will assure that the sanction ot be used in trivial situations, ill be invoked only when the vio- has compromised the integrity study. It further requires the issioner to consider the avail- y and probable effectiveness of sanctions as an alternative to alification. It would not, howev- reclude disqualification without warning.

pointed out in the preamble to roposed regulations, the provi- for disqualification are not to be reted as either the exclusive or ry administrative action for non- iance with good laboratory prac- Disqualification is designed to le FDA with an enforcement tool is more efficient and effective a study-by-study review when it es apparent that a testing facili- ot capable of producing accurate lid test results. The disqualifica- of a nonclinical testing facility e reserved for the the rare case the rejection of a particular is an inadequate regulatory re- e. The testing facility and/or the or of the nonclinical laboratory may also be prosecuted for viola- of Federal criminal laws, includ- ction 301(e) of the Federal Food, and Cosmetic Act (failure to a report required under certain sections of the act, because a y erroneous or inadequate report not fulfill the statutory obliga- and 18 U.S.C. 1001 (submission of se report to the government), where the testing facility is not a direct statutory obligation to it information to FDA, and in loes not send data to the agency merely transmits them to the or, the facility is likely to be that FDA will be the ultimate ent. In such cases, it may be for aiding and abetting in the ion (18 U.S.C. 2) or for causing iolation to be made by a third

. Two comments stated that the alification regulation seemed to only to private firms.

s interpretation is incorrect. The nble to the proposed regulations s clear the policy that the good atory practice regulations are to to any institution that generates herwise prepares safety data for ission to FDA. Included in that ition, to the extent that they pre- safety data to be submitted to in support of petitions for regu- products, are, for example, vet- ry and medical clinics, universi- and State experimental stations, State and Federal Government re- h laboratories. Accordingly, dis- fication provisions apply equally ll facilities that prepare safety for submission to FDA. The lan-

guage regarding the intended use of sanctions is incorporated into § 58.202(c).

NOTICE OF AND OPPORTUNITY FOR HEAR- ING ON PROPOSED DISQUALIFICATION

234. Several comments stated that the disqualification process, as pro- posed, would violate due process, deny a formal hearing, and deny a right of appeal to the courts.

The Commissioner advises, and the revisions to § 58.202 make clear, that the disqualification procedure will not be invoked for minor violations of the regulation. In addition, § 58.204 pro- vides that a regulatory hearing may be conducted in accordance with 21 CFR Part 16. Such a hearing provides all the safeguards essential to due proc- ess. See also the FEDERAL REGISTER of 40 FR 40713 et seq. (preamble to Sub- part F of 21 CFR Part 2, recodified as 21 CFR Part 16—Regulatory Hearing Before the Food and Drug Administra- tion; section 201(y) of the act (21 U.S.C. 321(y)) (procedural require- ments of an "informal hearing"); Goldberg v. Kelly, 397 U.S. 254 (1970). Judicial review of final administrative action is provided by the Administra- tive Procedure Act (5 U.S.C. 701 et seq.). See also § 10.45 Court Review of final administrative action; exhaus- tion of administrative remedies (21 CFR 10.45); and 40 FR 40689–40691 (preamble to procedural regulations, § 2.11 (recodified as 21 CFR 10.45)).

235. Several comments expressed the concern that any regulatory hearing conducted under 21 CFR Part 16 should provide for the confidentiality of all data on which the hearing is based.

The Commissioner advises that § 16.60(a) (21 CFR 16.60(a)) provides adequate safeguards when required to maintain the confidentiality of com- mercial information.

236. One comment stated that if notice for such a hearing should be mailed to a facility, more than 3 days should be allowed for a facility to be able to prepare itself to come to a meeting.

The Commissioner finds that the provisions of § 16.22 (21 CFR 16.22) provide adequate flexibility for any party responding to a notice of oppor- tunity for a hearing. See also the com- ments addressed to 21 CFR 52.204, set out in the preamble to the proposed regulations on obligations of sponsors and monitors, published in the FEDER- AL REGISTER of September 27, 1977 (42 FR 49619).

237. One comment suggested that § 58.204 include a provision specifying that a sponsor be allowed to intervene in the hearing process when a notice of opportunity for a hearing has issued to a testing facility that is per-

forming studies under contract for the sponsor.

Inasmuch as the disqualification process in such a case is directed at the testing facility rather than the sponsor and inasmuch as the alleged violations involved would be those of the testing facility, the Commissioner finds that intervention by a sponsor (or, in many cases, multiple sponsors) would serve no useful purpose. As noted in the preamble to the proposed regulation (41 FR 51218), a sponsor who wishes to contest a finding that a particular study or studies is or are in- adequate will be provided an opportu- nity to do so by the procedures for denying or withdrawing the approval of an application for a research or marketing permit.

238. Concern was also expressed that a reasonable time be provided to allow a sponsor to conduct a new test prior to termination or withdrawal.

The Commissioner emphasizes that in those cases in which a safety deci- sion has been based on data that have subsequently been called into ques- tion, protection of the public requires that proceedings be instituted without delay. As previously noted, opportuni- ty to contest a finding that a particu- lar study is so inadequate that it will not support a claim of safety of a product will be provided by procedures set forth in other regulations, e.g., withdrawal of an NDA.

FINAL ORDER ON DISQUALIFICATION

239. Several comments stated that § 58.206 should provide specifically for appeal to the Federal courts following a final decision to disqualify by the Commissioner.

The Commissioner notes that the provisions of 21 CFR 16.120 and 10.45 adequately address this point. These regulations clearly state the provisions that apply to court review of final ad- ministrative action.

240. One comment suggested that § 58.206(b) be modified to require that sponsors be notified, when applicable, at the time of issuance of a final order to a testing facility.

The Commissioner advises that such notification, which is discretionary, is expressly provided for in § 58.213(b). Additionally, § 58.206(a) and (b) are re- vised to reflect the requirement that the Commissioner must make the find- ings required by § 58.202 before a final order disqualifying a nonclinical test- ing facility shall issue.

ACTIONS UPON DISQUALIFICATION

241. Several comments objected to the retroactive provisions of § 58.210- (a), which state that once a testing fa- cility has been disqualified, each appli- cation for a research or marketing permit, whether approved or not, that contains or relies upon any nonclinical

laboratory study conducted by the disqualified testing facility may be examined to determine whether these studies were or would be essential to a decision.

The Commissioner advises that calling into question studies performed by a subsequently disqualified testing facility does not represent a departure from prior FDA policy in other areas. FDA must make additional inquiries to establish safety any time a question is raised about data previously submitted, regardless of whether a disqualification procedure exists. Section 58.210(a) allows the person relying on the study in question to establish that the study was not affected by the circumstances that led to disqualification. The safety of the public would not be adequately protected were no such validation required when serious questions are raised regarding the adequacy of data upon which regulatory decisions are based.

Section 58.210 is revised by the addition of paragraph (b), which states that no nonclinical laboratory study begun after a facility has been disqualified will be considered in support of any application for a research or marketing permit unless the facility has been reinstated under § 58.219. This addition makes it clear that, in such a case, no subsequent information can be submitted for purposes of subsequent validation. If the facility is reinstated, however, the study might by acceptable to FDA. This provision does not relieve the applicant from any other requirement under FDA regulations that all data and information regarding clinical experience with the article in question be submitted to the agency.

242. Many comments regarding § 58.210 were based on the assumption that the disqualification process might be invoked for a minor violation of the good laboratory practice regulation and stated that calling studies into question based on a minor violation was unreasonable.

As previously discussed, § 58.202 is revised to make it clear that the disqualification process will be reserved for those situations in which lesser sanctions, e.g., rejection of individual studies, will not suffice. Because disqualification will be reserved for use in serious situations, the Commissioner finds that calling into question all studies done before or after disqualification is warranted.

PUBLIC DISCLOSURE OF INFORMATION UPON DISQUALIFICATION

243. Several comments said that proprietary or trade secret documents should not be released. Others urged that disqualification records not be disclosed.

The Commissioner advises that release of all such documents is governed by the provisions of the Freedom of Information Act (5 U.S.C. 552) and 21 CFR Part 20 and need not be separately dealt with in this regulation. Interested parties are referred specifically to Part 20—Public Information (21 CFR Part 20). Section 20.61 (21 CFR 20.61) deals with trade secrets and commercial information and § 20.64 (21 CFR 20.64) deals with investigatory records. The preamble to the public information regulations (39 FR 44602 et seq.) (since recodified as Part 20) discusses these issues at length.

244. One comment on § 58.213 stated that no notification of other government departments or agencies should issue until completion of the judicial process.

The Commissioner disagrees and finds that withholding notification until completion of the administrative process by the agency provides an adequate opportunity for a testing facility to be heard prior to the issuance of any such notification.

245. Another comment stated that because FDA is a Federal agency, notification of State agencies is outside FDA's jurisdiction.

The Commissioner points out that section 705(b) of the act (21 U.S.C. 375(b)) provides for dissemination of information regarding food, drugs, or devices in situations involving imminent danger to health or gross deception of the consumer. In addition, the Commissioner emphasizes that he proposes to notify the States only in those situations for which adequate cause has been established and for which a final order has been issued. Section 58.213(a) is amended to make it clear that such notification shall state that it is given because of the relationship between the testing facility and the person notified and that the Food and Drug Administration is not advising or recommending that any action be taken by the person notified. Additionally, § 58.213 is modified to make it clear that notification of disqualification may be sent by the Commissioner not only to other Federal agencies but to any other person known to have professional relations with the disqualified testing facility. This includes sponsors of studies being performed by the facility.

246. A comment suggested that the scope of notification should be limited to those nonclinical laboratory studies upon which the decision to disqualify was based.

The language of § 58.213 makes it clear that notification may be given at the discretion of the Commissioner whenever he believes that such disclosure would further the public interest or would promote compliance with the

good laboratory practice regula[...] The Commissioner finds that, [...] the expressed purpose of notifica[...] further limitation would be ina[...] priate.

ALTERNATIVE OR ADDITIONAL ACTION[...] DISQUALIFICATION

247. One comment on § 58.215 [...] gested that informal procedure[...] used prior to the institution of [...] formal procedures.

The Commissioner notes that [...] approach was discussed in the pr[...] ble to the proposed regulation [...] FR 51218. Because such informal [...] cedures have, in the past, double[...] time and expense of all involved [...] ties without discernible benefit[...] Commissioner has decided not to [...] vide for informal procedures in [...] regulations.

SUSPENSION OR TERMINATION OF [...] TESTING FACILITY BY A SPONSO[...]

248. Many comments on § 58.21[...] that the section seemed to be a [...] tempt on the part of FDA to pr[...] legal grounds for the unilateral b[...] ing of contracts between private [...] ties.

The Commissioner finds that [...] section, as written, was subject [...] great deal of misunderstan[...] Therefore, the section is revised. [...] Commissioner advises that nothi[...] Part 58 is intended to infringe up[...] alter the private contractual arr[...] ments between a sponsor and a [...] clinical testing facility. A sponsor [...] terminate a testing facility for re[...] of its own whether or not FDA [...] begun any action to disqualify th[...] cility. Where a sponsor has inde[...] ent grounds for suspending or t[...] nating studies performed for [...] sponsor by the facility under con[...] the fact that FDA has not itsel[...] qualified the facility may not be r[...] by the contract facility as a de[...] against the sponsor.

249. Several comments said not[...] tion within 5 days was impractica[...]

The Commissioner agrees, and [...] time period is extended to 15 wo[...] days.

250. A number of comments sai[...] notification requirement provid[...] sponsor with an unfair opportun[...] impugn a contract facility that [...] have no opportunity for response[...]

The Commissioner emphasizes [...] termination of a nonclinical testi[...] cility by a sponsor should be subj[...] the contract between the two pa[...] A nonclinical testing facility, [...] party to the contract, may pr[...] itself from unjust termination b[...] terms of its contract with the sp[...] Remedies for both parties to s[...] contract may be spelled out in th[...] tract and are governed by princip[...] contract law. The Commissione[...]

FEDERAL REGISTER, VOL. 43, NO. 247—FRIDAY, DECEMBER 22, 1978

emphasizes that the requirement a sponsor notify FDA when it has [inated] or suspended a testing fa- ' applies only to those cases in h an application for a research or [teting] permit has been submitted. [re] no application has been sub- ed, no notification is required.

EINSTATEMENT OF A DISQUALIFIED TESTING FACILITY

l. One comment on § 58.219 ex- [sed] concern that when read with [210], it was confusing.

e Commissioner finds that the ad- [n] of § 58.210(b) substantially clari- the status of studies conducted [re], during, and after disqualifica- and that further amendment is [cessary].

l. A typographical error in the last [nce] of § 58.219 has been correct- [The] last sentence now reads: "A [rmination] that a testing facility been reinstated is disclosable to [public] under Part 20 of this Chap-

CONFORMING AMENDMENTS

l. The Commissioner is adding to [evising] provisions in the regula- [s] regarding food and color addi- . new drugs for investigational new drug applications, OTC drug [ucts], antibiotic drugs, new animal applications, biological product [ses], and performance standards [electronic] products to incorporate [opriate] implementing provisions and cross references to, Part 58, [h] is being added by this docu- [t]. Each of the regulations requires [submission] of data which may in- [e] nonclinical laboratory studies. regulations are being revised to [ire], with respect to each nonclini- [aboratory] study contained as part [e] submitted information, either a [ment] that the study was conduct- [i] compliance with the good labora- practice regulations set forth in 58 of this chapter, or, if the study not conducted in compliance with . regulations, a statement that de- [es] in detail all differences be- [n] the practices used in the study those required in the regulations. revisions highlight the fact that [ough] studies not conducted in [pliance] with the regulations may [inue] to be submitted to FDA, the [en] of establishing that the non- [pliance] did not affect the quality [e] data submitted is on the person [nitting] the noncomplying study. [erefore], under the Federal Food, [g], and Cosmetic Act (secs. 406, 408, 502, 503, 505, 506, 507, 510, 512- 518-520, 701(a), 706, and 801, 52 . 1049-1053 as amended, 1055, 1058 [mended], 55 Stat. 851 as amended, [tat]. 463 as amended, 68 Stat. 511- as amended, 72 Stat. 1785-1788 as

amended, 76 Stat. 794 as amended, 82 Stat. 343-351, 90 Stat. 539-574 (21 U.S.C. 346, 346a, 348, 352, 353, 355, 356, 357, 360, 360b-360f, 360h-360j, 371(a), 376, and 381)) and the Public Health Service Act (secs. 215, 351, 354-360F, 58 Stat. 690, 702 as amended, 82 Stat. 1173-1186 as amended (42 U.S.C. 216, 262, 263b-263n)) and under authority delegated to him (21 CFR 5.1), the Commissioner amends Chapter I of 21 CFR as follows:

SUBCHAPTER A—GENERAL

PART 16—REGULATORY HEARING BEFORE THE FOOD AND DRUG ADMINISTRATION

1. Part 16 is amended in § 16.1 by redesignating paragraph (b)(30) as paragraph (c) and by adding new paragraph (b)(30), to read as follows:

§ 16.1 Scope.

* * * * *

(b) * * *
(30) Section 58.204(b) of this chapter, relating to disqualifying a nonclinical laboratory testing facility.

(c) Any other provision in the regulations in this chapter under which a party who is adversely affected by regulatory action is entitled to an opportunity for a hearing, and no other procedural provisions in this part are by regulation applicable to such hearing.

———

2. Part 58 is added to read as follows:

PART 58—GOOD LABORATORY PRACTICE FOR NONCLINICAL LABORATORY STUDIES

AUTHORITY: Secs. 406, 408, 409, 502, 503, 505, 506, 507, 510, 512-516, 518-520, 701(a), 706, and 801, Pub. L. 717, 52 Stat. 1049-1053 as amended, 1055, 1058 as amended, 55 Stat. 851 as amended, 59 Stat. 463, as amended, 68 Stat. 511-517 as amended, 72 Stat. 1785-1788 as amended, 76 Stat. 794 as amended, 82 Stat. 343-351, 90 Stat. 539-574 (21 U.S.C. 346, 346a, 348, 352, 353, 355, 356, 357, 360, 360b-360f, 360h-360j, 371(a), 376, and 381); secs. 215, 351, 354-360F, Pub. L. 410, 58 Stat. 690, 702 as amended, 82 Stat. 1173-1186 as amended (42 U.S.C. 216, 262, 263b-263n).

Subpart A—General Provisions

§ 58.1 Scope.

This part prescribes good laboratory practices for conducting nonclinical laboratory studies that support or are intended to support applications for research or marketing permits for products regulated by the Food and Drug Administration, including food and color additives, animal food additives, human and animal drugs, medical devices for human use, biological products, and electronic products. Compliance with this part is intended to assure the quality and integrity of the safety data filed pursuant to sections 406, 408, 409, 502, 503, 505, 506, 507, 510, 512-516, 518-520, 706, and 801 of the Federal Food, Drug, and Cosmetic Act and sections 351 and 354-360F of the Public Health Service Act.

§ 58.3 Definitions.

As used in this part, the following terms shall have the meanings specified:

(a) "Act" means the Federal Food, Drug, and Cosmetic Act, as amended (secs. 201-902, 52 Stat. 1040 et seq., as amended (21 U.S.C. 321-392)).

(b) "Test article" means any food additive, color additive, drug, biological product, electronic product, medical device for human use, or any other article subject to regulation under the act or under sections 351 and 354-360F of the Public Health Service Act.

(c) "Control article" means any food additive, color additive, drug, biological product, electronic product, medical device for human use, or any other article other than a test article that is administered to the test system in the course of a nonclinical laboratory study for the purpose of establishing a basis for comparison with the test article.

(d) "Nonclinical laboratory study" means any in vivo or in vitro experiment in which a test article is studied prospectively in a test system under laboratory conditions to determine its safety. The term does not include studies utilizing human subjects or clinical studies or field trials in animals. The term does not include basic exploratory studies carried out to determine whether a test article has any potential utility or to determine physical or chemical characteristics of a test article.

(e) "Application for research or marketing permit" includes:

(1) A color additive petition, described in Part 71 of this chapter.

(2) A food additive petition, described in Parts 171 and 571 of this chapter.

(3) Data and information regarding a substance submitted as part of the procedures for establishing that a substance is generally recognized as safe for use, which use results or may reasonably be expected to result, directly or indirectly, in its becoming a component or otherwise affecting the characteristics of any food, described in §§ 170.35 and 570.35 of this chapter.

(4) Data and information regarding a food additive submitted as part of the procedures regarding food additives permitted to be used on an interim basis pending additional study, described in § 180.1 of this chapter.

(5) A "Notice of Claimed Investigational Exemption for a New Drug," described in Part 312 of this chapter.

(6) A "new drug application," described in Part 314 of this chapter.

(7) Data and information regarding an over-the-counter drug for human use, submitted as part of the procedures for classifying such drugs as generally recognized as safe and effective and not misbranded, described in Part 330 of this chapter.

(8) Data and information regarding a prescription drug for human use submitted as part of the procedures for classifying such drugs as generally recognized as safe and effective and not misbranded, to be described in this chapter.

(9) Data and information regarding an antibiotic drug submitted as part of the procedures for issuing, amending, or repealing regulations for such drugs, described in Part 430 of this chapter.

(10) A "Notice of Claimed Investigational Exemption for a New Animal Drug," described in Part 511 of this chapter.

(11) A "new animal drug application," described in Part 514 of this chapter.

(12) Data and information regarding a drug for animal use submitted as part of the procedures for classifying such drugs as generally recognized as safe and effective and not misbranded, to be described in this chapter.

(13) An "application for a biological product license," described in Part 601 of this chapter.

(14) An "application for an investigational device exemption," described in Part 812 of this chapter.

(15) An "Application for Premarket Approval of a Medical Device," described in section 515 of the act.

(16) A "Product Development Protocol for a Medical Device," described in section 515 of the act.

(17) Data and information regarding a medical device submitted as part of the procedures for classifying such devices, described in section 513 of the act.

(18) Data and information regarding a medical device submitted as part of the procedures for establishing, amending, or repealing a performance standard for such devices, described in section 514 of the act.

(19) Data and information regarding an electronic product submitted as part of the procedures for obtaining an exemption from notification of a radiation safety defect or failure of compliance with a radiation safety performance standard, described in Subpart D of Part 1003 of this chapter.

(20) Data and information regarding an electronic product submitted as part of the procedures for establishing, amending, or repealing a standard for such product, described in section 358 of the Public Health Service Act.

(21) Data and information regarding an electronic product submitted as part of the procedures for obtaining a variance from any electronic product performance standard as describe[d] § 1010.4 of this chapter.

(22) Data and information regar[ding] an electronic product submitte[d] part of the procedures for grant[ing] amending, or extending an exemp[tion] from any electronic product perfo[rm]ance standard, as described in § 10[10] of this chapter.

(f) "Sponsor" means:

(1) A person who initiates and [sup]ports, by provision of financial [or] other resources, a nonclinical lab[ora]tory study;

(2) A person who submits a noncl[ini]cal study to the Food and Drug [Ad]ministration in support of an app[lica]tion for a research or marke[ting] permit; or

(3) A testing facility, if it both i[niti]ates and actually conducts the stud[y].

(g) "Testing facility" means a pe[rson] who actually conducts a nonclin[ical] laboratory study, i.e., actually use[s the] test article in a test system. "Tes[ting] facility" includes any establishm[ent] required to register under section [510] of the act that conducts nonclin[ical] laboratory studies and any consul[ting] laboratory described in section 70[4 of] the act that conducts such stu[dies]. "Testing facility" encompasses [only] those operational units that are b[eing] or have been used to conduct nonc[lini]cal laboratory studies.

(h) "Person" includes an indivi[dual,] partnership, corporation, associa[tion,] scientific or academic establishm[ent,] government agency, or organizati[onal] unit thereof, and any other l[egal] entity.

(i) "Test system" means any ani[mal,] plant, microorganism, or subp[art] thereof to which the test or contr[ol ar]ticle is administered or added [for] study. "Test system" also include[s ap]propriate groups or components o[f a] system not treated with the tes[t or] control articles.

(j) "Specimen" means any mat[erial] derived from a test system for ex[ami]nation or analysis.

(k) "Raw data" means any labo[ra]tory worksheets, records, memora[nda,] notes, or exact copies thereof, tha[t are] the result of original observations [and] activities of a nonclinical labora[tory] study and are necessary for the re[con]struction and evaluation of the re[port] of that study. In the event that e[xact] transcripts of raw data have been [pre]pared (e.g., tapes which have [been] transcribed verbatim, dated, and [veri]fied accurate by signature), the e[xact] copy or exact transcript may be su[bsti]tuted for the original source as [raw] data. "Raw data" may include p[hoto]graphs, microfilm or micro[film] copies, computer printouts, mag[netic] media, including dictated observati[ons,] and recorded data from automate[d in]struments.

"Quality assurance unit" means person or organizational element, >t the study director, designated sting facility management to per- the duties relating to quality as- :ce of nonclinical laboratory stud-

) "Study director" means the indi- :l responsible for the overall con- of a nonclinical laboratory study.

"Batch" means a specific quanti- : lot of a test or control article has been characterized according 8.105(a).

) Applicability to studies performed nder grants and contracts.

:en a sponsor conducting a non- :al laboratory study intended to ibmitted to or reviewed by the and Drug Administration utilizes ervices of a consulting laboratory, actor, or grantee to perform an rsis or other service, it shall y the consulting laboratory, con- or, or grantee that the service is of a nonclinical laboratory study must be conducted in compliance the provisions of this part.

5 Inspection of a testing facility.

A testing facility shall permit an orized employee of the Food and : Administration, at reasonable 5 and in a reasonable manner, to :ct the facility and to inspect (and e case of records also to copy) all ·ds and specimens required to be .tained regarding studies within :cope of this part. The records in- :ion and copying requirements not apply to quality assurance records of findings and problems, actions recommended and taken. The Food and Drug Administra- will not consider a nonclinical lab- >ry study in support of an applica- for a research or marketing iit if the testing facility refuses to iit inspection. The determination a nonclinical laboratory study not be considered in support of an ication for a research or market- >ermit does not, however, relieve ipplicant for such a permit of any :ation under any applicable stat- >r regulation to submit the results ie study to the Food and Drug Ad- stration.

Subpart B—Organization and Personnel

:9 Personnel.

Each individual engaged in the iuct of or responsible for the su- ision of a nonclinical laboratory y shall have education, training, experience, or combination there- :o enable that individual to per- i the assigned functions.

) Each testing facility shall main- a current summary of training

and experience and job description for each individual engaged in or supervis- ing the conduct of a nonclinical labo- ratory study.

(c) There shall be a sufficient number of personnel for the timely and proper conduct of the study ac- cording to the protocol.

(d) Personnel shall take necessary personal sanitation and health precau- tions designed to avoid contamination of test and control articles and test systems.

(e) Personnel engaged in a nonclini- cal laboratory study shall wear cloth- ing appropriate for the duties they perform. Such clothing shall be changed as often as necessary to pre- vent microbiological, radiological, or chemical contamination of test sys- tems and test and control articles.

(f) Any individual found at any time to have an illness that may adversely affect the quality and integrity of the nonclinical laboratory study shall be excluded from direct contact with test systems, test and control articles and any other operation or function that may adversely affect the study until the condition is corrected. All person- nel shall be instructed to report to their immediate supervisors any health or medical conditions that may reasonably be considered to have an adverse effect on a nonclinical labora- tory study.

§ 58.31 Testing facility management.

For each nonclinical laboratory study, testing facility management shall:

(a) Designate a study director as de- scribed in § 58.33, before the study is initiated.

(b) Replace the study director promptly if it becomes necessary to do so during the conduct of a study, and document and maintain such action as raw data.

(c) Assure that there is a quality as- surance unit as described in § 58.35.

(d) Assure that test and control arti- cles or mixtures have been appropri- ately tested for identity, strength, purity, stability, and uniformity, as ap- plicable.

(e) Assure that personnel, resources, facilities, equipment, materials, and methodologies are available as sched- uled.

(f) Assure that personnel clearly un- derstand the functions they are to per- form.

(g) Assure that any deviations from these regulations reported by the quality assurance unit are communi- cated to the study director and correc- tive actions are taken and document- ed.

§ 58.33 Study director.

For each nonclinical laboratory study, a scientist or other professional

of appropriate education, training, and experience, or combination thereof, shall be identified as the study direc- tor. The study director has overall re- sponsibility for the technical conduct of the study, as well as for the inter- pretation, analysis, documentation and reporting of results, and repre- sents the single point of study control. The study director shall assure that:

(a) The protocol, including any change, is approved as provided by § 58.120 and is followed.

(b) All experimental data, including observations of unanticipated re- sponses to the test system are accu- rately recorded and verified.

(c) Unforeseen circumstances that may affect the quality and integrity of the nonclinical laboratory study are noted when they occur, and corrective action is taken and documented.

(d) Test systems are as specified in the protocol.

(e) All applicable good laboratory practice regulations are followed.

(f) All raw data, documentation, pro- tocols, specimens, and final reports are transferred to the archives during or at the close of the study.

§ 58.35 Quality assurance unit.

(a) A testing facility shall have a quality assurance unit composed of one or more individuals who shall be responsible for monitoring each study to assure management that the facili- ties, equipment, personnel, methods, practices, records, and controls are in conformance with the regulations in this part. For any given study the quality assurance unit shall be entire- ly separate from and independent of the personnel engaged in the direction and conduct of that study.

(b) The quality assurance unit shall:

(1) Maintain a copy of a master schedule sheet of all nonclinical labo- ratory studies conducted at the testing facility indexed by test article and containing the test system, nature of study, date study was initiated, cur- rent status of each study, name of the sponsor, name of the study director, and status of the final report.

(2) Maintain copies of all protocols pertaining to all nonclinical laboratory studies for which the unit is responsi- ble.

(3) Inspect each phase of a nonclini- cal laboratory study periodically and maintain written and properly signed records of each periodic inspection showing the date of the inspection, the study inspected, the phase or seg- ment of the study inspected, the person performing the inspection, findings and problems, action recom- mended and taken to resolve existing problems, and any scheduled date for re-inspection. For studies lasting more than 6 months, inspections shall be conducted every 3 months. For studies

lasting less than 6 months, inspections shall be conducted at intervals adequate to assure the integrity of the study. Any significant problems which are likely to affect study integrity found during the course of an inspection shall be brought to the attention of the study director and management immediately.

(4) Periodically submit to management and the study director written status reports on each study, noting any problems and the corrective actions taken.

(5) Determine that no deviations from approved protocols or standard operating procedures were made without proper authorization and documentation.

(6) Review the final study report to assure that such report accurately describes the methods and standard operating procedures, and that the reported results accurately reflect the raw data of the nonclinical laboratory study.

(7) Prepare and sign a statement to be included with the final study report which shall specify the dates inspections were made and findings reported to management and to the study director.

(c) The responsibilities and procedures applicable to the quality assurance unit, the records maintained by the quality assurance unit, and the method of indexing such records shall be in writing and shall be maintained. These items including inspection dates, the study inspected, the phase or segment of the study inspected, and the name of the individual performing the inspection shall be made available for inspection to authorized employees of the Food and Drug Administration.

(d) A designated representative of the Food and Drug Administration shall have access to the written procedures established for the inspection and may request testing facility management to certify that inspections are being implemented, performed, documented, and followed-up in accordance with this paragraph.

(e) All records maintained by the quality assurance unit shall be kept in one location at the testing facility.

Subpart C—Facilities

§ 58.41 General.

Each testing facility shall be of suitable size, construction, and location to facilitate the proper conduct of nonclinical laboratory studies. It shall be designed so that there is a degree of separation that will prevent any function or activity from having an adverse effect on the study.

§ 58.43 Animal care facilities.

(a) A testing facility shall have a sufficient number of animal rooms or areas, as needed, to assure proper: (1)

Separation of species or test systems, (2) isolation of individual projects, (3) quarantine of animals, and (4) routine or specialized housing of animals.

(b) A testing facility shall have a number of animal rooms or areas separate from those described in paragraph (a) of this section to ensure isolation of studies being done with test systems or test and control articles known to be biohazardous, including volatile substances, aerosols, radioactive materials, and infectious agents.

(c) Separate areas shall be provided for the diagnosis, treatment, and control of laboratory animal diseases. These areas shall provide effective isolation for the housing of animals either known or suspected of being diseased, or of being carriers of disease, from other animals.

(d) When animals are housed, facilities shall exist for the collection and disposal of all animal waste and refuse or for safe sanitary storage of waste before removal from the testing facility. Disposal facilities shall be so provided and operated as to minimize vermin infestation, odors, disease hazards, and environmental contamination.

(e) Animal facilities shall be designed, constructed, and located so as to minimize disturbances that interfere with the study.

§ 58.45 Animal supply facilities.

There shall be storage areas, as needed, for feed, bedding, supplies, and equipment. Storage areas for feed and bedding shall be separated from areas housing the test systems and shall be protected against infestation or contamination. Refrigeration shall be provided for perishable supplies or feed.

§ 58.47 Facilities for handling test and control articles.

(a) As necessary to prevent contamination or mixups, there shall be separate areas for:

(1) Receipt and storage of the test and control articles.

(2) Mixing of the test and control articles with a carrier, e.g., feed.

(3) Storage of the test and control article mixtures.

(b) Storage areas for the test and/or control article and test and control mixtures shall be separate from areas housing the test systems and shall be adequate to preserve the identity, strength, purity, and stability of the articles and mixtures.

§ 58.49 Laboratory operation areas.

(a) Separate laboratory space shall be provided, as needed, for the performance of the routine procedures required by nonclinical laboratory studies, including specialized areas for performing activities such as aseptic sur-

gery, intensive care, necropsy, hi[...]gy, radiography, and handling of hazardous materials.

(b) Separate space shall be pro[...] for cleaning, sterilizing, and main[...]ing equipment and supplies during the course of the study.

§ 58.51 Specimen and data storage [...]ties.

Space shall be provided for arch[...] limited to access by authorized pe[...]nel only, for the storage and retr[...] of all raw data and specimens [...] completed studies.

§ 58.53 Administrative and pers[...] facilities.

(a) There shall be space provide[...] the administration, supervision, direction of the testing facility.

(b) Separate space shall be pro[...] for locker, shower, toilet, and wa[...] facilities, as needed.

Subpart D—Equipment

§ 58.61 Equipment design.

Automatic, mechanical, or elect[...] equipment used in the gener[...] measurement, or assessment of [...] and equipment used for facility [...]ronmental control shall be of a[...] priate design and adequate capac[...] function according to the protoco[...] shall be suitably located for oper[...] inspection, cleaning, and mainten[...]

§ 58.63 Maintenance and calibratio[...] equipment.

(a) Equipment shall be adequ[...] inspected, cleaned, and mainta[...] Equipment used for the gener[...] measurement, or assessment of [...] shall be adequately tested, calib[...] and/or standardized.

(b) The written standard oper[...] procedures required [...] § 58.81(b)(11) shall set forth in [...] cient detail the methods, mate[...] and schedules to be used in the [...] tine inspection, cleaning, mainten[...] testing, calibration and/or stan[...] zation of equipment, and shall sp[...] remedial action to be taken in [...] event of failure or malfunctio[...] equipment. The written standard [...] ating procedures shall designat[...] person responsible for the per[...] ance of each operation, and cop[...] the standard operating proce[...] shall be made available to labor [...] personnel.

(c) Written records shall be [...] tained of all inspection, mainten[...] testing, calibrating and/or stand[...] ing operations. These records, co[...] ing the date of the operation, sh[...] scribe whether the maintenance [...] ations were routine and followe[...] written standard operating [...] dures. Written records shall be k[...] nonroutine repairs performed [...]

ment as a result of failure and
ction. Such records shall docu-
the nature of the defect, how
hen the defect was discovered,
ny remedial action taken in re-
: to the defect.

ubpart E—Testing Facilities Operation

Standard operating procedures.

A testing facility shall have
rd operating procedures in writ-
tting forth nonclinical labora-
tudy methods that management
sfied are adequate to insure the
y and integrity of the data gen-
in the course of a study. All de-
is in a study from standard oper-
procedures shall be authorized
: study director and shall be do-
ted in the raw data. Significant
es in established standard oper-
procedures shall be properly au-
ed in writing by management.
Standard operating procedures
be established for, but not limit-
the following:
nimal room preparation.
nimal care.
Receipt, identification, storage,
ing, mixing, and method of sam-
of the test and control articles.
est system observations.
aboratory tests.
Handling of animals found mori-
or dead during study.
Necropsy of animals or postmor-
amination of animals.
Collection and identification of
iens.
Histopathology.
Data handling, storage, and re-
l.
Maintenance and calibration of
nent.
Transfer, proper placement, and
fication of animals.
Each laboratory area shall have
liately available laboratory man-
and standard operating proce-
relative to the laboratory proce-
being performed, e.g., toxicol-
iistology, clinical chemistry, he-
ogy, teratology, necropsy. Pub-
literature may be used as a sup-
nt to standard operating proce-

A historical file of standard oper-
procedures, and all revisions
of, including the dates of such re-
s, shall be maintained.

Reagents and solutions.

reagents and solutions in the lab-
ry areas shall be labeled to indi-
identity, titer or concentration,
ge requirements, and expiration
Deteriorated or outdated rea-
and solutions shall not be used.

§ 58.90 Animal care.

(a) There shall be standard operating procedures for the housing, feeding, handling, and care of animals.

(b) All newly received animals from
outside sources shall be placed in quarantine until their health status has
been evaluated. This evaluation shall
be in accordance with acceptable veterinary medical practice.

(c) At the initiation of a nonclinical
laboratory study, animals shall be free
of any disease or condition that might
interfere with the purpose or conduct
of the study. If, during the course of
the study, the animals contract such a
disease or condition, the diseased animals shall be isolated. If necessary,
these animals may be treated for disease or signs of disease provided that
such treatment does not interfere with
the study. The diagnosis, authorizations of treatment, description of
treatment and each date of treatment
shall be documented and shall be retained.

(d) Warm-blooded animals, excluding suckling rodents, used in laboratory procedures that require manipulations and observations over an extended period of time or in studies
that require the animals to be removed from and returned to their
home cages for any reason (e.g., cage
cleaning, treatment, etc.), shall receive
appropriate identification (e.g., tattoo,
toe clip, color code, ear tag, ear punch,
etc.). All information needed to specifically identify each animal within
an animal-housing unit shall appear
on the outside of that unit.

(e) Animals of different species shall
be housed in separate rooms when
necessary. Animals of the same species, but used in different studies,
should not ordinarily be housed in the
same room when inadvertent exposure
to control or test articles or animal
mixup could affect the outcome of
either study. If such mixed housing is
necessary, adequate differentiation by
space and identification shall be made.

(f) Animal cages, racks and accessory
equipment shall be cleaned and sanitized at appropriate intervals.

(g) Feed and water used for the animals shall be analyzed periodically to
ensure that contaminants known to be
capable of interfering with the study
and reasonably expected to be present
in such feed or water are not present
at levels above those specified in the
protocol. Documentation of such analyses shall be maintained as raw data.

(h) Bedding used in animal cages or
pens shall not interfere with the purpose or conduct of the study and shall
be changed as often as necessary to
keep the animals dry and clean.

(i) If any pest control materials are
used, the use shall be documented.
Cleaning and pest control materials

that interfere with the study shall not
be used.

Subpart F—Test and Control Articles

§ 58.105 Test and control article characterization.

(a) The identity, strength, purity,
and composition or other characteristics which will appropriately define
the test or control article shall be determined for each batch and shall be
documented before the initiation of
the study. Methods of synthesis, fabrication, or derivation of the test and
control articles shall be documented
by the sponsor or the testing facility.
In those cases where marketed products are used as control articles, such
products will be characterized by their
labeling.

(b) The stability of each test or control article shall be determined by the
testing facility or by the sponsor
before initiation or a nonclinical laboratory study. If the stability of the
test and control articles cannot be determined before initiation of a study,
standard operating procedures shall be
established and followed to provide for
periodic re-analysis of each batch.

(c) Each storage container for a test
or control article shall be labeled by
name, chemical abstract number or
code number, batch number, expiration date, if any, and, where appropriate, storage conditions necessary to
maintain the identity, strength,
purity, and composition of the test or
control article. Storage containers
shall be assigned to a particular test
article for the duration of the study.

(d) For studies of more than 4 weeks'
duration, reserve samples from each
batch of test and control articles shall
be retained for the period of time provided by § 58.195.

§ 58.107 Test and control article handling.

Procedures shall be established for a
system for the handling of the test
and control articles to ensure that:

(a) There is proper storage.

(b) Distribution is made in a manner
designed to preclude the possibility of
contamination, deterioration, or
damage.

(c) Proper identification is maintained throughout the distribution
process.

(d) The receipt and distribution of
each batch is documented. Such documentation shall include the date and
quantity of each batch distributed or
returned.

§ 58.113 Mixtures of articles with carriers.

(a) For each test or control article
that is mixed with a carrier, tests by
appropriate analytical methods shall
be conducted:

(1) To determine the uniformity of
the mixture and to determine, periodi-

cally, the concentration of the test or control article in the mixture.

(2) To determine the stability of the test and control articles in the mixture. If the stability cannot be determined before initiation of the study, standard operating procedures shall be established and followed to provide for periodic re-analysis of the test and control articles in the mixture.

(b) For studies of more than 4 weeks' duration a reserve sample of each test or control carrier article mixture shall be taken and retained for the period of time provided by § 58.195.

(c) Where any of the components of the test or control article carrier mixture has an expiration date, that date shall be clearly shown on the container. If more than one component has an expiration date, the earliest date shall be shown.

Subpart G—Protocol for and Conduct of a Nonclinical Laboratory Study

§ 58.120 Protocol.

(a) Each study shall have an approved written protocol that clearly indicates the objectives and all methods for the conduct of the study. The protocol shall contain but shall not necessarily be limited to the following information:

(1) A descriptive title and statement of the purpose of the study.

(2) Identification of the test and control articles by name, chemical abstract number or code number.

(3) The name of the sponsor and the name and address of the testing facility at which the study is being conducted.

(4) The proposed starting and completion dates.

(5) Justification for selection of the test system.

(6) Where applicable, the number, body weight range, sex, source of supply, species, strain, substrain, and age of the test system.

(7) The procedure for identification of the test system.

(8) A description of the experimental design, including the methods for the control of bias.

(9) A description and/or identification of the diet used in the study as well as solvents, emulsifiers and/or other materials used to solubilize or suspend the test or control articles before mixing with the carrier. The description shall include specifications for acceptable levels of contaminants that are reasonably expected to be present in the dietary materials and are known to be capable of interfering with the purpose or conduct of the study if present at levels greater than established by the specifications.

(10) The route of administration and the reason for its choice.

(11) Each dosage level, expressed in milligrams per kilogram of body weight or other appropriate units, of the test or control article to be administered and the method and frequency of administration.

(12) Method by which the degree of absorption of the test and control articles by the test system will be determined if necessary to achieve the objectives of the study.

(13) The type and frequency of tests, analyses, and measurements to be made.

(14) The records to be maintained.

(15) The date of approval of the protocol by the sponsor and the signature of the study director.

(16) A statement of the proposed statistical methods to be used.

(b) All changes in or revisions of an approved protocol and the reasons therefor shall be documented, signed by the study director, dated, and maintained with the protocol.

§ 58.130 Conduct of a nonclinical laboratory study.

(a) The nonclinical laboratory study shall be conducted in accordance with the protocol.

(b) The test systems shall be monitored in conformity with the protocol.

(c) Specimens shall be identified by test system, study, nature, and date of collection. This information shall be located on the specimen container or shall accompany the specimen in a manner that precludes error in the recording and storage of data.

(d) Records of gross findings for a specimen from postmortem observations shall be available to a pathologist when examining that specimen histopathologically.

(e) All data generated during the conduct of a nonclinical laboratory study, except those that are generated as direct computer input, shall be recorded directly, promptly, and legibly in ink. All data entries shall be dated on the day of entry and signed or initialed by the person entering the data. Any change in entries shall be made so as not to obscure the original entry, shall indicate the reason for such change, and shall be dated and signed or identified at the time of the change. In computer driven data collection systems, the individual responsible for direct data input shall be identified at the time of data input. Any change in computer entries shall be made so as not to obscure the original entry, shall indicate the reason for change, and shall be dated and the responsible individual shall be identified.

Subparts H–I—[Reserved

Subpart J—Records and Rep

§ 58.185 Reporting of nonclinical tory study results.

(a) A final report shall be pr for each nonclinical laboratory and shall include, but not nece be limited to, the following:

(1) Name and address of the performing the study and the d which the study was initiate completed.

(2) Objectives and procedures in the approved protocol, inc any changes in the original prot

(3) Statistical methods employ analyzing the data.

(4) The test and control identified by name, chemical ab number or code number, st purity, and composition or oth propriate characteristics.

(5) Stability of the test and articles under the conditions of istration.

(6) A description of the m used.

(7) A description of the test used. Where applicable, the report shall include the num animals used, sex, body weight source of supply, species, stra substrain, age, and procedure u identification.

(8) A description of the dosage regimen, route of admi tion, and duration.

(9) A description of all cir stances that may have affect quality or integrity of the data.

(10) The name of the study di the names of other scientists fessionals, and the names of all visory personnel, involved i study.

(11) A description of the trans tions, calculations, or operation formed on the data, a summa analysis of the data, and a sta of the conclusions drawn fro analysis.

(12) The signed and dated rep each of the individual scient other professionals involved study.

(13) The locations where all mens, raw data, and the final are to be stored.

(14) The statement prepare signed by the quality assurance described in § 58.35(b)(7).

(b) The final report shall be by the study director.

(c) Corrections or additions to report shall be in the form amendment by the study direct amendment shall clearly identi part of the final report that i added to or corrected and the for the correction or additio ll be signed and dated by the son responsible.

190 Storage and retrieval of records and data.

) All raw data, documentation, pro-ols, specimens, and final reports erated as a result of a nonclinical oratory study shall be retained.

) There shall be archives for order-torage and expedient retrieval of aw data, documentation, protocols, :imens, and interim and final re-ts. Conditions of storage shall mini-e deterioration of the documents specimens in accordance with the uirements for the time period of ir retention and the nature of the uments or specimens. A testing fa-y may contract with commercial ives to provide a repository for all erial to be retained. Raw data and :imens may be retained elsewhere vided that the archives have specif-ference to those other locations.

) An individual shall be identified esponsible for the archives.

) Only authorized personnel shall er the archives.

) Material retained or referred to the archives shall be indexed by article, date of study, test system, nature of study.

195 Retention of records.

) Record retention requirements forth in this section do not super- the record retention require-its of any other regulations in this pter.

) Except as provided in paragraph of this section, documentation rec-s, raw data and specimens pertain-to a nonclinical laboratory study required to be made by this part ll be retained in the archive(s) for chever of the following periods is rtest:

) A period of at least 2 years fol-ing the date on which an applica-i for a research or marketing mit, in support of which the results the nonclinical laboratory study e submitted, is approved by the d and Drug Administration. This uirement does not apply to studies porting notices of claimed investi-ional exemption for new drugs D's) or applications for investiga-al device exemptions (IDE's), rec-s of which shall be governed by the visions of paragraph (b)(2) of this :ion.

) A period of at least 5 years fol-ing the date on which the results he nonclinical laboratory study are mitted to the Food and Drug Ad-istration in support of an applica-i for a research or marketing mit.

) In other situations (e.g., where nonclinical laboratory study does result in the submission of the

study in support of an application for a research or marketing permit), a period of at least 2 years following the date on which the study is completed, terminated, or discontinued.

(c) Wet specimens, samples of test or control articles, samples of test or control article carrier mixtures and specially prepared material (e.g., histochemical, electron microscopic, blood mounts, teratological preparation, and uteri from dominant lethal mutagenesis tests), which are relatively fragile and differ markedly in stability and quality during storage, shall be retained only as long as the quality of the preparation affords evaluation. In no case shall retention be required for longer periods than those set forth in paragraphs (a) and (b) of this section.

(d) The master schedule sheet, copies of protocols, and records of quality assurance inspections, as required by § 58.35(c) shall be maintained by the quality assurance unit as an easily accessible system of records for the period of time specified in paragraphs (a) and (b) of this section.

(e) Summaries of training and experience and job descriptions required to be maintained by § 58.29(b) may be retained along with all other testing facility employment records for the length of time specified in paragraphs (a) and (b) of this section.

(f) Records and reports of the maintenance and calibration and inspection of equipment, as required by § 58.63(b) and (c), shall be retained for the length of time specified in paragraph (b) of this section.

(g) If a facility conducting nonclinical testing goes out of business, all raw data, documentation, and other material specified in this section shall be transferred to the archives of the sponsor of the study. The Food and Drug Administration shall be notified in writing of such a transfer.

Subpart K—Disqualification of Testing Facilities

§ 58.200 Purpose.

(a) The purposes of disqualification are: (1) To permit the exclusion from consideration of completed studies that were conducted by a testing facility which has failed to comply with the requirements of the good laboratory practice regulations until it can be adequately demonstrated that such noncompliance did not occur during, or did not affect the validity or acceptability of data generated by, a particular study; and (2) to exclude from consideration all studies completed after the date of disqualification until the facility can satisfy the Commissioner that it will conduct studies in compliance with such regulations.

(b) The determination that a nonclinical laboratory study may not be

considered in support of an application for a research or marketing permit does not, however, relieve the applicant for such a permit of any obligation under any other applicable regulation to submit the results of the study to the Food and Drug Administration.

§ 58.202 Grounds for disqualification.

The Commissioner may disqualify a testing facility upon finding all of the following:

(a) The testing facility failed to comply with one or more of the regulations set forth in this part (or any other regulations regarding such facilities in this chapter);

(b) The noncompliance adversely affected the validity of the nonclinical laboratory studies; and

(c) Other lesser regulatory actions (e.g., warnings or rejection of individual studies) have not been or will probably not be adequate to achieve compliance with the good laboratory practice regulations.

§ 58.204 Notice of and opportunity for hearing on proposed disqualification.

(a) Whenever the Commissioner has information indicating that grounds exist under § 58.202 which in his opinion justify disqualification of a testing facility, he may issue to the testing facility a written notice proposing that the facility be disqualified.

(b) A hearing on the disqualification shall be conducted in accordance with the requirements for a regulatory hearing set forth in Part 16 of this chapter.

§ 58.206 Final order on disqualification.

(a) If the Commissioner, after the regulatory hearing, or after the time for requesting a hearing expires without a request being made, upon an evaluation of the administrative record of the disqualification proceeding, makes the findings required in § 58.202, he shall issue a final order disqualifying the facility. Such order shall include a statement of the basis for that determination. Upon issuing a final order, the Commissioner shall notify (with a copy of the order) the testing facility of the action.

(b) If the Commissioner, after a regulatory hearing or after the time for requesting a hearing expires without a request being made, upon an evaluation of the administrative record of the disqualification proceeding, does not make the findings required in § 58.202, he shall issue a final order terminating the disqualification proceeding. Such order shall include a statement of the basis for that determination. Upon issuing a final order the Commissioner shall notify the testing facility and provide a copy of the order.

FEDERAL REGISTER, VOL. 43, NO. 247—FRIDAY, DECEMBER 22, 1978

§ 58.210 Actions upon disqualification.

(a) Once a testing facility has been disqualified, each application for a research or marketing permit, whether approved or not, containing or relying upon any nonclinical laboratory study conducted by the disqualified testing facility may be examined to determine whether such study was or would be essential to a decision. If it is determined that a study was or would be essential, the Food and Drug Administration shall also determine whether the study is acceptable, notwithstanding the disqualification of the facility. Any study done by a testing facility before or after disqualification may be presumed to be unacceptable, and the person relying on the study may be required to establish that the study was not affected by the circumstances that led to the disqualification, e.g., by submitting validating information. If the study is then determined to be unacceptable, such data such be eliminated from consideration in support of the application; and such elimination may serve as new information justifying the termination or withdrawal of approval of the application.

(b) No nonclinical laboratory study begun by a testing facility after the date of the facility's disqualification shall be considered in support of any application for a research or marketing permit, unless the facility has been reinstated under § 58.219. The determination that a study may not be considered in support of an application for a research or marketing permit does not, however, relieve the applicant for such a permit of any obligation under any other applicable regulation to submit the results of the study to the Food and Drug Administration.

§ 58.213 Public disclosure of information regarding disqualification.

(a) Upon issuance of a final order disqualifying a testing facility under § 58.206(a), the Commissioner may notify all or any interested persons. Such notice may be given at the discretion of the Commissioner whenever he believes that such disclosure would further the public interest or would promote compliance with the good laboratory practice regulations set forth in this part. Such notice, if given, shall include a copy of the final order issued under § 58.206(a) and shall state that the disqualification constitutes a determination by the Food and Drug Administration that nonclinical laboratory studies performed by the facility will not be considered by the Food and Drug Administration in support of any application for a research or marketing permit. If such notice is sent to another Federal Government agency, the Food and Drug Administration will recommend that the agency also consider whether or not it should

accept nonclinical laboratory studies performed by the testing facility. If such notice is sent to any other person, it shall state that it is given because of the relationship between the testing facility and the person being notified and that the Food and Drug Administration is not advising or recommending that any action be taken by the person notified.

(b) A determination that a testing facility has been disqualified and the administrative record regarding such determination are disclosable to the public under Part 20 of this chapter.

§ 58.215 Alternative or additional actions to disqualification.

(a) Disqualification of a testing facility under this subpart is independent of, and neither in lieu of nor a precondition to, other proceedings or actions authorized by the act. The Food and Drug Administration may, at any time, institute against a testing facility and/or against the sponsor of a nonclinical laboratory study that has been submitted to the Food and Drug Administration any appropriate judicial proceedings (civil or criminal) and any other appropriate regulatory action, in addition to or in lieu of, and prior to, simultaneously with, or subsequent to, disqualification. The Food and Drug Administration may also refer the matter to another Federal, State, or local government law enforcement or regulatory agency for such action as that agency deems appropriate.

(b) The Food and Drug Administration may refuse to consider any particular nonclinical laboratory study in support of an application for a research or marketing permit, if it finds that the study was not conducted in accordance with the good laboratory practice regulations set forth in this part, without disqualifying the testing facility that conducted the study or undertaking other regulatory action.

§ 58.217 Suspension or termination of a testing facility by a sponsor.

Termination of a testing facility by a sponsor is independent of, and neither in lieu of nor a precondition to, proceedings or actions authorized by this subpart. If a sponsor terminates or suspends a testing facility from further participation in a nonclinical laboratory study that is being conducted as part of any application for a research or marketing permit that has been submitted to any Bureau of the Food and Drug Administration (whether approved or not), it shall notify that Bureau in writing within 15 working days of the action; the notice shall include a statement of the reasons for such action. Suspension or termination of a testing facility by a sponsor does not relieve it of any obligation under any other applicable reg-

ulation to submit the results of study to the Food and Drug Adm tration.

§ 58.219 Reinstatement of a disqua testing facility.

A testing facility that has been qualified may be reinstated as a ceptable source of nonclinical la tory studies to be submitted to Food and Drug Administration i Commissioner determines, upon evaluation of the submission of testing facility, that the facility adequately assure that it will con future nonclinical laboratory st in compliance with the good la tory practice regulations set fort this part and, if any studies are rently being conducted, that the ity and integrity of such studies not been seriously compromised. qualified testing facility that wish be so reinstated shall present in ing to the Commissioner reasons it believes it should be reinstated a detailed description of the corre actions it has taken or intends to to assure that the acts or omis which led to its disqualification not recur. The Commissioner ma dition reinstatement upon the te facility being found in compl with the good laboratory practice ulations upon an inspection. If a ing facility is reinstated, the Cor sioner shall so notify the testing ty and all organizations and pe who were notified, under § 58.2 the disqualification of the testir cility. A determination that a te facility has been reinstated is di ble to the public under Part 20 o chapter.

PART 71—COLOR ADDITIVE PETITIONS

3. Part 71 is amended:
a. § 71.1 by adding new para (g), to read as follows:

§ 71.1 Petitions.

*   *   *   *   *

(g) If nonclinical laboratory s are involved, petitions filed wit Commissioner under section 706 the act shall include with resp each nonclinical study contain the petition, either a statement the study was conducted in comp with the good laboratory practic ulations set forth in Part 58 o chapter, or, if the study was no ducted in compliance with such lations, a statement that descri detail all differences between the tices used in the study and tho quired in the regulations.

Friday
September 4, 1987

Part VI

Department of Health and Human Services

Food and Drug Administration

21 CFR Part 58
Good Laboratory Practice Regulations;
Final Rule

blank page

DEPARTMENT OF HEALTH AND HUMAN SERVICES

Food and Drug Administration

21 CFR Part 58

[Docket No. 83N-0142]

Good Laboratory Practice Regulations

AGENCY: Food and Drug Administration.
ACTION: Final rule.

SUMMARY: The Food and Drug Administration (FDA) is issuing a final rule that amends the regulations that specify good laboratory practice (GLP) for nonclinical laboratory studies. The amendments clarify, delete, or amend several provisions of the GLP regulations to reduce the regulatory burden on testing facilities. The changes will also achieve a substantial reduction in the paperwork burden imposed upon the regulated industries by the current regulations. Significant changes are made in the provisions respecting quality assurance, protocol preparation, test and control article characterization, and retention of specimens and samples based on FDA's experience in implementing the regulations. The agency has determined that the changes will not compromise the objective of the GLP regulations, which is to assure the quality and integrity of the safety data submitted in support of the approval of regulated products.

EFFECTIVE DATE: October 5, 1987.

FOR FURTHER INFORMATION CONTACT: Paul D. Lepore. Office of Regulatory Affairs (HFC-230), Food and Drug Administration. 5600 Fishers Lane. Rockville, MD 20857, 301-443-2390.

SUPPLEMENTARY INFORMATION:

Background

In the Federal Register of October 29, 198\. (49 FP 43530), FDA published a proposal to amend the agency's regulations in 21 CFR Part 58, which prescribe good laboratory practice for conducting nonclinical laboratory studies (the GLP regulations). The proposal was the result of an evaluation of the GLP regulations and of the data obtained by the agency's inspection program to assess laboratory compliance with the regulations. The evaluation led the agency to conclude that some of the provisions of the regulations could be revised to permit nonclinical testing laboratories greater flexibility in conducting nonclinical laboratory studies without compromising public protection. FDA invited comments on all aspects of the proposal and provided 60 days for interested persons to submit comments,

views, data, and information on the need to revise any other provisions of Part 58.

Comments

FDA received 33 comments: 19 from manufacturers of articles regulated by FDA, 4 from associations, 8 from foreign or domestic testing or consulting laboratories, and 2 from individuals within FDA. The majority of these comments endorsed the proposed changes. Many of the comments suggested additional revisions to the GLP regulations or modifications to the proposed changes. A summary of the comments received by FDA during the comment period and the agency's response to them follows.

General

1. One comment urged FDA to initiate training procedures for its field personnel so that the regulated community would obtain maximum benefit from the revisions to the GLP regulations.

FDA agrees that agency field personnel who conduct inspections of nonclinical testing laboratories need to understand the specific requirements of the GLP regulations to follow appropriate inspectional practices and procedures. FDA has, to date, conducted 17 training courses at its National Center for Toxicological Research in Jefferson, AR, to provide training in good laboratory practice and the associated laboratory inspection techniques. FDA intends to continue to provide such training for its personnel.

2. Eight comments urged FDA to encourage the Environmental Protection Agency (EPA) to adopt similar revisions to its good laboratory practice standards. The comments noted that unless EPA amends its good laboratory practice standards to conform them to FDA's GLP regulations, nonclinical laboratories will still be required to comply with EPA's more stringent requirements. Therefore, regardless of any changes that FDA makes in its regulations, laboratories will not benefit from the revisions unless the EPA regulations are similarly revised.

FDA recognizes that certain nonclinical laboratories that are subject to FDA's regulations are also subject to the good laboratory practice standards established by EPA under the Federal Insecticide, Fungicide and Rodenticide Act (7 U.S.C. 135 et seq.) and the Toxic Substances Control Act (15 U.S.C. 2600 et seq.). When this final rule becomes effective, some of the provisions of the GLP regulations will differ from the good laboratory practice standards established by EPA. FDA has consulted

with EPA officials respecting the changes to FDA's regulations effec by this final rule and will cooperat with EPA when that agency propo revise its regulations.

Scope

3. Except for editorial changes t § 58.1. FDA did not propose to cha the scope of Part 58. One comment however, urged the agency to revi: § 58.1 further to make clear that ba release safety tests performed on specific batches of biological prod intended for use in clinical trials a tests to establish the basic safety have been conducted are subject t GLP regulations.

FDA declines to change final § 5 the ground that the studies describ the comment are within the curren scope of Part 58. The animal tests performed with an investigational biological product prior to licensin including the batch release safety are intended to establish the safet the product. Accordingly, any suc would constitute a nonclinical laboratory study as defined in § 5: Because such test would also be intended to support a marketing application for a product regulatec FDA, it would be subject to the GI regulations.

Definitions

4. Four comments endorsed FDA proposal to change the definition control article" in § 58.3(c) to exc from the definition feed and water administered to control groups of system. One comment, however, expressed concern that, by relaxin essential standards, the proposed change would compromise the qua the animal test.

FDA does not agree that excludi feed and water from the definition "control article" will compromise quality. The regulations will conti contain provisions adequate to co the use of feed and water in a nonclinical laboratory study. For example, § 58.31(e) requires management to assure that materi available as scheduled, § 58.45 pr for proper feed storage, § 58.81(b)(requires the preparation of standa operating procedures for animal c: (e.g., nutrition), § 58.90(g) requires periodic analysis of feed and wate interfering contaminants, and fina § 58.120(a)(7) (formerly § 58.120(a] requires the protocol to contain a description or an identification of diet, including specifications for acceptable levels of contaminants Other sections of the regulations a

to feed or water that is used as a
r for the test or control article. For
ple, § 58.31(d) requires
gement to assure that test and
ol article mixtures have been
priately tested. § 58.47 requires
testing facility include storage
that are adequate to preserve the
ty, strength, purity, and stability of
mixtures, and § 58.113 requires the
atory to conduct appropriate
ses for uniformity of the mixture,
ll as concentration and stability of
st article in the mixture. As
ssed at length in the preamble to
oposal (49 FR 43531), the
dment to § 58.3(c) will mean that
ed and water provided to the
ol groups of a test system will not
bject to certain provisions of the
itions, e.g., those requiring control
es to be characterized and tested
ability (§ 58.105), retained as
ve samples (§ 58.195), or
ntable with respect to use
107). As discussed above, however,
gulations will continue to require
ovision of adequate supplies of
nd water, a description of the
proper storage, and use
ntability procedures as directed by
rotocol, and standard operating
dures. Further, only feed and water
n to be free from unacceptable
mination may be used in a study.
the reasons above, FDA concludes
he change to § 58.3(c) will not
romise the quality of the animal
nd that the term "control article"
d be reserved for the discrete
ances/articles and vehicles other
feed and water administered to
s of the test system to provide a
of comparison with the test article.
our comments on proposed
(d) endorsed FDA's proposal to
laboratories to conduct several
riments using the same test article
a single, comprehensive protocol.
comment, however, expressed
rn that by amending the definition
onclinical laboratory study," FDA
inadvertently encourage
atories to establish protocols that
e too brief to assure the quality and
rity of safety data developed
gh a study conducted under the
col, or (2) do not describe study
edures in sufficient detail for such
rance, because lengthy "umbrella
cols" may be difficult to administer
track during the amendment
ess.
der the revised definition in
3(d), a single "umbrella" protocol
be used for concurrent testing of
than one test article using a single
non procedure, e.g., mutagenicity

testing, or for a battery of studies of one
test article conducted in several test
systems. Section 58.120 requires that
each study have an approved written
protocol that clearly indicates the
objectives and all methods for the
conduct of the study, and § 58.33
requires the study director to assure that
the protocol is approved and followed.

FDA notes that the changed definition
of "nonclinical laboratory study" does
not require any laboratory to establish
"umbrella" protocols—it only allows it
as an option. The agency recognizes that
a longer, more complex protocol might
be more difficult to manage than a
simpler one; however, using an
"umbrella" protocol should be more
efficient than using several closely
related protocols. The quality or
accuracy of test data and procedures
should not be compromised, while the
paperwork burden should be reduced. In
any event, the laboratory remains
responsible for assuring that the validity
of any study that it conducts is not
adversely affected due to an inadequate
protocol.

6. One comment urged FDA to revise
further the definition of nonclinical
laboratory study to define the terms
"study initiation" and "study
termination."

FDA recognizes that differing words
and phrases are used within the GLP
regulations to denote dates respecting
significant events that occur during a
laboratory study. For this reason, the
agency agrees that it may be useful to
add to the regulations definitions of the
terms "study initiation" and "study
completion."

FDA advises that the study initiation
date represents the date on which the
study director has completed plans in
preparation for the technical conduct of
a study (see § 58.33) and on which,
under § 58.31(e), management is
required to make certain that personnel,
resources, facilities, equipment,
materials, and methodologies for the
study are available as scheduled. On the
study initiation date, the study is
entered on the master schedule sheet
(see § 58.35(b)(1)). After this date, any
protocol changes are to be made only in
accordance with the procedure
described in § 58.120(b). Accordingly,
FDA is adding new § 58.3(o) to define
"study initiation" to mean the date the
protocol is signed by the study director.

The study completion date is the date
on which the study director signs the
final report (see final § 58.185(b)). On
the study completion date, the study
director is required to make certain that
raw data, documentation, protocols,
specimens, and final reports are

transferred to the archives (see
§ 58.33(f)), and under § 58.35, the quality
assurance unit may retire the study from
the master schedule sheet. This date
also specifies the beginning of the
record retention period under
§ 58.195(b). After the study completion
date, final reports may be amended only
in accordance with the procedure
described in § 58.185(c). Accordingly,
FDA is adding new § 58.3(p) to define
"study completion" to mean the date the
final report is signed by the study
director. As a necessary conforming
amendment, FDA is also amending
§ 58.185(b) to provide that the final
report shall contain the dated signature
of the study director.

FDA advises that the phrase "close of
the study" as used in § 58.33(f) refers to
the study completion date. Also, the
terms "terminate" and "discontinue" as
used in § 58.195(b)(3) are used in their
ordinary senses to mean stop, cease,
break off, or give up, denoting that a
study has been ended before the
planned study completion date. For
these reasons, FDA believes that these
terms do not need any further definition
to make clear their meanings.

7a. Three comments urged FDA to
expand the definition of "raw data"
under § 58.3(k) to provide that the
computer record of "hand-recorded data
entered into the computer verbatim and
verified" could be substituted for the
original source as raw data.

FDA does not agree that the computer
record of hand-recorded data may be
considered as raw data. Individuals who
enter data from a laboratory study into
the computer commonly do not have any
knowledge of the conditions under
which the data were collected and may
not understand the data originator's
notations that regularly are included on
the hand-recorded data sheets. The
probability of error in data entry is
greatly increased under these
circumstances.

7b. One comment urged the agency to
revise § 58.3(k) to make clear that the
term "raw data" as it pertains to the
findings of the histopathological
examinations refers only to the signed
and dated final report of the pathologist.

FDA does not agree that it needs to
amend the definition of raw data
relative to the findings of
histopathological examinations. In
pertinent part, § 58.3(k) defines raw data
as laboratory worksheets, records,
memoranda, notes, or exact copies
thereof, that are the result of original
observations and activities and are
necessary for the reconstruction and
evaluation of the final report. Although
the notes taken by a pathologist during

histopathological examination of slides
are indeed the result of original
observations, these notes are not
necessary for the reconstruction and
evaluation of the final report. The final
report is evaluated by an analysis of the
pathology syndrome as described in the
pathologist's report. which is required
under § 58.185(a)(12). Further. because
§ 58.190(a) requires histopathological
blocks. tissues. and slides to be retained
as specimens. the final report can be
reconstructed by verification of the
pathology findings by, e.g.. a second
pathologist or by a team of pathologists.

The pathologist's interim notes,
therefore, which are subject to frequent
changes as the pathologist refines the
diagnosis, are not raw data because
they do not contribute to study
reconstruction. Accordingly, only the
signed and dated final report of the
pathologist comprises raw data
respecting the histopathological
evaluation of tissue specimens.

Testing Facility Management

8. One comment objected to FDA's
proposal to delete the provision in
§ 58.31(b) that requires testing facility
management to document as "raw data"
the replacement of a study director. The
comment argued that it could be difficult
to retrieve such documentation if data
were transferred to another location
after completion of a study.

FDA proposed to delete the
requirement in § 58.31(b) to document
study director replacement as "raw
data" in the agency's belief that other
provisions of the GLP regulations
adequately require documentation of
this event. The agency continues to
believe that the requirement in
§ 58.31(b) is redundant to such other
provisions and is not necessary to
assure the quality and integrity of the
safety data developed through a study
conducted by a laboratory. For example.
§ 58.35(b)(1) provides that the master
schedule sheet shall contain the name of
the study director. Thus. replacement of
the study director would necessitate an
updating of the master schedule sheet.
The master schedule sheet itself is "raw
data" because it is a record that is the
result of laboratory activities and is
necessary for the reconstruction of the
study. Also. § 58.120(b) requires that any
change in an approved protocol and the
reason or reasons for the change are to
be documented. Because § 58.120(a)(11)
of the final rule (previously
§ 58.120(a)(15)) requires that the
protocol contain the dated signature of
the study director. replacement of the
study director would constitute such a
change. Other provisions of the
regulations require the quality assurance

unit to retain the master schedule sheet
and copies of protocols as an easily
accessible system of records for a
specified period after completion of the
study (§ 58.195(d)). Therefore. it should
not be difficult to identify the term of
each study director even if records are
transferred elsewhere after study
completion.

Study Director

9. Several comments objected to
§ 58.33 in its entirety on the grounds that
(1) the regulation does not clearly define
the responsibility of the study director
and (2) the wording of the regulation
implies that the study director must be
technically competent in all areas of a
study. One comment argued that the
study director should be responsible
only for "coordinating" the technical
conduct. interpretation. analysis.
documentation. and reporting of results.

FDA discussed at length in the
preamble to the GLP final rule the intent
of § 58.33 and the requirements
applicable to the individual who is
designated the study director for any
study (43 FR 59986. 59995: December 22.
1978). As discussed in that preamble. the
study director represents the single.
fixed point of responsibility for overall
conduct of each study. Although
"coordination" of the pieces of a study
logically is part of the study director's
responsibilities. to limit his or her
responsibilities to mere "coordination"
would compromise public protection if
another person were not such
designated fixed point and would add
an unnecessary burden if FDA were to
require a laboratory to employ an
additional person to provide such a
point. The study director is charged with
the technical conduct of a study.
including interpretation. analysis.
documentation. and reporting of results.
FDA does not intend. however. that the
individual is to be technically competent
in all areas of a study. FDA's
inspectional experiences have
demonstrated that if responsibility for
proper study conduct is not assigned to
one person. a potential exists for the
issuance of conflicting instructions and
improper protocol implementation.

FDA concludes that the comments did
not provide any new data or information
to negate the agency's original
determinations and that it should retain
§ 58.33 as it was established in the
December 22. 1978. final rule.

10. One comment objected to FDA's
proposal to delete the phrase "and
verified" from § 58.33(b), which
currently requires that the study director
assure that all experimental data are
"verified" as well as accurately
recorded. The comment argued that

removing the study director's obligati
to assure that the data have been
verified would dilute the responsibili
of the study director for proper condu
of the study.

FDA has carefully reevaluated its
proposal to remove the phrase "and
verified" from § 58.33(b). FDA propos
to delete the phrase from the regulati
in response to arguments from
management of testing facilities that
they misinterpreted the provision.
apparently believing that it required
study director personally to witness
each data observation. FDA did not
intend the provision to so require.
Rather. the agency regards the study
director as responsible for assuring t
all experimental data are verified.

Ordinarily the verification require
§ 58.33(b) is obtained by the individu
collecting the data, using data
verification procedures described in
protocol or in the specific standard
operating procedure. and by the
individual's supervisor as part of the
supervisory quality control procedur
"Verified" is used to describe the stu
director's responsibility to assure the
accurate recording of data. Verificati
in this sense does not require the stu
director to observe every data collec
event but does require the study dire
to make certain that the study condu
procedures designated in the protoco
for a study and the standard operatir
procedures established for a study ar
followed. For all these reasons. FDA
decided to retain the phrase "and
verified" to confirm the need for data
verification in nonclinical laboratory
studies.

11. Two comments urged that the
study director's responsibility for
assuring placement in the archives o
the study records specified under
§ 58.33(f) be transferred to testing
facility management under § 58.31. T
comments argued that raw data and
documentation are under the immedi
control of facility management rathe
than the study director.

The change suggested by the
comments would conflict with the st
director's responsibility identified in
§ 58.33 of representing the single poi
of study control (see paragraph 9 of
preamble). The archived materials o
each study. including raw data and
documentation. constitute lasting pr
of study validity. The transfer of suc
materials to the archives is a critical
step in study control that assures th
the archived materials are complete
adequate for study reconstruction.
Indeed. these revised regulations fur
emphasize the study director's contr
function by defining the study

npletion date as the date the final port is signed and dated by the study ector; after this date, the study ector assures that all required terials are transferred to the archives. nsequently, FDA declines to accept e change proposed by the comments.

ality Assurance Unit

12. FDA received eight comments arding § 58.35(a), which sets forth the mposition and function of the Quality surance Unit (QAU). One comment dorsed FDA's proposal to substitute hich" for the current phrase. mposed of one or more individuals o." to make clear the personnel who n perform quality assurance duties. e comment suggested, however, that e agency use the term "function" or ctivity" in § 58.35(a) in place of the rrent term "unit" to make it clear that ality assurance monitoring need not performed by individuals of a rmanently staffed unit. Four mments disagreed strongly with the oposed change to § 58.35(a), arguing at the change inappropriately implies at assurance of a well-conducted dy does not require special training experience. The comments also serted that verifiable data are oduced as a result of the current quirement for an independent, fixed d permanently staffed quality surance unit. Other comments argued at individuals should not monitor dies similar to their own work. Some mments requested further clarification the required composition of the QAU.

FDA does not believe that entification of the quality assurance it as a "function" or "activity" would rve to clarify the composition or nction of the QAU. FDA never has tended that the QAU necessarily has be a separate entity or a permanently affed "unit" (see 43 FR 59996). Quality assurance unit" has become an cepted term to describe those dividuals responsible for quality. surance as described in § 58.35. FDA so does not agree that the proposed vision to § 58.35(a) implies that surance of a well-conducted study es not require special training or perience on the part of individuals onitoring the conduct and reporting of nonclinical laboratory study. FDA ntinues to believe that well-qualified d trained personnel are essential to ality assurance under the GLP gulations and that one of anagement's most important sponsibilities in maintaining effective ality assurance is to provide an dequate number of such personnel § 58.31 (c) and (e)).

FDA concludes from the comments that the requirements set forth in § 58.35(a) have been misinterpreted in some instances to mean that the regulations require that the QAU be composed of individuals whose sole duties are in quality assurance. In fact. the agency intends only that quality assurance activities be separated from study direction and conduct activities; that is, a trained and qualified person who works on one study can perform quality assurance duties on any study in which he or she is not involved. FDA's reason for requiring separation of quality assurance functions from study conduct functions is fundamental—to assure that quality assurance personnel can act candidly, without bias or a real or perceived conflict of interest. In effecting the separation required by the GLP regulations, FDA was aware that many small laboratories could not afford the operation of a permanently staffed QAU. For this reason, the agency concluded that the separation of functions on a study-by-study basis as permitted in the existing and revised regulations would provide effective quality assurance. The agency's intent in defining the composition and function of the QAU was discussed at length in the preamble to the current GLP regulations (see 43 FR 59996). FDA believes that the change now being made more clearly reflects the agency's original intent.

13. One comment recommended that FDA delete the word "sheet" from the term "master schedule sheet" in § 58.35(b)(1) on the ground that there are methods for maintaining a master schedule other than use of an actual "sheet."

FDA acknowledges that current technology allows for various methods for maintaining a master schedule, ranging from sophisticated computerized procedures to procedures whereby such information is contained in written records. Regardless of the method utilized, however, the master schedule information is "raw data" within the meaning of the GLP regulations and copies of the master schedule are required to be retained in the study archives in accordance with § 58.195(b). The agency is, therefore, retaining the term "master schedule sheet" to emphasize that the master schedule constitutes raw data subject to agency inspection and that the records must be retained.

14. One comment suggested that FDA delete the current provision in § 58.35(b)(1), which requires a laboratory to include the name of the sponsor of each study on the master schedule sheet for all studies conducted

at the facility. The comment urged the agency to allow sponsor identification by code.

FDA agrees that including the sponsors' name on the master schedule sheet is not essential either for the conduct of management's functions listed in § 58.31 or for the conduct of proper quality assurance under § 58.35. Sponsor identification by code is an adequate procedure, provided that the name of any sponsor is made available to FDA upon request. Accordingly, FDA is amending § 58.35(b)(1) to read "* · * identity of the sponsor * * *."

15. One comment suggested that FDA further revise § 58.35(b)(1) to allow the master schedule sheet to be indexed by study number rather than test article on the ground that multiple studies may be performed on each test article.

FDA recognizes that a nonclinical laboratory may have in progress sever studies on each test article that is liste on the master schedule sheet. The agency concludes, however, that indexing by study number alone and n by test article would be inappropriate. The master schedule sheet is the mechanism through which the QAU ca assure management that the facilities are adequate and that there are sufficient numbers of qualified personnel available to accomplish the scheduled work (see 43 FR 59997). In addition, § 58.31(e) requires management to assure that study materials (e.g., test articles) are available as scheduled. The use of stu numbers rather than test articles as index terms would, therefore, frustrate major purpose of the master schedule sheet and impede the conduct of an important management function.

16. Three comments endorsed FDA's proposal to delete the current requirement under § 58.35(b)(1) that th status of the final report be a distinct entry on the master schedule sheet. On comment, however, objected to the proposal on the basis that frequently there are delays in completing the fina report, and the study status often is different from the expected date of completion of the report.

FDA believes that the comment that objected to the proposal misconstrued the purpose of revising § 58.35(b)(1). Section 58.35(b)(1) currently requires that the master schedule sheet contain separate headings for the "current status" of the study and for the "statu of the final report" of the study. FDA considers the preparation of the final report to be a study event, current sta of which should be reflected on the master schedule sheet. Preparation of the final report is similar to other stud events (e.g., test article-mixture preparation, test system dosing, in life observations) that are listed under "current status." Under this revision to § 58.35(b)(1), the master schedule sheet would contain the same information respecting the final report as is required under the current GLP regulations, but the information would be included only under the "current status" heading. The agency advises that "expected date of completion of a final report" is not a "status" entry. Such information has not been required in the past nor is this information being required now.

17. FDA proposed to revise § 58.35(b)(3) to provide specifically that the QAU need only inspect "each nonclinical laboratory study" on a schedule adequate to assure the integrity of the study. Four comments recommended that the agency further revise the regulation to substitute the word "studies" for the phrase "each nonclinical laboratory study" alleging that inspection of multiple short duration studies can result in expenditure of significant time and effort with little derived benefit. Two comments argued that inspection of short duration studies conducted repeatedly at the same facility by the same personnel is not necessary and suggested that, in lieu of requiring inspection of each study, standard operating procedures be developed to determine the inspection frequency of various types of studies. These comments also recommended that such inspections be used to demonstrate compliance of other similar studies conducted in the same time frame.

FDA does not agree with the comments. The quality of each nonclinical laboratory study submitted to the agency in support of an application for a research or marketing permit for a product regulated by FDA is critical to a determination of the safety of the product. The principle of quality assurance advanced in the GLP regulations is to inspect studies to identify and correct problems in a timely fashion. FDA is convinced that such problems can be detected only through a program of vigorous inspection of each study. This does not mean, however, that every phase of every study needs to be inspected by the QAU (see paragraph 18 of this preamble).

18. Two comments disagreed with FDA's proposal to modify the current requirement in § 58.35(b)(3) that the QAU inspect each phase of a study at specified intervals. The comments argued that elimination of specified quality assurance inspection intervals may result in decreased compliance by

cost-cutting laboratories. Another comment urged FDA to identify the critical phases of a nonclinical laboratory study to be inspected by the QAU.

Section 58.35(b)(3) currently provides that the QAU is to inspect at periodic intervals each phase of a nonclinical laboratory study. For studies lasting more than 6 months, the inspections are to be conducted every 3 months. For studies lasting less than 6 months, the inspections are to be conducted at intervals adequate to assure the integrity of the study (including each phase at least once). The term "each phase" was intended to emphasize the need for repeated surveillance so that the QAU observes at least once during the course of the study each critical operation. The term "periodic" was included in the regulation to indicate the need for more than one inspection of certain repetitive, continuing operations. In light of current information, however, FDA does not believe that such a rigid schedule is essential to assure study quality. The agency has learned through its inspection program that the quality of toxicology testing is much higher than that envisioned in 1976 when FDA proposed to establish the GLP regulations (see 41 FR 51206, 51207–51208; November 19, 1976).

Contemporary concepts of quality assurance emphasize the effectiveness of thorough, in-depth inspections of study processes (i.e., all operations required to accomplish a study phase) in place of quick, spot checks of individual operations with a study. Thorough examination of personnel, facilities, equipment, standard operating procedures, data collection procedures, raw data books, and other features associated with a study phase can achieve more effective quality assurance than does a more superficial observation of the conduct of the same study phase in a series of studies.

The agency has concluded that the QAU's inspection schedule should take into account the need for inspection of each study on a schedule adequate to assure the integrity of the study being monitored. The change in § 58.35(b)(3) permits the QAU to exercise reasonable flexibility and judgment so that inspections can be scheduled to best achieve the goal of assuring that studies are properly conducted. The agency advises, however, that each study, no matter how short, needs to be inspected in-process at least once. Further, across a series of studies all phases should be inspected in order to assure the integrity of the studies. For these reasons, FDA does not believe that the regulation as

revised will result in decreased compliance with the GLP regula that the term "critical phase" ne be defined.

19. Eight comments addressed proposal to delete § 58.35(b)(4). currently requires the QAU to s periodic written status reports t management and to the study d One of the comments supported proposal on the basis that the p of § 58.35(b)(3), which require r management on problems likely study integrity, are adequate to study quality. Several comment questioned whether deleting § 58.35(b)(4) would provide any practical benefit to testing facili noting that management is likel continue to expect periodic repo comment noted that elimination requirement for status reports c allow management to disregard assurance problems if managen only marginally supportive of g laboratory practice and quality assurance. Three comments arg by deleting the provision. FDA inappropriately place on the QA than on the study director the o of determining what constitutes problem likely to affect the inte study. One comment argued tha because management is require involved in corrective actions u provisions of §§ 58.31(g) and 58 management must be kept infor respecting the status and the pr any study inspected by the QA Another comment suggested tha delete the requirement that stat reports be provided to the study but retain the requirement that management receive such repor

After careful consideration o comments, FDA has concluded § 58.35(b)(4) should be retained current form. The agency propo delete the paragraph based on tentative conclusion that routin of unremarkable findings by th are not essential to study quali comments, however, are persua that such reports are necessary demonstrate to management th QAU is functioning properly. It necessary that the study direct continue to receive such report that individual is responsible fo aspects of any study being cond a facility, including GLP compli

20. One comment suggested t lieu of deleting the entire parag § 58.35(b)(4) be modified by rer the requirement for status repo each study, thereby reducing pa

In accord with the conclusion in paragraph 19 of this preambl ieves that it is inappropriate to ninate the requirement for status orts for each study. Without status orts on each study, there would not adequate assurances concerning the lity of the ongoing studies.

1. One comment urged FDA to revise rent § 58.35(b)(5) to hold the QAU ponsible only to (1) determine that wn deviations from approved tocols or standard operating cedures were not made without per documentation, and (2) require horization for anticipated deviations. comment argued that the current rding of the regulation implies that QAU shall have prior knowledge of deviations and shall approve such riations, which is not the case in most tances.

DA would not consider a laboratory have violated § 58.35(b)(5) if a riation was not authorized in advance ause it was unanticipated. For mple. as recognized in the preamble he final rule (43 FR 59998). a fire in facility would necessitate immediate ion. Also, as discussed at length at 43 59998. FDA does not intend that the U is responsible for authorizing any riations from the protocol or standard rating procedures. rather it is ponsible for detecting any such riations by its inspection and audit cedures. The revision suggested by comment would remove the countability of the QAU for detecting riations and would undermine the uirements for quality assurance.

2. FDA proposed to revise current 8.35(b)(7) to provide that the tement which the QAU prepares to company the final report would be uired to identify the phases of the dy inspected and the number of pections conducted. Eleven nments objected to the proposed ange to the regulation. which rently requires only that the tement by the QAU specify the dates t inspections of the study were made d the dates that findings were orted to management and to the dy director. Most of the comments re concerned with the additional orting burden imposed by the posed revision. Some of the mments argued that the information ught by FDA through the proposed rision in § 58.35(b)(7) is currently uired under the provisions of § 58.35 (3) and (c). These comments pointed t that requiring additional cumentation would be duplicative d contrary to the stated purposes of proposed revisions to the GLP gulations.

FDA has carefully reevaluated its posal to require that the QAU

statement identify the phases of a study inspected and the number of inspections conducted. The agency is persuaded that the proposal would have provided information redundant to that which is available under other provisions of the GLP regulations. Accordingly. FDA has decided not to change § 58.35(b)(7) and is retaining this provision in its current form. Proposed § 58.35(b)(6) (as § 58.35(b)(7) would have been renumbered) is not being adopted.

Animal Care Facilities

23. One comment objected to the proposed modification to § 58.43(c). which would delete the current requirement that all laboratory facilities include separate areas for the diagnosis. treatment. and control of laboratory animal diseases. The comment noted that animal health is a major problem in toxicology testing. For this reason. the comment argued that any relaxation of the current requirements in the GLP regulations for animal care is ill-advised and contradictory to FDA's responsibility for the welfare of animals.

As discussed in the preamble to the proposed amendments to the GLP regulations (49 FR 43532). a laboratory may elect to dispose of diseased animals. thereby obviating the need for dedicated areas for such animals. FDA believes that it is not cost-effective to require separate areas in every case and has concluded that the decision concerning appropriate separation of animals should be made by the study director in consultation with other scientific personnel. The agency does not believe that providing for dedicated laboratory areas as appropriate compromises FDA's continuing commitment to animal welfare, which is specifically dealt with in § 58.90. Indeed. FDA believes that the GLP regulations foster quality animal testing under defined conditions so that fewer animals are required to establish product safety.

Animal Supply Facilities

24. FDA proposed to revise § 58.45 to permit laboratories to store perishable supplies or feed by methods most appropriate to the characteristics of the materials. One comment urged FDA to amend § 58.45 further to permit storage of animal feed in rooms housing small animals or small groups of animals that are isolated from other studies.

FDA declines to amend § 58.45 as recommended by the comment. In developing the current GLP regulations. the agency carefully considered whether animal feed or bedding might be stored within any test areas. FDA concluded at that time that storage areas needed for feed and bedding should be separate

from the areas housing the test system to preclude mixups and contamination of test article-carrier mixtures and inadvertent exposure of the test system to potentially interfering contaminants. FDA continues to believe that separate storage areas for feed and animal bedding should be required and the comment did not provide any data to counter this belief. FDA advises. however, that § 58.45 does not preclude holding of limited quantities of test or control article-feed mixtures for short periods of time in properly constructed and labeled containers in the animal rooms.

25. One comment objected to the proposed changes to § 58.47. arguing that it believed the current requirement respecting facilities for handling test an control articles have resulted in fewer mixups.

Based on the comment. FDA has reconsidered the proposed revision of § 58.47. The agency agrees with the comment, in principle. Indeed. the agency intended the revision to be only editorial to simplify and to make clear that laboratories shall provide separate areas for receipt. mixing, and storage o test and control articles and their mixtures as necessary to prevent contamination or mixups. Inadvertently however. the proposed revision would have deleted an essential requirement good laboratory practice. i.e.. the need to provide storage areas for test and control article mixtures adequate to preserve the identity. strength. purity. and stability of the mixtures. For this reason. the agency has concluded that the regulation is appropriate as currently stated and existing § 58.47 is retained.

Maintenance and Calibration of Equipment

26. FDA proposed to revise § 58.63(b to provide that written standard operating procedures respecting maintenance and calibration of equipment would allow laboratories to discard faulty equipment as an alternative to the current provisions of § 58.63(b). which provide only for the repair of equipment that fails or malfunctions. Under FDA's proposal. a testing facility would need to specify remedial action in the event of equipment failure or malfunction only when remedial action is "appropriate" to the particular piece of equipment. Three comments recommended that FDA amend § 58.63(a) to require testin; calibration. and/or standardization of equipment in accord with written standard operating procedures. The comments argued that by changing

current § 58.63(a) in this fashion, FDA might delete § 58.63(b) in its entirety without substantively affecting the requirements of the GLP regulations applicable to laboratory equipment.

FDA advises that § 58.63(b) concerns not only setting forth in standard operating procedures the details of routine inspection, cleaning, maintenance, testing, calibration, and/or standardization of equipment, but it also concerns describing remedial actions to be taken when appropriate if the equipment fails or malfunctions. In promulgating existing § 58.63(b), FDA concluded that the specific features set forth in the regulations need to be included in equipment standard operating procedures because of the crucial role that properly used equipment plays in study conduct: a role which pervades every phase of a study and is vital to study quality and final report integrity. The comments did not provide any data that negate FDA's original determination. The agency continues to believe that specification of these features provides useful guidance to persons subject to the GLP regulations and concludes, therefore, that it would be inappropriate to delete § 58.63(b) from the regulations.

27. Section 58.63(b) requires in part that written standard operating procedures designate the person responsible for certain operations respecting equipment, i.e., routine inspection, cleaning, maintenance, testing, calibration, and/or standardization. Although FDA did not propose to revise these requirements of § 58.63(b), three comments stated that by requiring that standard operating procedures designate an individual responsible for performing each operation, FDA obligates testing facilities to change their standard operating procedures frequently. The comments further argued that the requirement to make a copy of the standard operating procedures available to laboratory personnel is redundant to the requirements set forth in § 58.81(c) and should be deleted.

The comments misconstrue the meaning of the word "person" as it is used in § 58.63(b). The second sentence of § 58.63(b) requires, in pertinent part, that "the written standard operating procedures shall designate the person responsible for the performance of each operation * * *." As explained in paragraph 120 of the preamble to the final rule (43 FR 60001), FDA adopted the term "person" as defined in § 58.3(h) rather than the originally proposed term "individual" to allow the standard

operating procedures to designate an organizational unit.

The agency agrees with the comment that § 58.63(b) is redundant to § 58.81(c) insofar as it provides for standard operating procedure availability. Accordingly, FDA is removing the phrase "and copies of the standard operating procedures shall be made available to laboratory personnel" from § 58.63(b).

28. Section 58.63(c) requires that a testing facility maintain written records of all equipment inspection, maintenance, testing, calibration, and/or standardizing operations. Two comments recommended that FDA delete § 58.63(c) in its entirety, arguing that if management is satisfied that the standard operating procedures are adequate, further requirements are unnecessary.

FDA does not agree with the comments. FDA carefully considered the necessity for maintaining the records specified in § 58.63(b) when the agency developed the current GLP regulations. FDA concluded that such records are necessary to reconstruct a study and to ensure the validity and integrity of the data that are obtained from a study (see paragraph 28 of this preamble and 43 FR 60001). The purpose of the record retention requirement is to provide documentation throughout the study of equipment function in accord with design specifications for the equipment, and equipment use in accord with standard operating procedures for maintenance and calibration of the equipment. FDA's experience in administering the GLP regulations has shown the benefit of maintaining such records. For example, through FDA's laboratory inspection program, the agency has been able to identify the precise period of time of equipment malfunction by an examination of the equipment records. Thus, § 58.63(c) has made it possible to disregard the data collected by a single defective piece of equipment without having to disregard all the data obtained from a specific study.

Standard Operating Procedures

29. Two comments on the requirements for standard operating procedures suggested that § 58.81(a) be amended to permit appropriate supervisory personnel to authorize deviations from the written standard operating procedures, arguing that the study director may not have the technical expertise to evaluate such deviations. One of the comments further argued for the change on the ground that the principles of good laboratory practice established by the Organization

for Economic Cooperation and Development of the World Health Organization do not require study director approval of all standard operating procedure deviations.

FDA does not accept the suggestion As discussed in paragraph 9 of this preamble, it is not necessary for the study director to be technically competent in all aspects of a study to assure that appropriate action is taken in response to any circumstances that may affect the quality and integrity of study. The study director, however, is to be aware of and authorize any deviations that could have an impact the study. FDA does not agree that the responsibility of the study director, as set forth in § 58.81(a), is inconsistent with the Organization for Economic Cooperation and Development principles of GLP. Chapter 2, section (Ref. 1, p. 27) of the Organization for Economic Cooperation and Development document defines the responsibilities of the study director part, as "ensur[ing] that the procedures specified in the study plan are followed and that authorization for any modification is obtained and documented together with the reason for them * * *." Accordingly, both the GLP regulations and the Organization for Economic Cooperation and Development principles of GLP require that the study director make certain specified procedures are followed a that all modifications to the procedures in the approved study plan (i.e., standard operating procedures) are documented and approved.

30. One comment recommended t FDA amend § 58.81(b) to delete all examples of standard operating procedures that FDA requires a test facility to establish on the ground th the list is not all-inclusive and may to misinterpretation.

FDA disagrees with the recommendation. The examples list § 58.81(b) are those minimal laborator procedures which the agency believe are essential to assuring the quality integrity of the data generated in th course of any study (see 41 FR 5121 FDA recognizes that circumstances necessitate establishment of addition standard operating procedures. The is not intended to be all-inclusive a should not be interpreted as such.

31. Three comments objected to t proposed deletion of examples of laboratory manuals and standard operating procedures required to be the laboratory under § 58.81(c). The comments argued that the listing of examples serves to spell out the agency's intent, helps laboratories

e which areas to cover, and that
ut such guidance a potential exists
isinterpretation of FDA's intent.
A does not agree. Section 58.81(c)
res that each laboratory area have
diately available standard
ting procedures relative to the
atory procedures being performed.
e the list in § 58.81(b), which
sents specific minimal
rements, the examples listed in
1(c) encompass very broad areas.
of the areas presented would
re the preparation of a group of
lard operating procedures to cover
uately the operations within that
Consequently, the agency
ludes that the list in § 58.81(c) is too
d to serve as useful guidance.

ents and Solutions

FDA received five comments that
nmended that the agency amend
3. Four comments stated that only
reagents and solutions used in the
uct of nonclinical laboratory
es need to be labeled in accord
the requirements of § 58.83, i.e., "to
ate identity, titer or concentration.
ge requirements, and expiration
' of the material. These comments
d that such a modification would
the laboratory flexibility in
ning the most suitable system to
e that deteriorated or outdated
nts and solutions are not used. The
comment recommended that FDA
t into the first sentence of § 58.83
hrase "as appropriate" to make the
ision read "all reagents and
ions in the laboratory areas shall
beled to indicate identity, titer or
entration, storage requirements,
as appropriate, expiration date."
comment argued that it was not
ssary to have expiration dates on
in stable reagents and solutions
as water and saline.

)A declines to amend § 58.83. The
cy continues to believe, as
issed in paragraph 148 of the
mble to the final rule (43 FR 60003),
all reagents and solutions
itained in the laboratory area for
n the conduct of nonclinical
ratory studies should be labeled as
ired by § 58.83. Accordingly, the
I should include the identity, titer or
entration, storage requirements,
xpiration date. This label
mation is the minimum information
ssary to make clear to the
ratory personnel that the reagents
solutions are suitable for use in the
edures specified in the protocol and
otect against inadvertent mixups of
ents and solutions that are used in
linical laboratory studies with
e that are not intended for such use.

Further, FDA disagrees with the
comment that suggested that expiration
dating is not necessary for some
reagents and solutions. FDA believes
that expiration dates should be required
on all reagents and solutions, without
regard to their stability so that there is
no doubt about the suitability of the
materials for use in nonclinical
laboratory studies.

Animal Care

33. One comment recommended that
FDA delete § 58.90(a) on the basis that
requiring standard operating procedures
for housing, feeding, handling, and care
of animals is redundant to other
provisions of the GLP regulations.

FDA recognizes that § 58.81(b)(2)
requires testing facilities to establish
standard operating procedures for
animal care. Section 58.90(a), however,
expressly specifies that standard
operating procedures shall also cover
animal housing, feeding, and handling.
The agency believes that these items are
essential features for providing
adequate humane treatment of animals
and, therefore, is retaining § 58.90(a).

34. Two comments objected to the
proposed amendment of the
requirements under §§ 58.90 (b) and (c)
to provide that animals may be isolated
rather than quarantined. The comments
argued that allowing laboratories to
develop isolation and health status
evaluation procedures in lieu of
quarantine may not provide adequate
assurance of test animal health because
such evaluations are done only on a
small randomly selected number of
animals and are not as reliable as an
adequate quarantine period. One
comment suggested that the term
"segregated" or the term "separated"
may more accurately reflect FDA's
intent in amending §§ 58.90 (b) and (c)
than does the term "isolated."

As discussed in the preamble to the
proposed amendments (49 FR 43533),
substitution of a requirement for
isolation and health status evaluations
in lieu of quarantine of newly received
animals will permit laboratories to
develop specific isolation and health
status evaluation procedures in concert
with the age, species, and class of
animals, and with the type of study to
be done. As used in current § 58.90(b),
the term "quarantine" connotes a rigid
set of prestudy procedures which a
facility is obligated to follow, including
a mandatory holding period, specified
diagnostic procedures, and the use of
specialized facilities and animal care
practices. The agency has concluded
that isolation and health status
evaluations should provide adequate
precautions against entry of unhealthy

animals into a study. Health status
evaluations may be performed during
the prestudy acclimation period. Under
§ 58.90(b), the health status of each
animal is to be evaluated soon after
receipt. Section 58.90(c) prohibits the
entry of any diseased animal into the
study.

The agency has also concluded that a
devoted area equipped to provide
isolation of diseased animals is not
necessary in all cases. As discussed in
the preamble to the proposal (49 FR
43533), FDA believes that it should
allow certain options for handling
diseased animals, thereby permitting
increased flexibility in laboratory
operation. FDA agrees that the term
"segregate" or the term "separate"
rather than the term "isolate" could be
used in as much as each term connotes
"to set apart." Because "isolate" is a
term commonly used and understood in
contemporary veterinary medical
practice, however, FDA will continue to
use that term in these regulations.

35. One comment recommended that
FDA amend § 58.90(d) to permit
procedural means to provide
"appropriate identification" of
individual animals in a study.

FDA does not accept the
recommendation. The agency notes that
proper animal identification throughout
a study is essential to study integrity.
When animals are housed individually
in cages, procedural means can be used
to identify each individual animal, i.e., a
cage card may be used if it provides all
the information necessary to identify the
animal specifically, and the animal
handling standard operating procedures
specify detailed procedures for
preventing animal mixups. The agency
is not aware, however, of any
procedural means that could be used to
identify adequately individual animals
that are housed within a group in a cage.
FDA advises that § 58.90(d) does not
preclude identification by means other
than the examples enumerated in that
section provided that individual
identification can be maintained and
documented throughout a study.

*Test and Control Article
Characterization*

36. One comment urged the agency to
delete the requirement in § 58.105(b) for
stability determination of test and
control articles before initiation of the
study. The comment argued that
changing § 58.105(b) as recommended
would make it consistent with FDA's
proposed changes to § 58.105(a), which
would allow test and control article
characterization after completion of the
study.

FDA disagrees with the recommendation. As discussed in the preamble to the proposal (49 FR 43533), FDA has concluded that characterization of test and control articles need not be performed until initial toxicology studies with the test article show reasonable promise of the article's reaching the marketplace. In arriving at this conclusion. the agency considered that prior knowledge of the precise molecular structure of a test article is not vital to the conduct of a valid toxicology test. It is important. however. to know the strength. purity, and stability of a test or control article that is used in a nonclinical laboratory study.

A stability determination of a test or control article conducted after completion of the nonclinical laboratory test with the article does not provide any information about the continued strength of either the test or control article previously given to the test system. Determining that the test and control articles are stable for the duration of the study is fundamental to interpreting the results of the study. For this reason. it would be inappropriate to allow for stability determinations only after-the-fact.

The agency does believe. however. that the continued strength of the test or control article may adequately be determined either by stability testing before initiation of the study or through appropriate periodic analysis of each batch. Section 58.105(b) currently allows stability testing of the test and control articles through periodic analysis only if it is not possible to determine their stability before study initiation. Because experience has shown that it is adequate either to determine stability of the test and control articles before initiation of a study or by periodic analysis of the articles while the study is in progress. the agency on its own initiative is revising § 58.105(b) to provide facilities and sponsors the flexibility to use either approach. Therefore. § 58.105(b) has been revised to provide for determination of the stability of the test or control article either before study initiation or through periodic analysis of each batch according to established standard operating procedures.

37. One comment observed that § 58.105(a) provides that testing facilities may rely on the labeling of marketed products for purposes of characterizing such products used as control articles in a study. The comment recommended that the agency also revise § 58.105(b) to allow product labeling to serve as documentation of

stability for marketed products used as control articles.

FDA declines to adopt this recommendation. The only stability information typically included on the labeling of marketed products is the expiration date. The manufacturer's expiration dating on a marketed product is not adequately precise to provide data on the strength of the control article used throughout a nonclinical laboratory study. Lacking precise stability information respecting the control article could raise doubts concerning whether the test and control articles are comparable. Furthermore, mixing the marketed product with a carrier or use of the product in a manner not in accord with its labeling could alter the stability characteristics of the product. For these reasons, FDA concludes that determination of the stability in accordance with the procedures in § 58.105(b) still is appropriate for any products used as control articles.

38. Several comments recommended that FDA revise § 58.105(c) to remove the current requirement that storage containers be assigned to a test article for the duration of a nonclinical laboratory study. The comments argued that the requirement is unnecessary and is inconsistent with the Organization for Economic Cooperation and Development principles of GLP. One comment alleged that the statement is vague and questioned whether the provision would permit a storage container that a laboratory emptied during the conduct of. but before completion of. a study to be destroyed or reused. The same comment also questioned whether, as the test article is depleted during the conduct of a study, the provision would permit a laboratory to transfer the test article into a container smaller than that originally assigned to the article.

FDA does not believe that it would be appropriate to eliminate the storage container provision in § 58.105(c). FDA advises that the provision simply requires that each test article storage container be assigned to a test article at the beginning of a study and remain so assigned until the study is completed. terminated. or discontinued. The test article may not be transferred to different sized storage containers as a study progresses. nor may assigned storage containers be destroyed while a study is in progress.

The agency recognizes that it may be inconvenient for laboratories. especially small laboratories. to devote space to a test article container which is emptied before a study is completely terminated.

or discontinued. or which might be replaced by progressively smaller containers while a study is in prog Destroying or reusing or otherwise substituting for originally assigned identified test article storage conta however. could adversely affect th integrity of the study. FDA establi the current provision because the agency observed a lack of accountability for test materials. ostensibly due to the very acts proscribed by the regulation. i.e. transfer of test articles to different containers and destruction of emp containers during the progress of nonclinical laboratory studies.

FDA continues to believe that th requirement concerning assignmen storage containers is necessary to ensure the integrity of a study. For example. the mere act of transferri test article from one storage conta another introduces the opportunity contamination of the test article b other laboratory materials or for m ups with other laboratory material addition. during a transfer the test article would be exposed to air-bo contaminants as well as to moistu either of which may compromise tl integrity of the test article.

FDA also believes that requiring containers be assigned to a test ar for the duration of a study is fully consistent with the Organization fi Economic Cooperation and Development principles of GLP. Cl 2. section 2.3 (Ref. 1. p. 31) of the Organization for Economic Cooper and Development document states "storage container(s) should carry identification information. earliest expiration date. and specific stora instructions." Although the Organi for Economic Cooperation and Development document does not f characterize storage instructions tl the Organization for Economic Cooperation and Development wo recommend be included on contai FDA concludes that its regulation identifies a "specific" storage instruction that is necessary to ass the integrity of the test article.

Mixtures of Articles with Carriers

39. Two comments argued that t requirements in § 58.113(a) to dete the uniformity, concentration, and stability of test and control article mixtures is unnecessarily burdens for short-term studies. One comme stated that such tests are not nece for test articles that are prepared a dispensed on the same day as use single-dose acute toxicity tests. Th comments suggested that the agen

se § 58.113(a) to require that tests of
tures of test and control articles
i be conducted only "for studies
r than short-term."

ection 58.113(a) requires, for all test
ontrol articles mixed with a carrier,
termination of the uniformity of the
ture and a periodic analysis of the
centration of the test or control
cle in the mixture. In addition,
.113(a)(2), as amended, requires a
rmination of the stability of the test
ontrol article in the mixture
quate to support time of use. Each of
e analyses is necessary to assure
the test system is exposed to the
or to the control article in the
unts specifically designated in the
tocol for a study. Determination of
uniformity of a mixture assures that
h member of the test system receives
intended dose of test or control
cle. Periodic analysis of the
centration of the test or control
cle serves as a spot check to assure
the test or control article mixture
been prepared properly and in
ord with the protocol and with
licable standard operating
cedures. Determining the stability of
test or control article in the mixture
ps to determine the period of time
ing which the test or control article
ture may be suitable for
ninistration to the test system. FDA
ieves that knowledge of the dose of
or control article used in any test is
ential for the proper evaluation of the
ults of that test. For these reasons,
A has concluded that the
uirements of § 58.113(a) should
tinue to apply to short-term tests as
ll as to studies other than short term.
s discussed in paragraph 36 of this
amble, however, FDA has decided to
w facilities and sponsors the
ibility to determine the stability of
t and control articles either before
dy initiation or through periodic
lysis of each batch according to
ablished standard operating
cedures. The agency believes that it
ppropriate to allow similar flexibility
th respect to determining the stability
mixtures of articles with carriers. For
s reason, on its own initiative, FDA
s revised § 58.113(a)(2) to allow
termination of the stability of test and
ntrol articles in the mixture either
fore study initiation or through
riodic analysis of the mixture
cording to established standard
erating procedures.

40. One comment recommended that
A deletes § 58.113(a)(1), alleging that
riodic determinations of the test or
ntrol article in mixtures is routine and
ovisions respecting such

determination would more appropriately
be included in § 58.35(b)(3).

The comment misconstrues the
function of the QAU (see especially
paragraphs 12 and 18 of this preamble).
Current § 58.35(b)(3), as well as
§ 58.35(b)(3) as revised by this final rule
assigns responsibility for periodic
inspection of laboratory operations to
the QAU. The QAU does not, indeed,
under § 58.35(a), the QAU may not
conduct any portion of a study. Rather,
it is responsible for assuring that a study
is conducted according to the protocol,
the standard operating procedures, and
the GLP regulations.

Protocol

41. A comment urged the agency to
delete the provision in current
§ 58.120(a)(14) (final § 58.120(a)(10)) that
requires that the protocol for each study
identify the records for the study to be
maintained. The comment argued that
the requirement in renumbered
§ 58.120(a)(10) is redundant to the same
requirements in §§ 58.33(f), 58.190(a),
and 58.195(b).

FDA recognizes that §§ 58.33(f),
58.190(a), and 58.195(b) address records
that the GLP regulations require a
testing facility to retain in all events. As
the authorized master plan for a study,
however, the protocol should identify all
records of the study to be maintained by
the testing facility to inform all study
participants fully of recordkeeping
obligations. The agency believes that
inclusion of this information in the
protocol is essential to ensure adequate
documentation of the conduct of the
study.

For these reasons, FDA declines to
remove § 58.120(a)(10) from the GLP
regulations.

42. One comment recommended that
FDA delete the provision in current
§ 58.120(a)(16) (final § 58.120(a)(12)) that
requires that the protocol for each study
include a statement of the proposed
statistical methods to be used in the
study. The comment argued that a
determination of the statistical methods
used to evaluate data can, in many
cases, be made only after data have
been reviewed.

FDA recognizes that circumstances
occasionally require a testing facility to
modify the proposed statistical methods
for analysis of the data in a given study.
Good scientific practice, however,
requires consideration of the statistical
analysis of a study as part of the design
of the study to assess whether the
objectives of the study can be met. FDA
concludes that the requirement is
appropriate.

*Conduct of a Nonclinical Laboratory
Study*

43. Section 58.130(c) provides that
materials derived from a test system for
examination or analysis (specimens) are
to be identified by test system, study,
nature, and the date of collection, and
requires that such identification is to be
located on the specimen container or is
to accompany the specimen in a manner
that precludes error in the recording and
storage of data. Five comments argued
that the requirements under § 58.130(c)
for specimen identification are overly
restrictive and suggested that
management should be responsible for
determining methods for identifying
specimens.

The identification requirements
specified in § 58.130(c) are designed to
preclude error during the conduct of a
nonclinical laboratory study. FDA has
reviewed the requirements and has
concluded that the identifying
information specified in the regulation is
the minimum information needed to
distinguish each specimen from all
others that have been collected in the
facility thereby protecting against
mixups and permitting orderly storage o
specimens. Such information also shows
whether the data collected on each
specimen are assigned to the correct
component of the test system.

FDA has always provided flexibility
in the methods to be used for identifying
the specimens in that the identifying
information can be encoded through, fo
example, the use of accession numbers
affixed to the specimen with the
numbers decoded in accompanying
information (see, e.g., paragraph 205 of
the preamble to the final rule (43 FR
60008)). The agency believes that the
requirements with respect to specimen
identification are the minimum
requirements necessary to prevent
mixup of specimens or lost specimen
identity.

44. Two comments disagreed with
FDA's proposal to revise § 58.130(d),
which currently requires that records of
gross findings for a specimen from
postmortem (necropsy) observations
shall, in all cases, be provided to a
pathologist during study of the
specimen. The proposed revision
provides only that such records
"should" be made available to the
pathologist. Four comments supported
the proposed change on the ground that
the resulting flexibility allows
"blinding" of the pathologist. One
comment suggested that § 58.130(d) be
deleted in its entirety or, alternatively,
modified by specifying that the records

are to be provided to the pathologist "if required by the protocol."

FDA established § 58.130(d) in the belief that the provision would increase the pathologist's ability to describe correctly microscopic findings and to relate such findings properly with the gross postmortem observations (41 FR 51214). FDA continues to believe that for most studies it is important for the pathologist to have available the records of gross findings when examining a specimen histopathologically. The agency recognizes, however, that for certain nonclinical laboratory studies it may be appropriate for pathologists to evaluate the histopathological specimens without being informed of the necropsy findings. The change made in the regulation will permit the study director, in concert with management, to determine the need for "blinding" in relation to the specific objectives of the study.

FDA advises that it continues to believe that for most studies it is a preferred practice not to use "blinding" in histopathological evaluation. For this reason, the agency is not deleting § 58.130(d) as suggested by one comment. The alternative phrase "if required by the protocol" suggested by the comment would serve the same purpose as the modification proposed by the agency. Therefore, the agency is adopting § 58.130(d) as proposed.

45. One comment on § 58.130(e) requested that FDA clarify the meaning of the term "automated data collection systems," stating that it is unclear whether § 58.130(e) applies to tabulation of source data from "hard" copy by automated systems.

As discussed in paragraph 7a. of this preamble, hand-recorded data collected during a study are "raw data." The subsequent entry of these hand-recorded data into a computer system, for example, by data processing clerks or through the use of optical readers, does not alter the status of the hand-recorded data as "raw data." The processing of these data by the computer, e.g., tabulation, is not "direct data input" and is, therefore, not encompassed by the requirements of § 58.130(e).

Reporting of Nonclinical Laboratory Study Results

46. Section 58.185(a) requires that a final report be prepared for each nonclinical laboratory study conducted and specifies the minimum information that shall be included in a final report. One comment urged that § 58.185(a) should be modified to require final reports to include the specified information only "where appropriate" to

make the requirements respecting final reports consistent with FDA's proposal to change § 58.120(a) respecting protocols. Alternatively, the comment urged that, at the least, § 58.185(a)(8), which requires that any final report shall include a description of the dosage, dosage regimen, route of administration, and duration, should be so modified. The comment argued that some of the information required under this paragraph is not applicable to certain studies involving medical devices.

FDA proposed to change § 58.120(a) in recognition of the fact that certain of the enumerated items are not necessary for the protocols for all studies. FDA has carefully reviewed each of the items listed in § 58.185(a) and has concluded that the information required by this section to be included in a final report for a nonclinical laboratory study is necessary to evaluate any such study. FDA notes that § 58.185 currently allows certain appropriate flexibilities where specified information is determined by the laboratory not to be relevant to a study. For example, § 58.185(a)(4) provides that test and control articles may be identified by "other appropriate characteristics" and § 58.185(a)(7) includes the term "where applicable." The agency believes that the information required by § 58.185(a)(8) is applicable to studies involving medical devices. For example, "dosage" and "dosage regimen" for devices may be expressed in units used per animal at a designated frequency of use. "Route of administration" may describe the means by which the device is used in relation to the animal. "Duration" would pertain to the period of time the device was tested in the animal. These clarifications should permit nonclinical laboratory studies on medical devices to be reported properly.

47. Section 58.185(a)(10) requires that each final report of a nonclinical laboratory study include the name of the study director, the names of other scientists or professionals, and the names of all supervisory personnel involved in the study. One comment recommended that FDA revise § 58.185(a)(10) to require identification only of the study director and of the principal scientists involved in the study.

FDA disagrees with this suggestion. Supervisors play an important role in the data collection process. They supervise those who perform the procedures, may recommend or actually provide training, and assure that data collection is carried out in accordance with the protocol, the standard operating procedures, and the GLP regulations. The names of all scientists,

professionals, and supervisors are needed to assure the accountability all of those individuals responsible the integrity of the study. The comm did not provide any data or informa to support its recommendation to re § 58.185(a)(10) and the agency concl that it should retain the current requirement.

48a. Section 58.185(a)(12) requires the final report of a nonclinical laboratory study include the signed dated reports of each of the individi scientists or other professionals whe were involved in the study. Two comments recommended that FDA revise § 58.185(a)(12) to allow for combined reports signed by the principal scientists. According to th comments, allowing for such combii reports would be consistent with th Organization for Economic Coopera and Development principles of GLP

The agency does not believe that combined report from scientists of different disciplines would be appropriate. Each individual scienti involved in a study has to be accountable for reporting data, information, and views within his o designated area of responsibility. Reports which combine the data, information, and views of more tha such person would obscure the individual's accountability for accu reporting. Furthermore, FDA believ that the comment has misinterprete standard that the Organization for Economic Cooperation and Development established respectin reports of persons involved in a nonclinical laboratory study. Chap section 2.3 (Ref. 1, p. 38) of the Organization for Economic Cooper and Development document provid that "if reports of principal scientis from co-operating disciplines are included in the final report, they sh sign and date them." It provides fui that the final report should contain "names of other principal personne having contributed reports to the fi report." These provisions do not im that combined reports would be appropriate. Rather, they support tl need for individual accountability. Organization for Economic Cooper and Development provisions are ei consistent with the provisions of § 58.185(a)(12).

48b. On its own initiative, FDA i amending § 58.185(b) to provide th final report shall contain the dated signature of the study director to conform the provision to the defini of "study completion date" being a by this final rule (see paragraph 6 preamble).

rage and Retrieval of Records and
ta

9. FDA proposed to revise § 58.190(a)
allow wet specimens and specimens
m mutagenicity tests to be discarded
er evaluation and recording. The
ency's proposal was supported by
eral comments. One comment.
wever. requested that the agency
rify the meaning of "mutagenicity
ts" as the term is used in § 58.190(a).
e comment asked whether. for
ample, in vitro cell transformation is
erpreted to be a mutagenicity test
der this provision.

FDA advises that. for the purpose of
8.190(a), the agency considers
utagenicity tests to be those tests
signed to assess the capacity of a test
control article to induce heritable
anges in a test system. Accordingly.
A considers in vitro cell
nsformation to be a mutagenicity test
the term is used in the GLP
gulations.

50. One comment recommended that
A further revise § 58.190(a) to assure
at the regulation makes clear that final
ports must be retained.
As discussed at length in the
eamble to the proposal (49 FR 43534).
A intended to exclude from the
tention requirements of the GLP
gulations specimens that are relatively
agile or contribute only in a minor way
safety evaluation. FDA did not intend
change the current requirement that
al reports of any nonclinical
boratory study are to be retained in
cordance with § 58.195. To preclude
y possible confusion regarding the
quirement that final reports must be
tained. FDA is rewording § 58.190(a)
read "all raw data. documentation.
otocols. final reports, and specimens
xcept those specimens obtained from
utagenicity tests and wet specimens of
ood. urine. feces, and biological fluids)
nerated as a result of a nonclinical
boratory study shall be retained."

etention of Records

51. FDA proposed to delete from
urrent § 58.195(c) the listed examples of
aterials that the agency believes need
be retained only so long as the quality
f the preparation affords evaluation.
ne comment noted that § 58.195(c)
hould be further revised to delete wet
pecimens to be consistent with the
roposed revisions in § 58.190(a) (see
aragraph 50 of this preamble).

FDA agrees with the comment and in
his final rule is conforming § 58.195(c)
revised § 58.190(a). In pertinent part.
58.195(c) now reads "Wet specimens
xcept those specimens obtained from
utagenicity tests and wet specimens of

blood. urine. feces. and biological
fluids). samples of test or control
articles. and specially prepared
material * * *."

52. Two comments suggested that
FDA should specifically provide in new
§ 58.195(g) that records to be retained
under the GLP regulations may be
retained as magnetic media.

Section 58.195(g). which is being
added to Part 58 by this final rule.
provides that records respecting a
nonclinical laboratory study may be
retained as original records or as true
copies such as photocopies. microfilm.
microfiche. or other accurate
reproductions of the original records.
Magnetic media could qualify as either
"original records" or "accurate
reproductions of the original records."
This conclusion is consistent with
§ 58.3(k). which includes "magnetic
media" within the meaning of "raw
data." Consequently. the agency does
not believe that it is necessary to change
new § 58.195(g) in order to achieve the
result desired by the comments.

Reference

The following reference has been
placed on display in the Dockets
Management Branch (HFA–305). Food
and Drug Administration. Rm. 4–62. 5600
Fishers Lane. Rockville. MD 20857. and
may be seen in that office between 9
a.m. and 4 p.m., Monday through Friday.

1. "Good Laboratory Practice in the Testing
of Chemicals." Organization for Economic
Cooperation and Development Publications.
Paris. France. 1982.

Economic Assessment

As announced in the proposal. FDA
has examined the economic
consequences of this final rule in
accordance with Executive Order. 12291
and the Regulatory Flexibility Act (Pub.
L 96–354). At that time. the agency
concluded that the changes would not
constitute a major rule as defined in the
Order and that no regulatory flexibility
analysis would be required. The agency
also certified that the revisions would
not have a significant impact on a
substantial number of small entities. The
agency has not received any new
information or comments that would
alter its previous determination.

Environmental Impact

The agency has determined under. 21
CFR 25.24(a)(10) that this action is of a
type that does not individually or
cumulatively have a significant effect on
the human environment. Therefore.
neither an environmental assessment
nor an environmental impact statement
is required.

Paperwork Reduction Act of 1980

Sections 58.35(b) (1). (3). and (6).
58.63(b). 58.90(c). 58.105(a). 58.120(a).
58.130(e). and 58.190 (a) and (e) of this
final rule contain collection of
information requirements. FDA
submitted a copy of the proposed rule
containing the same requirements to the
Office of Management and Budget
(OMB). These collection of information
requirements were approved under
OMB Control No. 0910–0203.

List of Subjects in 21 CFR Part 58

Laboratories.

Therefore. under the Federal Food.
Drug. and Cosmetic Act and under 21
CFR 5.11. Part 58 is amended as follows:

**PART 58—GOOD LABORATORY
PRACTICE FOR NONCLINICAL
LABORATORY STUDIES**

1. The authority citation for 21 CFR
Part 58 is revised to read as follows:

Authority: Secs. 306. 402(a). 406. 408. 409.
502. 503. 505. 506. 507. 510. 512–516. 518–520.
701(a). 706. 801. Pub. L. 717. 52 Stat. 1045–10
as amended. 1049–1053 as amended. 1055.
1058 as amended. 55 Stat. 851 as amended.
Stat. 463 as amended. 68 Stat. 511–517 as
amended. 72 Stat. 1785–1788 as amended. 76
Stat. 794 as amended. 82 Stat. 343–351. 90
Stat. 539–574 (21 U.S.C. 336. 342(a). 346. 346.
348. 352. 353. 355. 356. 357. 360. 360b–360f.
360f. 360h–360j. 371(a). 376. 381); secs. 215.
351. 354–360F. Pub. L. 410. 58 Stat. 690. 702 a
amended. 82 Stat. 1173–1186 as amended (4:
U.S.C. 216. 262. 263b–263n); 21 CFR 5.11.

2. In § 58.1 by designating the existin
text as paragraph (a) and adding new
paragraph (b). to read as follows:

§ 58.1 Scope.

* * * * *

(b) References in this part to
regulatory sections of the Code of
Federal Regulations are to Chapter I o
Title 21. unless otherwise noted.

3. In § 58.3 by removing the phrase "
this chapter" wherever it appears; by
revising paragraphs (c). (d). and (e)(8);
by removing and reserving paragraph
(e)(12); by replacing "in section 513 of
the act" with "in Part 860" in paragrap
(e)(17); by replacing "section 514 of the
act" with "in Part 861" in paragraph
(e)(18); and by adding new paragraphs
(o) and (p). to read as follows:

§ 58.3 Definitions.

* * * * *

(c) "Control article" means any food
additive. color additive. drug. biologic
product. electronic product. medical
device for human use. or any article
other than a test article. feed. or water
that is administered to the test system
the course of a nonclinical laboratory

study for the purpose of establishing a basis for comparison with the test article.

(d) "Nonclinical laboratory study" means in vivo or in vitro experiments in which test articles are studied prospectively in test systems under laboratory conditions to determine their safety. The term does not include studies utilizing human subjects or clinical studies or field trials in animals. The term does not include basic exploratory studies carried out to determine whether a test article has any potential utility or to determine physical or chemical characteristics of a test article.

(e) * * *

(8) Data and information about a substance submitted as part of the procedures for establishing a tolerance for unavoidable contaminants in food and food-packaging materials. described in Parts 109 and 509.

* * * * *

(12) [Reserved]

* * * * *

(o) "Study initiation date" means the date the protocol is signed by the study director.

(p) "Study completion date" means the date the final report is signed by the study director.

4. In § 58.31 by revising paragraph (b), to read as follows:

§ 58.31 Testing facility management.

* * * * *

(b) Replace the study director promptly if it becomes necessary to do so during the conduct of a study.

* * * * *

5. In § 58.35 by removing paragraph (e). by revising paragraphs (a) and (b) introductory text. (1) and (3). and by adding an OMB number at the end of the section to read as follows:

§ 58.35 Quality assurance unit.

(a) A testing facility shall have a quality assurance unit which shall be responsible for monitoring each study to assure management that the facilities. equipment. personnel. methods. practices. records. and controls are in conformance with the regulations in this part. For any given study. the quality assurance unit shall be entirely separate from and independent of the personnel engaged in the direction and conduct of that study.

(b) The quality assurance unit shall:
(1) Maintain a copy of a master schedule sheet of all nonclinical laboratory studies conducted at the testing facility indexed by test article and containing the test system. nature of study. date study was initiated. current status of

each study, identity of the sponsor. and name of the study director.

* * * * *

(3) Inspect each nonclinical laboratory study at intervals adequate to assure the integrity of the study and maintain written and properly signed records of each periodic inspection showing the date of the inspection. the study inspected. the phase or segment of the study inspected. the person performing the inspection. findings and problems, action recommended and taken to resolve existing problems. and any scheduled date for reinspection. Any problems found during the course of an inspection which are likely to affect study integrity shall be brought to the attention of the study director and management immediately.

* * * * *

(Collection of information requirements approved by the Office of Management and Budget under number 0910–0203.)

6. By revising § 58.41. to read as follows:

§ 58.41 General.

Each testing facility shall be of suitable size and construction to facilitate the proper conduct of nonclinical laboratory studies. It shall be designed so that there is a degree of separation that will prevent any function or activity from having an adverse effect on the study.

7. In § 58.43 by removing paragraph (e) and by revising the first sentence of paragraph (c) to read as follows:

§ 58.43 Animal care facilities.

* * * * *

(c) Separate areas shall be provided. as appropriate. for the diagnosis. treatment. and control of laboratory animal diseases. * * *

* * * * *

8. In § 58.45 by revising the last sentence. to read as follows:

§ 58.45 Animal supply facilities.

* * * Perishable supplies shall be preserved by appropriate means.

9. By revising § 58.49. to read as follows:

§ 58.49 Laboratory operation areas.

Separate laboratory space shall be provided. as needed. for the performance of the routine and specialized procedures required by nonclinical laboratory studies.

§ 58.53 [Removed]

10. By removing § 58.53.
Administrative and personnel facilities.

11. By revising § 58.61. to read as follows:

§ 58.61 Equipment design.

Equipment used in the generati measurement. or assessment of d equipment used for facility environmental control shall be of appropriate design and adequate capacity to function according to protocol and shall be suitably loc for operation. inspection. cleanin; maintenance.

12. In § 58.63 by revising parag and by adding an OMB numbert : end of the section. to read as follc

§ 58.63 Maintenance and calibratio equipment.

* * * * *

(b) The written standard opera procedures required under § 58.8 shall set forth in sufficient detail methods. materials. and schedule used in the routine inspection. cle maintenance. testing. calibration. standardization of equipment. an specify. when appropriate. remed action to be taken in the event of or malfunction of equipment. The written standard operating proce shall designate the person respon for the performance of each oper:

* * * * *

(Collection of information requiremer approved by the Office of Manageme Budget under number 0910–0203.)

13. In § 58.81 by revising the fir sentence of paragraph (c). to reac follows:

§ 58.81 Standard operating proced

* * * * *

(c) Each laboratory area shall l immediately available laboratory manuals and standard operating procedures relative to the labora procedures being performed. * *

* * * * *

14. In § 58.90 by revising parag (b) and (c) and by adding an OM number at the end of the section. as follows:

§ 58.90 Animal care.

* * * * *

(b) All newly received animals outside sources shall be isolated their health status shall be evalu accordance with acceptable vete medical practice.

(c) At the initiation of a nonclii laboratory study. animals shall b of any disease or condition that r interfere with the purpose or con the study. If. during the course of study. the animals contract such disease or condition. the disease animals shall be isolated. if nece These animals may be treated fo disease or signs of disease provic

treatment does not interfere with
udy. The diagnosis, authorizations
atment, description of treatment,
ach date of treatment shall be
nented and shall be retained.

*   *   *   *

ction of information requirements
ved by the Office of Management and
t under number 0910–0203.)

In § 58.105 by revising the first
nce of paragraph (a), by revising
graph (b), and by adding an OMB
er at the end of the section, to read
lows:

05   Test and control article
cterization.

The identity, strength, purity, and
osition or other characteristics
h will appropriately define the test
ntrol article shall be determined for
batch and shall be
mented. * * *
The stability of each test or control
e shall be determined by the
ig facility or by the sponsor either:
fore study initiation, or (2)
omitantly according to written
iard operating procedures, which
de for periodic analysis of each
l.

*   *   *   *

ction of information requirements
ved by the Office of Management and
t under number 0910–0203.)

In § 58.113 by revising paragraph
, to read as follows:

13   Mixtures of articles with carriers.
* * *

To determine the stability of the
ind control articles in the mixture
quired by the conditions of the
y either (i) before study initiation, or
oncomitantly according to written
iard operating procedures which
ide for periodic analysis of the test
control articles in the mixture.

*   *   *   *

. In § 58.120 by revising paragraph
nd by adding an OMB number at
nd of the section, to read as
ws:

120   Protocol.

) Each study shall have an approved
ten protocol that clearly indicates
bjectives and all methods for the
luct of the study. The protocol shall
ain, as applicable, the following
mation:
) A descriptive title and statement of
urpose of the study.
) Identification of the test and
rol articles by name, chemical
ract number, or code number.
) The name of the sponsor and the
e and address of the testing facility
hich the study is being conducted.

(4) The number, body weight range,
sex, source of supply, species, strain,
substrain, and age of the test system.
(5) The procedure for identification of
the test system.
(6) A description of the experimental
design, including the methods for the
control of bias.
(7) A description and/or identification
of the diet used in the study as well as
solvents, emulsifiers, and/or other
materials used to solubilize or suspend
the test or control articles before mixing
with the carrier. The description shall
include specifications for acceptable
levels of contaminants that are
reasonably expected to be present in the
dietary materials and are known to be
capable of interfering with the purpose
or conduct of the study if present at
levels greater than established by the
specifications.
(8) Each dosage level, expressed in
milligrams per kilogram of body weight
or other appropriate units, of the test or
control article to be administered and
the method and frequency of
administration.
(9) The type and frequency of tests,
analyses, and measurements to be
made.
(10) The records to be maintained.
(11) The date of approval of the
protocol by the sponsor and the dated
signature of the study director.
(12) A statement of the proposed
statistical methods to be used.

*   *   *   *

(Collection of information requirements
approved by the Office of Management and
Budget under number 0910–0203.)

18. In § 58.130 by revising paragraphs
(d) and (e) and by adding an OMB
number at the end of the section, to read
as follows:

§ 58.130   Conduct of a nonclinical
laboratory study.
*   *   *   *

(d) Records of gross findings for a
specimen from postmortem observations
should be available to a pathologist
when examining that specimen
histopathologically.
(e) All data generated during the
conduct of a nonclinical laboratory
study, except those that are generated
by automated data collection systems,
shall be recorded directly, promptly, and
legibly in ink. All data entries shall be
dated on the date of entry and signed or
initialed by the person entering the data.
Any change in entries shall be made so
as not to obscure the original entry,
shall indicate the reason for such
change, and shall be dated and signed
or identified at the time of the change. In
automated data collection systems, the
individual responsible for direct data

input shall be identified at the time of
data input. Any change in automated
data entries shall be made so as not to
obscure the original entry, shall indicate
the reason for change, shall be dated,
and the responsible individual shall be
identified.

(Collection of information requirements
approved by the Office of Management and
Budget under number 0910–0203.)

19. In § 58.185 by revising paragraph
(b), to read as follows:

§ 58.185   Reporting of nonclinical
laboratory study results.
*   *   *   *

(b) The final report shall be signed
and dated by the study director.
20. In § 58.190 by revising paragraphs
(a) and (e) and by adding an OMB
number at the end of the section, to read
as follows:

§ 58.190   Storage and retrieval of records
and data.

(a) All raw data, documentation,
protocols, final reports, and specimens
(except those specimens obtained from
mutagenicity tests and wet specimens of
blood, urine, feces, and biological fluids)
generated as a result of a nonclinical
laboratory study shall be retained.

*   *   *   *   *

(e) Material retained or referred to in
the archives shall be indexed to permit
expedient retrieval.

(Collection of information requirements
approved by the Office of Management and
Budget under number 0910–0203.)

21. In § 58.195 by revising paragraph
(c), redesignating paragraph (g) as
paragraph (h), and adding new
paragraph (g), to read as follows:

§ 58.195   Retention of records.
*   *   *   *

(c) Wet specimens (except those
specimens obtained from mutagenicity
tests and wet specimens of blood, urine,
feces, and biological fluids), samples of
test or control articles, and specially
prepared material, which are relatively
fragile and differ markedly in stability
and quality during storage, shall be
retained only as long as the quality of
the preparation affords evaluation. In no
case shall retention be required for
longer periods than those set forth in
paragraphs (a) and (b) of this section.

*   *   *   *

(g) Records required by this part may
be retained either as original records or
as true copies such as photocopies,
microfilm, microfiche, or other accurate
reproductions of the original records.

*   *   *   *

§ 58.204 [Amended]

22. In § 58.204 *Notice of and opportunity for hearing on proposed disqualification* in paragraph (b) by removing "of this chapter."

§ 58.213 [Amended]

23. In § 58.213 *Public disclosure of information regarding disqualification* in paragraph (b) by removing "of this chapter."

§ 58.219 [Amended]

24. In § 58.219 *Reinstatement of a disqualified testing facility* by removing "of this chapter."

Frank E. Young,
Commissioner of Food and Drugs.
Otis R. Bowen.
Secretary of Health and Human Services.

Dated: August 10. 1987.

[FR Doc. 87–20375 Filed 9–3–87: 8:45 am]

BILLING CODE 4160-01-M

OECD PRINCIPLES OF GOOD LABORATORY PRACTICE

The publication, *The Organisation for Economic Co-operation and Development Principles of Good Laboratory Practice,* first appeared in 1982 in *Good Laboratory Practice in the Testing of Chemicals.* These principles, like the U.S. Food and Drug Administration and the U.S. Environmental Protection Agency GLP requirements, were promulgated as a means of assuring the quality and integrity of scientific research data.

In 1992 the OECD provided the *Series on Good Laboratory Practice and Compliance Monitoring* as reflected in Environment Monograph No. 45, the *OECD Principles of Good Laboratory Practice.* These publications continued to be enhanced, and after more than fifteen years of use, "Member countries considered that there was a need to review and update the the Principles of GLP to account for scientific and technical progress in the field of safety testing and the fact that safety testing was currently required in many more areas . . . " In 1995 an Expert Group was established, led by Germany with experts from Australia, Austria, Belgium, Canada, the Czech Republic, Denmark, Finland, France, Germany, Greece, Hungary, Ireland, Italy, Japan, Korea, the Netherlands, Norway, Poland, Portugal, the Slovak Republic, Spain, Sweden, Switzerland, the United Kingdom, the United States, and the International Organisation for Standardisation. Their work was completed in 1996, and the *Revised OECD Principles of GLP* were reviewed by the relevant policy bodies of the Organisation and were adopted by Council on November 26, 1997.

The following information is herewith reproduced for the convenience of the reader wishing to have on hand the items referenced in the checklist section of this book. It is not the complete work: The OECD Principles of Good Laboratory Practice. The complete text is available from:

OECD Environment Directorate,

Environmental Health and Safety Division

2 rue André-Pascal

75775 Paris Cedex 16

France

Fax: (33-1) 45 24 16 75

E-mail: ehscont@oecd.org

OECD Environmental Health and Safety Publications

Series on Principles of Good Laboratory Practice

and Compliance Monitoring

No. 1

OECD Principles of Good Laboratory Practice

(as revised in 1997)

Environment Directorate

Organisation for Economic Co-operation and Development

Paris 1998

FOREWORD

Chemicals control legislation in OECD Member countries is founded in a proactive philosophy of preventing risk by testing and assessing chemicals to determine their potential hazards. The requirement that evaluations of chemicals be based on safety test data of sufficient quality, rigour and reproducibility is a basic principle in this legislation. The Principles of Good Laboratory Practice (GLP) have been developed to promote the quality and validity of test data used for determining the safety of chemicals and chemicals products. It is a managerial concept covering the organisational process and the conditions under which laboratory studies are planned, performed, monitored, recorded and reported. Its principles are required to be followed by test facilities carrying out studies to be submitted to national authorities for the purposes of assessment of chemicals and other uses relating to the protection of man and the environment.

The issue of data quality has an important international dimension. If regulatory authorities in countries can rely on safety test data developed abroad, duplicative testing can be avoided and costs saved to government and industry. Moreover, common principles for GLP facilitate the exchange of information and prevent the emergence of non-tariff barriers to trade, while contributing to the protection of human health and the environment.

The OECD Principles of Good Laboratory Practice were first developed by an Expert Group on GLP established in 1978 under the Special Programme on the Control of Chemicals. The GLP regulations for non-clinical laboratory studies published by the US Food and Drug Administration in 1976 provided the basis for the work of the Expert Group, which was led by the United States and comprised experts from the following countries and organisations: Australia, Austria, Belgium, Canada, Denmark, France, the Federal Republic of Germany, Greece, Italy, Japan, the Netherlands, New Zealand, Norway, Sweden, Switzerland, the United Kingdom, the United States, the Commission of the European Communities, the World Health Organisation and the International Organisation for Standardisation.

Those Principles of GLP were formally recommended for use in Member countries by the OECD Council in 1981. They were set out (in Annex II) as an integral part of the Council Decision on Mutual Acceptance of Data in the Assessment of Chemicals, which states that "data generated in the testing of chemicals in an OECD Member country in accordance with OECD Test Guidelines* and OECD Principles of Good Laboratory Practice shall be accepted in other Member countries for purposes of assessment and other uses relating to the protection of man and the environment" [C(81)30(Final)].

After a decade and half of use, Member countries considered that there was a need to review and update the Principles of GLP to account for scientific and technical progress in the field of safety testing and the fact that safety testing was currently required in many more areas than was the case at the end of the 1970's. On the proposal of the Joint Meeting of the Chemicals Group and Management Committee of the Special Programme on the Control of Chemicals, another Expert Group was therefore established in 1995 to develop a proposal to revise the Principles of GLP. The Expert Group, which completed its work in 1996, was led by Germany and comprised experts from Australia, Austria, Belgium, Canada, the Czech Republic, Denmark, Finland, France, Germany, Greece, Hungary, Ireland, Italy, Japan, Korea, the Netherlands, Norway, Poland, Portugal, the Slovak Republic, Spain, Sweden, Switzerland, the United Kingdom, the United States and the International Organisation for Standardisation.

---

*OECD Guidelines for the Testing of Chemicals, 1981 and continuing series.

The Revised OECD Principles of GLP were reviewed in the relevant policy bodies of the Organisation and were adopted by Council on 26th November, 1997 [C(97)186/Final], which formally amended Annex II of the 1981 Council Decision. This publication, the first in the OECD series on Principles of Good Laboratory Practice and Compliance Monitoring, contains the Principles of GLP as revised in 1997 and, in Part Two, the three OECD Council Acts related to the Mutual Acceptance of Data.

SECTION II

GOOD LABORATORY PRACTICE PRINCIPLES

1. Test Facility Organisation and Personnel

1.1 *Test Facility Management's Responsibilities*

1. Each test facility management should ensure that these Principles of Good Laboratory Practice are complied with, in its test facility.

2. At a minimum it should:

   a) ensure that a statement exists which identifies the individual(s) within a test facility who fulfil the responsibilities of management as defined by these Principles of Good Laboratory Practice;

   b) ensure that a sufficient number of qualified personnel, appropriate facilities, equipment, and materials are available for the timely and proper conduct of the study;

   c) ensure the maintenance of a record of the qualifications, training, experience and job description for each professional and technical individual;

   d) ensure that personnel clearly understand the functions they are to perform and, where necessary, provide training for these functions:

   e) ensure that appropriate and technically valid Standard Operating Procedures are established and followed, and approve all original and revised Standard Operating Procedures;

   f) ensure that there is a Quality Assurance Programme with designated personnel and assure that the quality assurance responsibility is being performed in accordance with these Principles of Good Laboratory Practice;

   g) ensure that for each study an individual with the appropriate qualifications, training, and experience is designated by the management as the Study Director before the study is initiated. Replacement of a Study Director should be done according to established procedures, and should be documented.

   h) ensure, in the event of a multi-site study, that, if needed, a Principal Investigator is designated, who is appropriately trained, qualified and experienced to supervise the delegated phase(s) of the study. Replacement of a Principal Investigator should be done according to established procedures, and should be documented.

i) ensure documented approval of the study plan by the Study Director;

j) ensure that the Study Director has made the approved study plan available to the Quality Assurance personnel;

k) ensure the maintenance of an historical file of all Standard Operating Procedures;

l) ensure that an individual is identified as responsible for the management of the archive(s);

m) ensure the maintenance of a master schedule;

n) ensure that test facility supplies meet requirements appropriate to their use in a study;

o) ensure for a multi-site study that clear lines of communication exist between the Study Director, Principal Investigator(s), the Quality Assurance Programme(s) and study personnel;

p) ensure that test and reference items are appropriately characterised;

q) establish procedures to ensure that computerised systems are suitable for their intended purpose, and are validated, operated and maintained in accordance with these Principles of Good Laboratory Practice.

3. When a phase(s) of a study is conducted at a test site, test site management (if appointed) will have the responsibilities as defined above with the following exceptions: 1.1.2 g), i), j) and o).

1.2 *Study Director's Responsibilities*

1. The Study Director is the single point of study control and has the responsibility for the overall conduct of the study and for its final report.

2. These responsibilities should include, but not be limited to, the following functions. The Study Director should:

a) approve the study plan and any amendments to the study plan by dated signature;

b) ensure that the Quality Assurance personnel have a copy of the study plan and any amendments in a timely manner and communicate effectively with the Quality Assurance personnel as required during the conduct of the study;

c) ensure that study plans and amendments and Standard Operating Procedures are available to study personnel;

d) ensure that the study plan and the final report for a multi-site study identify and define the role of any Principal Investigator(s) and any test facilities and test sites involved in the conduct of the study;

e) ensure that the procedures specified in the study plan are followed, and assess and document the impact of any deviations from the study plan on the quality and integrity of the study, and take appropriate corrective action if necessary; acknowledge deviations from Standard Operating Procedures during the conduct of the study;

f) ensure that all raw data generated are fully documented and recorded;

g) ensure that computerised systems used in the study have been validated;

h) sign and date the final report to indicate acceptance of responsibility for the validity of the data and to indicate the extent to which the study complies with these Principles of Good Laboratory Practice;

i) ensure that after completion (including termination) of the study, the study plan, the final report, raw data and supporting material are archived.

1.3 *Principal Investigator's Responsibilities*

The Principal Investigator will ensure that the delegated phases of the study are conducted in accordance with the applicable Principles of Good Laboratory Practice.

1.4 *Study Personnel's Responsibilities*

1. All personnel involved in the conduct of the study must be knowledgeable in those parts of the Principles of Good Laboratory Practice which are applicable to their involvement in the study.

2. Study personnel will have access to the study plan and appropriate Standard Operating Procedures applicable to their involvement in the study. It is their responsibility to comply with the instructions given in these documents. Any deviation from these instructions should be documented and communicated directly to the Study Director, and/or if appropriate, the Principal Investigator(s).

3. All study personnel are responsible for recording raw data promptly and accurately and in compliance with these Principles of Good Laboratory Practice, and are responsible for the quality of their data.

4. Study personnel should exercise health precautions to minimise risk to themselves and to ensure the integrity of the study. They should communicate to the appropriate person any relevant known health or medical condition in order that they can be excluded from operations that may affect the study.

2. Quality Assurance Programme

2.1 *General*

1. The test facility should have a documented Quality Assurance Programme to assure that studies performed are in compliance with these Principles of Good Laboratory Practice.

2. The Quality Assurance Programme should be carried out by an individual or by individuals designated by and directly responsible to management and who are familiar with the test procedures.

3. This individual(s) should not be involved in the conduct of the study being assured.

2.2 *Responsibilities of the Quality Assurance Personnel*

1. The responsibilities of the Quality Assurance personnel include, but are not limited to, the following functions. They should:

    a) maintain copies of all approved study plans and Standard Operating Procedures in use in the test facility and have access to an up-to-date copy of the master schedule;

    b) verify that the study plan contains the information required for compliance with these Principles of Good Laboratory Practice. This verification should be documented;

    c) conduct inspections to determine if all studies are conducted in accordance with these Principles of Good Laboratory Practice. Inspections should also determine that study plans and Standard Operating Procedures have been made available to study personnel and are being followed.

       Inspections can be of three types as specified by Quality Assurance Programme Standard Operating Procedures:

       - Study-based inspections

       - Facility-based inspections

       - Process-based inspections

       Records of such inspections should be retained.

    d) inspect the final reports to confirm that the methods, procedures, and observations are accurately and completely described, and that the reported results accurately and completely reflect the raw data of the studies;

    e) promptly report any inspection results in writing to management and to the Study Director, and to the Principal Investigator(s) and the respective management, when applicable;

f) prepare and sign a statement, to be included with the final report, which specifies types of inspections and their dates, including the phase(s) of the study inspected, and the dates inspection results were reported to management and the Study Director and Principal Investigator(s), if applicable. The statement would also serve to confirm that the final report reflects the raw data.

3. Facilities

3.1 *General*

1. The test facility should be of suitable size, construction and location to meet the requirements of the study and to minimise disturbance that would interfere with the validity of the study.

2. The design of the test facility should provide an adequate degree of separation of the different activities to assure the proper conduct of each study.

3.2 *Test System Facilities*

1. The test facility should have a sufficient number of rooms or areas to assure the isolation of test systems and the isolation of individual projects, involving substances or organisms known to be or suspected of being biohazardous.

2. Suitable rooms or areas should be available for the diagnosis, treatment and control of diseases, in order to ensure that there is no unacceptable degree of deterioration of test systems.

3. There should be storage rooms or areas as needed for supplies and equipment. Storage rooms or areas should be separated from rooms or areas housing the test systems and should provide adequate protection against infestation, contamination, and/or deterioration.

3.3 *Facilities for Handling Test and Reference Items*

1. To prevent contamination or mix-ups, there should be separate rooms or areas for receipt and storage of the test and reference items, and mixing of the test items with a vehicle.

2. Storage rooms or areas for the test items should be separate from rooms or areas containing the test systems. They should be adequate to preserve identity, concentration, purity, and stability, and ensure safe storage for hazardous substances.

3.4 *Archive Facilities*

Archive facilities should be provided for the secure storage and retrieval of study plans, raw data, final reports, samples of test items, and specimens. Archive design and archive conditions should protect contents from untimely deterioration.

3.5 *Waste Disposal*

Handling and disposal of wastes should be carried out in such a way as not to jeopardise the integrity of studies. This includes provision for appropriate collection, storage and disposal facilities, and decontamination and transportation procedures.

4. Apparatus, Material, and Reagents

1. Apparatus, including validated computerised systems, used for the generation, storage and retrieval of data, and for controlling environmental factors relevant to the study should be suitably located and of appropriate design and adequate capacity.

2. Apparatus used in a study should be periodically inspected, cleaned, maintained, and calibrated according to Standard Operating Procedures. Records of these activities should be maintained. Calibration should, where appropriate, be traceable to national or international standards of measurement.

3. Apparatus and materials used in a study should not interfere adversely with the test systems.

4. Chemicals, reagents, and solutions should be labelled to indicate identity (with concentration if appropriate), expiry date and specific storage instructions. Information concerning source, preparation date and stability should be available. The expiry date may be extended on the basis of documented evaluation or analysis.

5. Test Systems

5.1 *Physical/Chemical*

1. Apparatus used for the generation of physical/chemical data should be suitably located and of appropriate design and adequate capacity.

2. The integrity of the physical/chemical test systems should be ensured.

5.2 *Biological*

1. Proper conditions should be established and maintained for the storage, housing, handling and care of biological test systems, in order to ensure the quality of the data.

2. Newly received animal and plant test systems should be isolated until their health status has been evaluated. If any unusual mortality or morbidity occurs, this lot should not be used in studies and, when appropriate, should be humanely destroyed. At the experimental starting date of a study, test systems should be free of any disease or condition that might interfere with the purpose or conduct of the study. Test systems that become diseased or injured during the course of a study should be isolated and treated, if necessary to maintain the integrity of the study. Any diagnosis and treatment of any disease before or during a study should be recorded.

3. Records of source, date of arrival, and arrival condition of test systems should be maintained.

4. Biological test systems should be acclimatised to the test environment for an adequate period before the first administration/application of the test or reference item.

5. All information needed to properly identify the test systems should appear on their housing or containers. Individual test systems that are to be removed from their housing or containers during the conduct of the study should bear appropriate identification, wherever possible.

6. During use, housing or containers for test systems should be cleaned and sanitised at appropriate intervals. Any material that comes into contact with the test system should be free of contaminants at levels that would interfere with the study. Bedding for animals should be changed as required by sound husbandry practice. Use of pest control agents should be documented.

7. Test systems used in field studies should be located so as to avoid interference in the study from spray drift and from past usage of pesticides.

6. Test and Reference Items

6.1 *Receipt, Handling, Sampling and Storage*

1. Records including test item and reference item characterisation, date of receipt, expiry date, quantities received and used in studies should be maintained.

2. Handling, sampling, and storage procedures should be identified in order that the homogeneity and stability are assured to the degree possible and contamination or mix-up are precluded.

3. Storage container(s) should carry identification information, expiry date, and specific storage instructions.

6.2 *Characterisation*

1. Each test and reference item should be appropriately identified (e.g., code, Chemical Abstracts Service Registry Number [CAS number], name, biological parameters).

2. For each study, the identity, including batch number, purity, composition, concentrations, or other characteristics to appropriately define each batch of the test or reference items should be known.

3. In cases where the test item is supplied by the sponsor, there should be a mechanism, developed in co-operation between the sponsor and the test facility, to verify the identity of the test item subject to the study.

4. The stability of test and reference items under storage and test conditions should be known for all studies.

5. If the test item is administered or applied in a vehicle, the homogeneity, concentration and stability of the test item in that vehicle should be determined. For test items used in field studies (e.g., tank mixes), these may be determined through separate laboratory experiments.

6.   A sample for analytical purposes from each batch of test item should be retained for all studies except short-term studies.

7.   Standard Operating Procedures

7.1   A test facility should have written Standard Operating Procedures approved by test facility management that are intended to ensure the quality and integrity of the data generated by that test facility. Revisions to Standard Operating Procedures should be approved by test facility management.

7.2   Each separate test facility unit or area should have immediately available current Standard Operating Procedures relevant to the activities being performed therein. Published text books, analytical methods, articles and manuals may be used as supplements to these Standard Operating Procedures.

7.3   Deviations from Standard Operating Procedures related to the study should be documented and should be acknowledged by the Study Director and the Principal Investigator(s), as applicable.

7.4   Standard Operating Procedures should be available for, but not be limited to, the following categories of test facility activities. The details given under each heading are to be considered as illustrative examples.

1.   *Test and Reference Items*

Receipt, identification, labelling, handling, sampling and storage.

2.   *Apparatus, Materials and Reagents*

a)   *Apparatus*

Use, maintenance, cleaning and calibration.

b)   *Computerised Systems*

Validation, operation, maintenance, security, change control and back-up.

c)   *Materials, Reagents and Solutions*

Preparation and labelling.

3.   *Record Keeping, Reporting, Storage, and Retrieval*

Coding of studies, data collection, preparation of reports, indexing systems, handling of data, including the use of computerised systems.

4.   *Test System (where appropriate)*

a)   Room preparation and environmental room conditions for the test system.

b)   Procedures for receipt, transfer, proper placement, characterisation, identification, and care of the test system.

c) Test system preparation, observations and examinations, before, during and at the conclusion of the study.

d) Handling of test system individuals found moribund or dead during the study.

e) Collection, identification and handling of specimens including necropsy and histopathology.

f) Siting and placement of test systems in test plots.

5. *Quality Assurance Procedures*

Operation of Quality Assurance personnel in planning, scheduling, performing, documenting and reporting inspections.

8. Performance of the Study

8.1 *Study Plan*

1. For each study, a written plan should exist prior to the initiation of the study. The study plan should be approved by dated signature of the Study Director and verified for GLP compliance by Quality Assurance personnel as specified in Section 2.2.1.b., above. The study plan should also be approved by the test facility management and the sponsor, if required by national regulation or legislation in the country where the study is being performed.

2. a) Amendments to the study plan should be justified and approved by dated signature of the Study Director and maintained with the study plan.

b) Deviations from the study plan should be described, explained, acknowledged and dated in a timely fashion by the Study Director and/or Principal Investigator(s) and maintained with the study raw data.

3. For short-term studies, a general study plan accompanied by a study specific supplement may be used.

8.2 *Content of the Study Plan*

The study plan should contain, but not be limited to the following information:

1. *Identification of the Study, the Test Item and Reference Item*

a) A descriptive title;

b) A statement which reveals the nature and purpose of the study;

c) Identification of the test item by code or name (IUPAC; CAS number, biological parameters, etc.);

d) The reference item to be used.

2. *Information Concerning the Sponsor and the Test Facility*

    a) Name and address of the sponsor;

    b) Name and address of any test facilities and test sites involved;

    c) Name and address of the Study Director;

    d) Name and address of the Principal Investigator(s), and the phase(s) of the study delegated by the Study Director and under the responsibility of the Principal Investigator(s).

3. *Dates*

    a) The date of approval of the study plan by signature of the Study Director. The date of approval of the study plan by signature of the test facility management and sponsor if required by national regulation or legislation in the country where the study is being performed.

    b) The proposed experimental starting and completion dates.

4. *Test Methods*

    Reference to the OECD Test Guideline or other test guideline or method to be used.

5. *Issues (where applicable)*

    a) The justification for selection of the test system;

    b) Characterisation of the test system, such as the species, strain, substrain, source of supply, number, body weight range, sex, age and other pertinent information;

    c) The method of administration and the reason for its choice;

    d) The dose levels and/or concentration(s), frequency, and duration of administration/application;

    e) Detailed information on the experimental design, including a description of the chronological procedure of the study, all methods, materials and conditions, type and frequency of analysis, measurements, observations and examinations to be performed, and statistical methods to be used (if any).

6. *Records*

    A list of records to be retained.

8.3   *Conduct of the Study*

1.   A unique identification should be given to each study. All items concerning this study should carry this identification. Specimens from the study should be identified to confirm their origin. Such identification should enable traceability, as appropriate for the specimen and study.

2.   The study should be conducted in accordance with the study plan.

3.   All data generated during the conduct of the study should be recorded directly, promptly, accurately, and legibly by the individual entering the data. These entries should be signed or initialled and dated.

4.   Any change in the raw data should be made so as not to obscure the previous entry, should indicate the reason for change and should be dated and signed or initialled by the individual making the change.

5.   Data generated as a direct computer input should be identified at the time of data input by the individual(s) responsible for direct data entries. Computerised system design should always provide for the retention of full audit trails to show all changes to the data without obscuring the original data. It should be possible to associate all changes to data with the persons having made those changes, for example, by use of timed and dated (electronic) signatures. Reason for changes should be given.

9.   Reporting of Study Results

9.1   *General*

1.   A final report should be prepared for each study. In the case of short term studies, a standardised final report accompanied by a study specific extension may be prepared.

2.   Reports of Principal Investigators or scientists involved in the study should be signed and dated by them.

3.   The final report should be signed and dated by the Study Director to indicate acceptance of responsibility for the validity of the data. The extent of compliance with these Principles of Good Laboratory Practice should be indicated.

4.   Corrections and additions to a final report should be in the form of amendments. Amendments should clearly specify the reason for the corrections or additions and should be signed and dated by the Study Director.

5.   Reformatting of the final report to comply with the submission requirements of a national registration or regulatory authority does not constitute a correction, addition or amendment to the final report.

9.2   *Content of the Final Report*

The final report should include, but not be limited to, the following information:

1.   *Identification of the Study, the Test Item and Reference Item*

    a)   A descriptive title;

    b)   Identification of the test item by code or name (IUPAC, CAS number, biological parameters, etc.);

    c)   Identification of the reference item by name;

    d)   Characterisation of the test item including purity, stability and homogeneity.

2.   *Information Concerning the Sponsor and the Test Facility*

    a)   Name and address of the sponsor;

    b)   Name and address of any test facilities and test sites involved.

    c)   Name and address of the Study Director;

    d)   Name and address of the Principal Investigator(s) and the phase(s) of the study delegated, if applicable;

    e)   Name and address of scientists having contributed reports to the final report.

3.   *Dates*

Experimental starting and completion dates.

4.   *Statement*

A Quality Assurance Programme statement listing the types of inspections made and their dates, including the phase(s) inspected, and the dates any inspection results were reported to management and to the Study Director and Principal Investigator(s), if applicable. This statement would also serve to confirm that the final report reflects the raw data.

5.   *Description of Materials and Test Methods*

    a)   Description of methods and materials used;

    b)   Reference to OECD Test Guideline or other test guideline or method.

6)   *Results*

    a)   A summary of results;

    b)   All information and data required by the study plan;

c) A presentation of the results, including calculations and determinations of statistical significance;

d) An evaluation and discussion of the results and, where appropriate, conclusions.

7) *Storage*

The location(s) where the study plan, samples of test and reference items, specimens, raw data and the final report are to be stored.

10. Storage and Retention of Records and Materials

10.1 The following should be retained in the archives for the period specified by the appropriate authorities:

a) The study plan, raw data, samples of test and reference items, specimens, and the final report of each study;

b) Records of all inspections performed by the Quality Assurance Programme, as well as master schedules;

c) Record of qualifications, training, experience and job descriptions of personnel;

d) Records and reports of the maintenance and calibration of apparatus;

e) Validation documentation for computerised systems;

f) The historical file of all Standard Operating Procedures;

g) Environmental monitoring records.

In the absence of a required retention period, the final disposition of any study materials should be documented. When samples of test and reference items and specimens are disposed of before the expiry of the required retention period for any reason, this should be justified and documented. Samples of test and reference items and specimens should be retained only as long as the quality of the preparation permits evaluation.

10.2 Material retained in the archives should be indexed so as to facilitate orderly storage and retrieval.

10.3 Only personnel authorised by management should have access to the archives. Movement of material in and out of the archives should be properly recorded.

10.4 If a test facility or an archive contracting facility goes out of business and has no legal successor, the archive should be transferred to the archives of the sponsor(s) of the study(s).

Guidance for Industry

Good Laboratory Practice Regulations
Management Briefings

Post Conference Report

U.S. Department of Health and Human Services
Food and Drug Administration
Office of Regulatory Affairs
August 1979
(Minor editorial and formatting changes made November 1998)

MANAGEMENT BRIEFINGS
ON THE GOOD LABORATORY PRACTICE REGULATIONS
POST CONFERENCE REPORT

On May 1, 2, and 3, 1979, FDA conducted half-day briefing sessions in Washington, Chicago and San Francisco on the Good Laboratory Practice Regulations. The purpose of the sessions was to provide the regulated industry with information to understand and comply with the regulations. The program included speakers from FDA as well as representatives from the American Association for Accreditation of Laboratory Animal Care (Dr. J. W. Ward), the National Association of Life Science Industries (Mr. D. P Neilsen and Dr. H. C. Brown, Jr.), and the Society of Toxicology (Dr. R. B. Forney). Attendance at the three sessions was estimated at 800 persons affiliated with some 149 sponsor laboratories, 68 contractor laboratories, 19 university laboratories and 10 government laboratories. Some three hundred questions were posed, many of which were answered by the panelists during the question and answer portion of the sessions. At the sessions, the agency announced its intention to make available to the registrants and other interested persons a post conference report which would include the substance of all the answers to the questions posed at the conferences, including those questions which were not responded to because of time limitations.

INTRODUCTION

The questions received pertained to general and specific issues concerning the provisions of the GLPs, inspectional procedures, and FDA's enforcement policies. Many of the questions and their answers have been consolidated to eliminate redundancy and to focus more sharply on the issues.

For further information contact:

> Bioresearch Monitoring Coordinator (HFC-230)
> U.S. Food and Drug Administration
> 5600 Fishers Lane
> Rockville, MD 20857

QUESTIONS AND ANSWERS

I. THE GLP REGULATIONS—GENERAL

1. Do the GLPs require the establishment of Technical Operation Manuals?

   No.

2. If a laboratory is accredited by AAALAC (American Association for Accreditation of Laboratory Animal Care), does this serve as assurance of meeting the GLP requirements for animal care and facilities?

   AAALAC accreditation does not substitute for Agency inspection nor does it guarantee automatic compliance with the applicable GLP sections. It is of value, however, in that it demonstrates that the facility has favorably passed a peer group review.

3. Results of the quality assurance unit inspections are not routinely available to an Agency investigator. However, the conforming amendments require that GLP deviations are to be reported in detail with each submission to the FDA. Are we required to send the contents of the quality assurance unit inspection report to the FDA?

   No. The GLP compliance statement in the conforming amendments to the GLPs was included for several reasons:

   (a) to provide an orderly transition across the effective date of the regulations. It was understood that applications for research and marketing permits submitted to the Agency for some period of time after the GLP effective date of June 20, 1979, would contain final reports of nonclinical studies begun and completed prior to the effective date, begun prior to the effective date and completed thereafter, and begun and completed after the effective date. Studies begun and completed prior to the effective date are not required to comply with the GLPs and accordingly, the conforming amendments require that differences be noted. Similar considerations apply to studies begun prior to and completed after the effective date, although in these studies, those portions underway as of the effective date are required to comply.

   (b) to provide for the submission of final reports of studies which were not required to comply with the GLPs but which otherwise, contribute to safety evaluation. The GLPs do not apply to safety studies conducted by independent investigators studying regulated products. Such studies are not sponsored by the product manufacturer, nor is there any intention to submit the results to the Agency. The study results are published in the open literature. The sponsor is required to submit the study to the Agency but could in no way control the research. If the sponsor wishes to use the data in support of the application, the conforming amendments provide a mechanism by which the sponsor can prove that the study was not compromised. A similar situation exists for preliminary exploratory safety studies done by the sponsor.

(c) to foster GLP compliance attitudes by management. The conforming amendment causes management to act responsively to all cases of GLP non-compliance and to take prompt corrective actions.

With these purposes in mind, the conforming amendments require a brief statement of overall GLP compliance and need not contain the Quality Assurance Unit findings. The Quality Assurance Unit findings should cover short-term GLP deviations which are promptly corrected. The conforming amendments statement should cover those systematic GLP deviations which have occurred throughout the study.

4.   Who provides the GLP compliance statement required by the conforming amendments?

This statement is provided by the applicant for the research or marketing permit.

5.   What is the degree of compliance with GLPs which the FDA will require for INDs submitted after June 20, 1979, but which includes toxicology studies initiated before June 20, 1979, and completed after June 20, 1979?

Those portions of the studies underway as of the effective date will have to be done in accord with the applicable provisions of the GLPs.

6.   Do nonclinical laboratory studies completed prior to June 20, 1979, but submitted as part of an IND or NDA subsequent to that date fall under the conforming amendments?

These studies would not have to have been conducted under the GLPs but the conforming amendments statement of compliance is required.

7.   How many members of the National Association of Life Science Industries (NALSI) come under the GLPs? How can the membership list be obtained?

The Agency has not compiled such a list. A membership list is available from NALSI, 1747 Pennsylvania Avenue, NW, Suite 300, Washington, D.C. 20006. All members who conduct nonclinical laboratory studies are subject to the GLPs.

8.   Should a contract laboratory ask a sponsor if the article they are testing is subject to FDA regulations? Should these studies then be listed as a separate master list of studies to comply with the GLP regulations?

Contract laboratories should ask sponsors to identify studies which are associated with FDA regulated products, although the GLPs place this responsibility on the sponsor. A separate listing of such studies, apart from the firm's master list of all studies undertaken by the firm will satisfy the requirement of the GLPs.

9.   What impact have the GLP regulations had on the cost of performing toxicology studies?

The president of a large contracting laboratory has stated that three years ago a chronic rat study could be done for about $80,000; and that the current cost is closer $250,000. He estimated that half of the increased cost is due to GLPs, 30% to larger numbers of test animals per study on present day protocols and 20% to inflation. The Agency has not developed cost estimates.

SUBPART A: GENERAL PROVISIONS

58.1   SCOPE

58.3   DEFINITIONS

58.10   APPLICABILITY TO STUDIES PERFORMED UNDER GRANTS AND
CONTRACTS

58.15   INSPECTION OF A TESTING FACILITY

1.   Are short-term microbiological screening tests and microbiological preservative stability research and development covered by the GLPs?

Microbiological preservative stability research, development and quality control tests are not covered by the GLPs. However, microbiological tests conducted to establish the toxicological profile of an article are covered.

2.   Does the Agency intend to audit analytical data collected on a test article?

Yes, insofar as it contributes to the evaluation of a nonclinical laboratory study.

3.   Does the Agency intend to audit draft final protocols and draft final reports?

The regulations do not require that such materials be retained, however, if draft reports are available, they may be audited in order to help the Agency follow the process from raw data to final report.

4.   Explain why the GLPs apply to "microorganisms or subparts thereof." How are microorganisms currently used by FDA in assessment of safety?

For certain products, FDA does request that microbial tests be done for the purpose of obtaining information on potential neoplastic and mutagenic activity. Likewise, microsomal preparations (subparts thereof) are used as activating systems for certain in vitro tests. When this happens, the tests should be done in accord with the GLPs.

5.   Do the GLPs apply to engineering/electronic testing laboratories that perform functionality tests on medical devices?

No.

6.   Is a licensed manufacturer of human biological products subject to continuing GLP inspection?

The GLPs apply to safety studies submitted to the Agency in order to obtain the license. They do not apply to such studies conducted for the purpose of obtaining batch release of licensed biologicals.

7. Will nonclinical studies in support of medical devices which do not come in contact with man (e.g., stopcocks, a gas machine, a urine bag) be subject to the GLP regulations?

If the medical device application for a research or marketing permit does not require the submission of safety data for approval, then the GLPs do not apply.

8. If a test article is produced by microbial fermentation, are tests run on the bacteria, such as pathogenicity or virulence covered by the GLPs?

No.

9. Are studies performed for label purposes as required by the Federal Hazardous Substances Act considered to be nonclinical laboratory studies under the GLPs?

No.

10. When an application for Premarket Approval for a Class III Device is scrutinized, would a GLP audit by FDA become a criterion for premarket approval?

Safety data are required for Class III Devices and such data are to be collected under the GLPs, but an FDA audit will not automatically become part of the premarket approval mechanisms.

11. Are Class I, II and III Devices regulated products within the meaning of the GLPs?

Yes.

12. Are data contained in a 510(k) notification subject to the GLPs?

No.

13. How do the GLPs apply to the testing of electromechanical medical devices (non-animal work)?

It is presumed that the question refers to engineering tests and in vitro tests of such devices conducted to assess functionality. In these cases, the GLPs do not apply.

14. Please elaborate on the preamble statement (43 FR 59989) that studies involving "diagnostic products" and "medical devices, which do not come in contact with or are implanted in man" are not within the scope of the GLPs.

Failure of diagnostic products or medical devices, which do not come in contact with man or are not implanted does pose a safety hazard. This is also true for implantable devices.

Tests to establish the reliability of these articles are functionality tests, not safety tests. The GLPs cover implantable devices, which may cause adverse tissue reactions or may have components, which leach into the tissues and cause a toxic response.

15. Is an in vitro study to quantitate the amounts of residual proteolytic enzyme on a soft contact lens (the enzyme is used to clean the lens) a safety study which is covered by the GLPs?

No, the enzyme is part of the lens manufacturing process and its analysis would be covered by the GMPs and not the GLPs. If, however, the proteolytic enzyme is sold as a means of cleaning lenses after purchase by a person, the enzyme is an accessory to a medical device and the safety studies supporting the use of the enzyme would be subject to the GLPs.

16. Do engineering laboratory tests done on components of implantable medical devices fall under the GLPs?

No.

17. Are safety tests conducted on biological products exempt from the GLPs?

Two kinds of safety tests are performed on human biological products. Those which are performed by the manufacturer prior to licensing, and those performed post licensing. The tests performed prior to licensing establish the basic safety profile of the product and they are covered by the GLPs. The safety tests performed post licensing are part of the required quality control assays which permit the release of each batch of product. These tests are not covered by the GLPs. Safety testing of interstate biological products for use in animals is not covered by the GLPs since these products are not regulated by FDA.

18. Do the GLPs apply to veterinary drug and biological manufacturers even when the end products are strictly for veterinary use?

The GLPs apply to animal drugs used on a prescription basis but they do not apply to interstate veterinary biologicals since these products are regulated by USDA. Intrastate veterinary biologicals, which are considered to be new animal drugs, are also covered by the GLPs.

19. If an organization has separate divisions for basic research and for toxicological safety testing, will the basic research division be subject to inspection under the GLPs?

No, as long as the basic research division is not providing any service function for the safety-testing unit.

20. Do the GLP requirements apply to an equal degree to acute, medium-term, and long-term studies?

The GLPs apply equally to all nonclinical laboratory studies. It should be recognized, however, that short-term (less than 6 months) studies need not be inspected as frequently as long term (more than 6 months) studies by the quality assurance unit.

21. Are preliminary protocol development or design studies that employ laboratory animals covered by the GLPs?

   No, these are preliminary studies.

22. If an acute oral toxicity study, a 90-day oral toxicity study, and a two-year chronic study are done, is only the two year study required to be done under GLPs?

   No. Each study, regardless of its duration or complexity should be considered in terms of its purpose. A study which is conducted for the purpose of estimating the safety of a product in humans or animals and which will be submitted to FDA is covered under the GLPs. This includes acute oral toxicity studies as well as 90-day oral toxicity studies and two-year chronic studies. In early phases of research, acute studies are often used to select the most promising product from a group of candidate products. In this sense acute studies are exploratory or screening in nature and would be exempted from the GLPs. There are also special situations where a 90-day oral toxicity study or even a chronic oral toxicity study may be exempted from GLPs. For example, a multinational company may want to develop Product A for a very specific foreign market. The company has no intention of ever applying to FDA for an investigational or marketing permit for Product A. Long-term safety studies with Product A for the purpose of foreign registration would be exempted from GLPs.

23. Will you please define a range-finding study and will such studies be inspected?

   A range-finding study is conducted to gather information such as dose range or toxicological end point to permit the more proper design of a subsequent nonclinical laboratory study. Such studies, which are usually short-term, are preliminary exploratory studies which are exempt from the GLPs if properly labeled as "range-finding" or "preliminary pilot study" or similar designation. These studies will usually not serve as the basis of inspection, but may be reviewed to determine whether the operation of a facility is in compliance with the GLPs. Although the studies are exempt from the GLPs, they must still be submitted to the Agency as part of the respective application for a research or marketing permit.

24. Does the Agency agree that the GLPs are applicable to safety studies intended for submission to the Agency in support of the approval of a regulated product and that they are not applicable to preliminary exploratory studies, screening studies, and range-finding studies whose purpose is to develop or improve the experimental design of a planned nonclinical laboratory study?

   Yes.

25. Many toxicological studies are conducted on products or formulations which are comprised entirely of materials which are known to be safe. Such studies are intended to be a quality control measure to determine lack of product integrity or to detect adulteration. Do the GLPs apply to such studies?

No. The Agency considers such studies to be quality control studies which are not subject to the GLPs.

26. Does a food manufacturer's laboratory which conducts only microbiological screening studies have to comply with the GLPs?

    Generally no. The GLPs apply to safety studies intended for submission to the Agency in support of product approval. Food microbiology studies are quality control studies not subject to the regulations.

27. Do the GLPs apply to laboratories which perform routine sterility analyses on marketable medical devices which have been treated with gas for the purpose of sterilization?

    No.

28. Are studies of approved drugs or devices undertaken for physician education, advertising or pharmaceutical marketing purposes subject to the GLPs?

    No.

29. Do the GLPs apply to safety substantiation studies conducted on over-the-counter drugs which are covered by a final monograph?

    No.

30. It is not clear whether a laboratory involved solely in chemical analysis support of a non-clinical laboratory study would be required to comply with the GLPs. Can this be clarified?

    Yes. Analytical laboratories must comply with the GLPs to the extent that they provide data which support the nonclinical laboratory study. Only those portions of the laboratory, those procedures and those personnel involved are required to be in compliance with the GLPs.

31. What is FDA's position regarding the testing of "medical foods" according to GLP requirements?

    By "medical foods," it is assumed that you mean either diets which complement human therapy, or dietary products used for nutritional purposes. Such products usually do not require an application for a research or marketing permit and therefore they do not fall under the scope of the GLPs. If an application is required, the safety tests would be within the scope.

32. How do previous GLP inspections prior to these new regulations affect our being accredited by AAALAC?

    Not at all. AAALAC accreditation deals with animal care practices and is a process which is independent from FDA's GLP inspections.

33. What about the special problems university laboratories have with complying to the GLPs? Are these laboratories expected to comply to the same degree as industry laboratories?

In crafting the final order, the Agency was cognizant of the problems of university laboratories and certain changes were made which would simplify compliance for all laboratories without frustrating the intent of the GLPs. All laboratories are expected to comply to the same degree since product safety decisions are of equal importance regardless of the size or of the organizational structure of the laboratory doing the study.

34. Are analytical laboratories which perform support characterization of a substance subject to GLP inspection? If so, when and under what circumstances?

Yes, the laboratories are subject to inspection at the request of the headquarters bureau which is evaluating the nonclinical laboratory studies on that substance. The kind of inspection will be a data audit which will include only those records, personnel and portions of the laboratory which collected the data on that substance.

35. Does the definition of nonclinical laboratory study include electrical safety of medical devices or evaluation of 'safe" operation of equipment, i.e., fail-safe studies for a critical device?

No, functionality studies do not fall within the scope of the GLPs.

36. Do metabolism studies come under the scope of the GLPs?

For drugs and feed additives used in food producing animals, metabolism studies come under the GLPs. In these cases, the studies are intended to define the tissue residues of toxicological concern as well as to estimate tissue depletion. Such studies on other regulated products are usually conducted as part of the pharmacological evaluation and would not be covered. However, metabolism studies on food additives are covered.

37. Does the FDA have a list of laboratories which do and do not comply with the GLPs?

No, but the Agency maintains a list of the laboratories which have been inspected. Copies of individual inspection reports may be obtained as a Freedom of Information request.

38. Does the term "nonclinical laboratory study" include animal laboratory studies, which are designed for the explicit purpose of determining whether a test article has reasonable promise of clinical effectiveness, and in which observations bearing on clinical safety are only incidental or fragmentary, or at most, clearly secondary?

No.

39. With regard to the Submission of foreign toxicity data to the Agency, must a sponsor monitor and inspect the foreign laboratories and audit the final study report?

Not necessarily. The foreign laboratory would be considered a contract laboratory and the sponsor's responsibilities would be as set forth in question 40 (below).

40. If a sponsor company utilizes a contract laboratory, who is responsible for the GLP compliance of the contract laboratory? Should a sponsor have its own quality assurance unit to monitor contracted studies? If a contract laboratory has its own quality assurance unit, is it necessary for the sponsor to audit these studies also? How does a sponsor validate a report of a study performed at a contract lab?

The ultimate responsibility for assuring the quality and integrity of a nonclinical laboratory study rests with the person (sponsor) who submits the application for a research or marketing permit to the Agency. This responsibility can be discharged as follows:

Case 1. The contract laboratory has a fully functional quality assurance unit and is operating in conformance with the GLPs. In this case, the sponsor should assure itself that the contract facility has adequate personnel, facilities, equipment and standard operating procedures to perform the study properly. Likewise, the sponsor should examine the procedures used by the contract facility's quality assurance unit and make a determination that such procedures are adequate to obtain GLP compliance. Finally, the sponsor should review the final report (not audit since this has already been done by the contract facility) for consistency and accuracy.

Case II. The contract laboratory does not have a quality assurance unit and may or may not be operating in conformance with the other provisions of the GLPs. In this case, the sponsor must perform all quality assurance functions and take whatever steps are required to promote the GLP compliance of the contract facility. The final report will have to be audited since this has not been done by the contractor.

SUBPART B: ORGANIZATION AND PERSONNEL

58.29   PERSONNEL

58.31   TESTING FACILITY MANAGEMENT

58.33   STUDY DIRECTOR

58.35   QUALITY ASSURANCE UNIT

1.   Must an employee with a cold or the flu be removed from the study?

This decision is left to management. If an employee's disease can adversely affect the test system or the study results, the employee should be removed from the study until the employee is well.

2.   In view of the precautions being taken to adequately document diet preparation, the provision for quality assurance unit inspection of the procedure more than once on each study, what is the Agency's thinking on what is to be accomplished by retaining all samples for the period required?

Maintaining a reserve sample is necessary to provide independent assurance that the test system was exposed to the test article as specified in the protocol. If the results of the study raise questions about the composition of the test article, the reserve sample analysis may provide answers to the questions. The Agency is willing to accept a petition from industry to consider changing the reserve sample retention provisions as discussed elsewhere.

3.   Under what circumstances may QAU audit reports be inspected by FDA? Is there any requirement to maintain these reports or can they be discarded?

QAU audit reports as a matter of administrative policy are exempt from routine FDA inspection. FDA's access to QAU audit reports would be through the Courts should the subject matter of those reports be litigated. Since there is no FDA requirement that these reports be maintained, the disposition of these reports is up to the firm's management. FDA advises that such records not be destroyed without the firm seeking advice from its legal counsel.

4.   What are the quality assurance unit inspection requirements for acute and short-term studies?

For studies lasting less than 4 weeks, each final report should be reviewed by the quality assurance unit for accuracy. With regard to the in process phases (dose preparation, dose administration, in vivo observation and measurement, necropsy, etc.), a random sampling approach could be used so that over a series of studies each critical phase has been monitored. The random sampling approach should be statistically designed so that it is adequate for revealing GLP deviations. The approach and its justification should be made a part of the standard operating procedures of the quality assurance unit.

5. What constitutes proper quality assurance unit inspection of each phase of a nonclinical laboratory study?

A variety of procedures are acceptable for performing a quality assurance unit inspection. The GLPs do not mandate specific procedures. The development of an acceptable procedure should not necessarily be limited to but should consider the following:

(a) nonclinical laboratory studies lasting longer than 6 months should be inspected every 3 months; whereas, studies lasting less than 6 months should be inspected at suitable intervals,

(b) each phase of the study should be inspected,

(c) inspection reports are to be submitted to management and to the study director, and

(d) the purpose of the inspections is to identify significant problems which may affect study integrity and to determine that no changes from approved protocols or standard operating procedures were made without proper authorization.

The phases of a particular study will be determined by the nature of the study. For example, the phases of a typical feeding study include the following:

1. protocol development and approval

2. test article characterization

3. test article stability determination

4. test article-carrier mixture preparation

5. test article-carrier mixture sampling

6. test article-carrier mixture homogeneity determination

7. test system quarantine

8. test system allocation to housing

9. test article-carrier mixture distribution to test system

10. periodic measurements

  – animal observations

  – food consumption

  – body weights

  – blood sampling—hematology and clinical chemistry

11. necropsy—histopathology

12. statistical analyses and report preparation

The type of inspection will depend on the nature of the phase. Each phase must be inspected at least once during the study; the times selected for inspection should be those most likely to reveal problems before the quality of the data generated could be adversely affected.

6. Could you take a typical subacute 14-day study and define the phases?

Phases in a short term study (depending on the type) would include protocol preparation, dose preparation, animal allocation, test system dosage, animal observation, necropsy, data recording, data analysis and final report writing.

7. By what authority may the Agency examine master schedule sheets for studies which may never be used in support of an application for a research or marketing permit?

Studies that are not intended to be used to support an application for a research or marketing permit are not covered by the GLPs and need not appear on the master schedule sheet. If however, the studies are intended to be submitted, then they should be listed and can be inspected by the Agency under its authority to evaluate the results of studies designed to demonstrate product safety.

8. Are acute studies to be included on the master schedule sheet?

Yes, if they fall within the scope of the GLPs.

9. In regard to the master schedule sheet, can the "current status of each study" be satisfied by listing the starting date and completion date of the study? Can the "status of the final report" be satisfied by listing the estimated or actual date of issuance of the final report?

Although the GLPs do not specify entries for "current status of each study," dates alone would not be adequate. Suggested entries that are possible include "study proceeding according to protocol," "study proceeding according to protocol as amended on such-and-such date," "study terminated due to such-and-such," etc. Likewise, entries for the status of the final report might include "awaiting final hematology report," "data in statistical analysis," "first draft prepared," "draft under circulation for review and comment," etc.

10. In our laboratory, critical operations for all studies are carried out by the same individuals using essentially similar procedures. Would it be adequate for the quality assurance unit to inspect a set of representative operations for GLP and standard operating procedure compliance that would incorporate a good cross-section of studies?

No, but refer to the answer under question 4 above.

11. In reference to the quality assurance unit review of the final report, you have indicated that not all numbers have to be traced. Do you have in mind a standard which describes an acceptable level of accuracy, e.g., 90%, 99%, 99.9%, 99.99%?

The quality assurance unit review is to ensure that the final report accurately reflects the raw data. Inasmuch as final reports of certain long-term studies can encompass several hundred thousand observations, it would be a prodigious exercise for the quality assurance unit to verify and trace all raw data. Further, the Agency did not mean to require that the quality assurance unit review would include a check of the accuracy of the calculations used to arrive at the final report. This activity would be redundant since the contributing scientists would have already done so in preparing their reports. Rather, the review was expected to be of sufficient depth to reveal inaccuracies in the final report. Consequently, the Agency envisioned the development of a statistically based system whereby a random sample of the results in the final report is traced. The procedure should be made a part of the standard operating procedures.

The Agency has not established an acceptable level of accuracy of the trace.

12. Is the master schedule sheet intended to be prospective or historical? If it is historical, what is the required retention period?

The master schedule sheet is intended to include a listing of all nonclinical laboratory studies currently in progress as well as those which have been conducted during the terms specified in section 58.195 of the GLPs.

13. Does the master schedule sheet have to list studies on compounds for which no data has yet been submitted to the Agency?

Yes. The GLPs cover all nonclinical laboratory studies of Agency regulated products that support or are intended to support applications for research or marketing permits.

14. The GLPs state that the quality assurance unit should assure that the final report reflects the study results. Is it required that every final report be reviewed by the quality assurance unit?

Yes. This procedure helps to ensure the accuracy of the final report.

15. Does the quality assurance unit review of each final study report have to be reported to management?

Yes. The quality assurance unit must make periodic reports to management and the study director on each study. These reports should include the results of the final report review.

16. At our facility the quality assurance unit reports directly to the executive vice president of the company and not to the vice president of research and development. Is it necessary for us to formulate a separate quality assurance unit within the research and development department?

The GLPs require that the quality assurance unit director and the study director cannot be the same person. The quality assurance unit must report to a level of management that has the authority to effect the corrective action as indicated by the quality assurance unit inspection reports. How this is accomplished organizationally is a management prerogative.

17. Is it acceptable for the quality assurance unit to report to the management person who is also responsible for drug safety evaluation?

This is acceptable provided that the management person is not the study director for the studies being inspected by the quality assurance unit.

18. Is it permissible to have a pharmacologist in the research division serve as the director of the quality assurance unit?

The GLPs state that a person may not perform both quality assurance functions and study direction and conduct functions for the same study. Thus, a pharmacologist in a research division could serve as the director of the quality assurance unit as long as he or she did not otherwise participate in the studies under review by the quality assurance unit.

19. How is the requirement for a quality assurance unit to be interpreted when the testing facility is itself a quality assurance unit?

By definition, a testing facility could not be a quality assurance unit. A quality assurance unit which conducts nonclinical laboratory studies should make separate provision for the performance of the GLP quality assurance functions.

20. Is a member of the statistical department of a testing facility entitled to be a member of the quality assurance unit?

This decision rests with facility management but such a choice is acceptable.

21. Company A is conducting a study. Company B performs animal work for Company A to the extent of implanting test material, recovering test materials and tissues, and returning these to Company A for analysis and conclusions. Which company is designated as the testing facility, which company designates the study director, and which company does the study director work for?

In the cited example, Company A would be the study sponsor while Company B would be a contract laboratory performing a portion of a nonclinical laboratory study. Both companies would be considered testing facilities, but, since the GLPs require a single study director for each study, Company A would designate the study director. Company B would, no doubt, designate a participating scientist in charge of the animal work and would have the responsibility of submitting a participating scientist's report to Company A for inclusion into the final report.

22. Is it acceptable to have two study directors for a single study at the same time?

No. The regulations require a single point of study control, which has been vested in the study director.

23. Do the GLPs permit the designation of a "deputy" or "acting" study director to be in charge of a nonclinical laboratory study when the study director is out of town, on vacation, etc.?

Yes.

24. Must the study director personally verify all observations made during a nonclinical laboratory study?

No. The study director must assure that study procedures are adequate to ensure the collection of valid data.

25. A study is only as good as the people who perform it and most importantly as the person who directs it. What does the Agency do to assess the training and experience of toxicologists?

The assessment of the training and experience of personnel is a routine part of the GLP Compliance Program. Agency investigators collect summaries of training and experience for individuals participating in the study. These summaries are evaluated by the headquarters scientific review staff.

26. In view of the shortage of board certified pathologists, is it permissible to permit either non-veterinarians or non-board certified veterinary pathologists to conduct necropsies? Is certification required for a pathologist to participate in a nonclinical laboratory study?

The Agency recognizes the serious shortage of trained and certified pathologists as well as toxicologists. The GLPs require that personnel possess the appropriate combination of education, training and experience needed to do their jobs. Therefore, it is permissible to have non-veterinarians conduct necropsies provided their training and experience are adequate. The GLPs do not require board certification for either pathologists or toxicologists.

27. What does the agency consider to be the minimal acceptable educational requirements for someone appointed as "study director?"

Due to the wide range of nonclinical laboratory studies and the numerous combinations of education, training and experience which would be acceptable, the Agency did not specify minimal educational requirements for nonclinical laboratory study participants. The GLPs specify that the study director should have the appropriate mixture of education, training and experience to permit the performance of the assigned functions.

28. Will I, as the director of a contract pathology laboratory, be required to have a quality assurance unit and to store slides, blocks, wet tissues, etc. in the archives?

The GLPs require that the quality assurance functions be performed. In your case, either you or the sponsor must have a quality assurance unit. Again, either you, the sponsor, or a separate commercial facility will have to store slides, blocks, wet tissues, etc., and the archives will have to specify the storage location.

SUBPART C: FACILITIES

58.41 GENERAL

58.43 ANIMAL CARE FACILITIES

58.45 ANIMAL SUPPLY FACILITIES

58.47 FACILITIES FOR HANDLING TEST AND CONTROL ARTICLES

58.49 LABORATORY OPERATION AREAS

58.51 SPECIMEN AND DATA STORAGE FACILITIES

58.53 ADMINISTRATIVE AND PERSONNEL FACILITIES

1. Would there be any criticism of a laboratory where animals of the same species, used concurrently in 6-8 short term eye or dermal irritation studies, were housed in the same room, assuming there is sufficient spatial separation?

    No. This procedure would be acceptable provided that precautions were taken to prevent animal and experimental mix-ups and cross-contamination.

2. What is the relationship between the FDA and the USDA inspection of animal facilities?

    The USDA inspection is directed towards ensuring the humane care of animals used in research whereas the FDA inspection is directed towards ensuring the quality of data obtained from safety experiments that involve animals.

3. We feel that storage of test article-diet mixtures in animal rooms in well-labeled, vermin proof containers will lead to fewer errors than storage in a central common area. Is this permissible in light of section 58.47(b)?

    Yes. Section 58.47(b) requires separate areas for test article diet mixtures, which need not be a separate common area or a separate room. In the cited example, each animal room could have a separate area devoted to feed storage.

4. Is it necessary to provide space for the isolation of diseased animals if they are immediately removed from the study and sacrificed?

    No. The intent of the regulations is to ensure that diseased animals are handled in a manner that will not adversely impact on the nonclinical laboratory study.

5. Is it acceptable for a nonclinical laboratory to quarantine all newly arrived animals for the required period and then begin the study in the same area?

    Yes.

SUBPART D: EQUIPMENT

58.61   EQUIPMENT DESIGN

58.63   MAINTENANCE AND CALIBRATION OF EQUIPMENT

1.   Regarding GLP required standard operating procedures for preventive maintenance, is it expected that detailed instructions be prepared for each piece of laboratory equipment? Can the standard operating procedures refer to an equipment manual for detailed instructions as appropriate?

Specific standard operating procedures are required for each piece of equipment. These procedures can incorporate verbatim the instructions contained in the equipment manuals.

2.   In order to calibrate a scale used to weigh large farm animals is it necessary to use a set of standard weights similar to those used for laboratory animal scales only much, much heavier?

In this case, calibration and maintenance of a periodic nature can be performed by a manufacturer's representative and the records should reflect these operations. Additionally, calibration can be accomplished through use of secondary standards.

SUBPART E: TESTING FACILITIES OPERATION

58.81   STANDARD OPERATING PROCEDURES

58.83   REAGENTS AND SOLUTIONS

58.90   ANIMAL CARE

1.   Is there a published tolerance regarding the amount of copper in water on the basis of species?

The Agency is not aware of any.

2.   With regard to section 58.90(c), does "separate" mean a separate air supply as well as space?

Yes, insofar as it is required to ensure effective isolation of the disease.

3.   There are many common reagents used in safety studies (e.g. glucose, sodium chloride, etc.). Do the GLPs intend that these reagents be labeled with storage conditions and expiration dates?

Yes. It is of utmost importance that outdated and deteriorated reagents not be used in the study.

4.   What are the environmental requirements for large animal (cattle/horses) safety studies?

Guidance on this matter can be obtained by contacting the appropriate preclearance division within the Bureau of Veterinary Medicine.

5.   How long do animal care records (cage cards, vendor information, etc.) need to be retained?

These records should be retained in the archives for the terms specified in section 58.195.

6.   Does approximate age of the test system need to be listed on the cage cards?

No.

7.   Why can't textbooks and manufacturer's literature be used as standard operating procedures?

Textbooks and manufacturer's literature are not necessarily complete and it is highly unlikely that such materials could be used without modifications to more precisely fit a laboratory's needs. These materials may be used, however, as supplements to and references for standard operating procedures.

8. In the absence of the "Guide for the Care of Laboratory Animals," what reference will FDA use in inspection of facilities for determining appropriate cage sizes, animal environment, animal facilities, veterinary care, and animal care practices?

    References to the guide and regulations promulgated by other agencies have been deleted from the final order on the GLPs. Nonetheless, these materials do provide guidance on the current state-of-the-art for animal care and they are helpful both to the laboratory and to the Agency in determining the adequacy of animal care practices.

9. Are expiration dates required on purchased chemicals and reagents present in the laboratory?

    Yes, expiration dates are required on such chemicals and reagents when they are used in a nonclinical laboratory study.

10. Are expiration dates required on prepared solutions made from purchased chemicals and reagents?

    Yes.

11. Are stability data required to substantiate the expiration dates of reagents and solutions?

    Not necessarily. It is sufficient to use scientific judgement coupled with literature documentation, manufacturer's literature or laboratory experience.

12. With respect to evaluating the effectiveness of reagents and solutions throughout their shelf life, what requirements are there on the certification of efficacy of the test reagents used to evaluate the effectiveness of the GLP reagents and solutions?

    Standard operating procedures for the analyses should provide such efficacy tests for reagents and solutions as the scientific literature, the manufacturer's literature, and the laboratory experience indicate are necessary.

13. What does the Agency expect in the area of analysis of feed and drinking water for known interfering contaminants?

    The GLPs require analysis for and control of contaminants known to be capable of interfering with the nonclinical laboratory study and which are reasonably expected to be present in the feed and water. Certain contaminants may affect study outcome by masking the effects of the test article, as was the case in recent toxicological studies of pentachlorophenol and diethylstilbestrol. In these studies the feeds used as carriers of the test article were found to contain varying quantities of pentachlorophenol and estrogenic activity. These contaminants invalidated the studies by producing erratic results. The use of positive and negative controls in these studies was insufficient to compensate for the variability in the concentration of the contaminants.

To implement this provision of the GLPs, the study director and associated scientists should consider each study in the light of its length, the expected toxicological endpoints and pharmacological activity of the test article, the test system, the route of administration, and other relevant factors to determine what contaminants could reasonably be expected to interfere. These considerations coupled with scientific literature, experience and anticipated levels of contamination should be used to determine which contaminants should be controlled and analyzed.

It is unlikely that a blanket analysis conducted either by feed manufacturers or water authorities would be sufficient. These analyses would either provide data on contaminants which would not be expected to interfere or neglect to provide data for certain interfering contaminants.

For acute studies in which the test article dosage is sufficiently high, in most instances, to overcome any effects from feed or water contaminants, the analytical requirement would be minimized.

14. Study directors are frequently unfamiliar with certain aspects of their studies (e.g. chemical analyses, histopathology, etc.). Is it appropriate for the study director to authorize all deviations from standard operating procedures?

   Yes. As the focal point for study direction and conduct, the study director must be made aware of and react positively to any deviation from a standard operating procedure. Where necessary, a study director should consult with other scientists to determine the impact of a deviation on the study.

15. Is it required that the quality assurance unit test the reagents used in a nonclinical laboratory study?

   Whatever testing is required by section 58.83 of the GLPs for reagents and solutions may be accomplished by those organizational units that normally conduct such testing. It need not be done by the quality assurance unit.

16. May reagent grade chemicals be used in a study on the basis of label analysis declaration?

   Yes, provided that the reagent is labeled with an expiration date.

17. If animals do not have some form of unique identification actually attached to the animal, is identification using only cage cards appropriate? If the test system is housed in individual cages, which are uniquely identified, must each and every animal be identified?

   Section 58.90(d) requires that animals which are to be removed from their home cages or which are to be observed over a long period of time have appropriate identification. Therefore, identification using only cage cards is not sufficient in most cases and each animal should be identified.

SUBPART F: TEST AND CONTROL ARTICLES

58.105 TEST AND CONTROL ARTICLE CHARACTERIZATION

58.107 TEST AND CONTROL ARTICLE HANDLING

58.113 MIXTURES OF ARTICLES WITH CARRIERS

1. Are laboratories required to go beyond shelf storage of reserve samples of test article-carrier mixtures to whatever methods (e.g., cryogenic temperatures), regardless of cost that will maximize stability? Does the Agency expect stability studies to determine optimum storage conditions for each sample?

    No, heroic measures need not be taken. Storage conditions should be consistent with the knowledge of the stability of the mixture under conditions of use and reasonable so as not to permit accelerated decomposition.

2. What are the details of the Agency's reserve sample retention policy?

    With regard to reserve sample retention, the GLPs provide as follows:

    Reserve samples are to be retained from each batch of test and control article prepared in accord with section 58.105(a) for all nonclinical laboratory studies lasting more than 4 weeks. For the purposes of these sections, the 4-week period includes initial dosing to the final *in vivo* observations. Only sufficient sample need be retained to permit meaningful reanalysis. The samples need be retained either for the terms specified in section 58.195 or for the useful life of the sample (dependent on the stability or the quality of the sample) whichever is shorter. Storage conditions should be those commonly accepted as minimizing the deterioration of sample quality and need not require exhaustive study to determine those which maximize stability. All batches of test and control article mixtures are to be retained even if they are prepared daily.

3. For medical devices, how can stability be demonstrated any more effectively than by the continued functioning of a device within specifications during an *in vivo* nonclinical study?

    The stated procedure is acceptable.

4. The cost of chemical assay development and assay of dosage forms prior to conducting acute studies far exceeds the cost of doing the experiment. Will data confirming the weighing, mixing and administration of the test article be considered sufficient?

    No. The test article must be sufficiently characterized to ensure that the same article is used in any further studies.

5.   Does FDA expect a firm to conduct long-term stability tests on test article-carrier mixtures which are used within a day of preparation?

The firm must determine the stability of the mixtures over the period of their use. The GLPs require retention of samples of all batches of test article-carrier mixtures for studies that last longer than 4 weeks. The regulations do not require stability studies on such samples. Samples placed in storage may be analyzed periodically to determine their useful storage life.

6.   Am I correct in assuming that the chemical testing done by the sponsor to characterize the test article is not covered by the GLPs when the test article is subsequently submitted to a contract laboratory as a blind sample for safety testing?

The GLPs do not cover the basic exploratory chemical tests done to derive the specifications of the test article. They do cover those chemical tests done on discrete batches of test article to determine identity, strength, purity and composition.

7.   Does the phrase "mixtures of articles and carriers" also refer to solutions and suspensions, e.g., a solution of a test article in distilled water?

Yes.

8.   For acute studies, is it necessary for the laboratory to analyze each batch of test article-carrier mixture prior to dosing the test system?

No. Uniformity of the mixture must be known and periodic batch analyses need to be done.

9.   Will dialogues such as this and recent inspection experience bring about substantive changes in the final regulations through FDA initiated proposed amendments? What changes are anticipated in the reserve sample retention requirements?

The Agency does not believe the initiative to change the GLPs rests with FDA. Petitions for change may be submitted to the Agency in accord with the 21 CFR 10.30. As was mentioned at the meeting, the Agency recognizes that the reserve sample retention requirements are extensive and expensive and a petition for change would be considered.

10.   What guidelines can be used by a laboratory or sponsor in deciding how frequently concentration analyses should be made?

The Agency has not established guidelines with regard to the frequency of periodic reanalysis of test article-carrier mixtures. Enough batches should be analyzed to assure that the test systems are being exposed to the quantities of test article in the specified protocol.

11.   How long must one retain samples of feed used in nonclinical laboratory studies and should they be frozen?

The sample retention period differs for the various regulated products and the periods are listed in section 58.195. Feed samples need not be frozen for storage.

12. What is the definition of carrier?

Carrier is the material with which the test article is mixed for administration to the test system. It can be feed, water, solvents and excipients depending on dosage form and route of administration.

13. Once stability of a given concentration of a test article-carrier mixture is substantiated, is it necessary to establish a stability profile for each batch at that concentration?

No. Stability need be determined only on a single batch of test article-carrier mixture; however, periodic reanalysis to determine concentration must be done.

14. In the course of a 14-C tissue residue study in the target animal, is it necessary to retain:

    a. a sample of the 14-C labeled drug,

    b. samples of the diet fed control and experimental animals,

    c.. samples of urine and feces after completion of the analyses,

    d. samples of collected tissues after completion of the analyses,

    e. if they must be retained, for how long?

    f. is similar sample retention necessary when doing "cold" tissue residue studies in target animals?

All samples listed in a–d and f above should be retained for the term listed in section 58.195.

15. If a battery of different tests on a substance is being conducted by different contractors, is it necessary to run replicate stability analyses from each and every contractor especially when long-term stability has been documented for the substance?

No. Once stability has been determined in accord with good science, it is not necessary to continually replicate the stability determination.

SUBPART G: PROTOCOL FOR AND CONDUCT
OF A NONCLINICAL LABORATORY STUDY

58.120 PROTOCOL

58.130 CONDUCT OR A NONCLINICAL LABORATORY STUDY

1. In as much as only wet tissues, blocks and slides are necessary to reconstruct the histopathologic aspects of a study by a third party, are written notes, tapes, etc., of the histopathologist's thought process in arriving at a final report legitimately considered "raw data" in the presence of a signed and dated final report? Does the Agency have the right to inspect the written notes from the pathologist?

    Raw data in this case, refers only to the signed and dated final report of the pathologist. Agency investigators may wish to examine the interim notes and reports in an attempt to reconstruct the study but not to second-guess the scientific process used to arrive at the final report. The GLPs do not require that these interim reports and notes be retained.

2. What is considered to be raw data in computer systems when the data is generated from dictated results?

    Transcribed dictation which has been proofread and corrected for typographical and transcription errors is raw data.

3. Do the GLPs require that the protocol be amended to reflect the actual starting date of the study?

    Yes, this is a critical piece of information which should be supplied by way of a formal protocol amendment.

4. It is said that raw data may be any verified exact copy of the original data. In a computerized data system where data is put directly on disc thence to tape, what documentation of the program performing this transfer is required to assure that the tape copy is exact?

    The standard operating procedures which cover computer operations should describe the computer program and the procedure used to assure the production of an exact tape copy.

5. If reformatting of data is done as part of the transfer described in question 4 above, is the new file not raw data even if all data is transferred intact although in a different organization?

    The Agency cannot precisely answer this question without further details of the new data format.

6. Are initials and dates on data printouts (e.g., scintillation counters, gas chromatographs), when these printouts include standards, sufficient documentation for standardization?

    Yes.

7. Is there a time limit for submission of the final report of a nonclinical laboratory study after its conclusion?

    Generally no. On occasion, for marketed products, the Agency may establish time frames for study conduct. Of course alarming findings on marketed products should be reported as soon as possible.

8. Is it permissible to list changes in a final report on a page which is appended to the original final report?

    Yes.

9. Does "studies in progress on June 20, 1979" refer to the phase of dosing of the test system or the phase post-dosing but not yet reported?

    The quotation pertains to all studies for which the final report has not yet been completed. Included are all post-dosing phases.

10. The final report requires a list of participants. Should this include technicians as well as people who perform support functions?

    The final report should include the name of the study director, the names of other scientists or professionals, and the names of all supervisory personnel involved in the study.

11. When an analysis protocol is developed for the first time by using standard scientific technique, who shall validate the protocol?

    The Agency does not *per se* validate protocols. Persons developing new protocols may submit them to the responsible bureau for review and comment prior to initiating a nonclinical laboratory study.

12. Why is the signature of the sponsor required on a protocol for routine acute testing when these procedures are published and sufficiently standardized by the industry? Would written standard operating procedures of the testing facility be sufficient to replace the protocol without the sponsor signature?

    One of the testing deficiencies found in the early Agency investigations of nonclinical studies was protocol changes that were made without informing the sponsor. The changes prejudiced the validity of the studies. Accordingly, the GLPs require that each study have a specific protocol which is attested to by the sponsor.

13. The identity of the individual collecting data entered into a computer can be recorded via the use of a code known only to the individual but directly identifying the individual; similarly the identity of the individuals witnessing or reviewing the data can be recorded. Is this acceptable?

Yes, this procedure is acceptable. The key to the code must be made available to Agency investigators. Do note, however, that the final GLPs do not require that data entries need be witnessed by a second person.

14. Does the following proposal on data entry to computer files satisfy the GLP intent?

> Data is entered through keyboard commands and stored in a "temporary" computer file with accompanying date, time, and analyst codes. The analyst may be technician level personnel. At the conclusion of a set of observations, no more than one day's worth, the data in the "temporary file" is reviewed by a scientist (this person may or may not be the same person who entered the original data) and "corrected" for any typing or entry errors. When it is determined that the data are correct, the data are transferred to a "permanent" computer file. Only authorized personnel may make changes to the "permanent" file.

> No audit trail is kept for changes to "temporary" file. All changes to permanent file are recorded in a change file with appropriate data, personnel code, comments regarding reason for change and original entry.

No. this method would permit unauthorized tampering with the temporary file before the raw data are transferred to the permanent file.

15. When should a protocol amendment issue? Should it be as soon as possible or could a list of all deviations from a protocol be prepared at the end of the study?

If the deviation from the protocol is intended to be permanent, the protocol should be amended as soon as possible. If the deviation is an error, it should be promptly corrected and noted in the raw data.

16. Section 58.120 describes a sixteen-part protocol and section 58.185 describes a fourteen part final report. Must all of these be included in protocols and reports for LD 50's and other short-term tests?

Yes.

17. Is a protocol required for routine research and experimentation?

Protocols are required for all studies covered by the GLPs.

18. If all raw data are not required in a final report, does this mean, for example, that weekly body weight or food intake averages can be in a report without the individual animal data?

The data appearing in a final report depends on the type of study and the kind of regulated product. Specific advice can be obtained by contacting the Agency bureau which has responsibility for the regulated product.

19. If a compound or formula is proprietary, must the final report describe its detailed composition or chemical structure?

   If the proprietary material is a commercially available article to be used as a control, the final report need only describe the trade or chemical name, the source and the manufacturer's batch number.

20. How does the requirement for "approval" of protocols apply to "in house" studies which are conducted in the laboratories of the actual "sponsor?" Who approves? What is an "approved" protocol?

   The word "approved" was retained in the final order to emphasize that a sponsor should have a mechanism for evaluation and approval of initial protocols and all amendments. The specifics of the mechanism can vary but a formal mechanism should be in place.

21. Must the protocol contain both the name and the code number of the test article?

   No, either designation is acceptable.

22. Section 58.120 states that the protocol shall contain the records to be maintained. Is this intended as a detailed list of each data form to be generated?

   No, in this case generalized statements would be satisfactory.

23. How much raw data must be entered into notebooks when performing well-documented routine tests?

   Basically, the GLPs define raw data as the immediate results of original observations. All such immediate results must be entered.

24. What is meant by the statement in section 58.120(a)(12) which pertains to the method by which the degree of absorption of the test and control articles by the test system will be determined?

   The GLPs do not mandate that absorption studies need be done, or which kind of study is satisfactory. The GLPs do require, however, that the protocol describe the method used if one is necessary to achieve the study objectives.

25. Please clarify the issue of having to provide reasons for all corrections to data entries. It seems unreasonable to require reasons for "obvious" error corrections such as misspellings, transposed numbers, and wrong year early in a calendar year.

It must be remembered that "raw data" is basically the results of original observations. Thus, the wrong year is not raw data and can be easily corrected. Misspellings may or may not be raw data whereas in all probability numbers are raw data. The Agency believes that it is sometimes difficult for a second party, such as the personnel in your quality assurance unit, to distinguish "obvious" errors. Consequently, the Agency insists that all corrections to raw data entries be justified.

26. How and to what extent is the selection of the test system to be justified in the protocol?

    Usually, the test system is selected after consideration of the state-of-the-art of toxicology testing in the area of interest. The protocol need not contain extensive justification.

27. Are we expected to label all specimens (e.g., serum, blood, urine, tissue slides) with their exact nature?

    Yes. Such information is useful in preventing mix-ups.

28. Why does "test system, study, nature and date of collection" have to be located on a specimen container? Can such information be coded?

    Specimen refers to any material derived from a test system for examination or analysis. Consequently, blood, tissues, urine, feces, etc., are considered to be specimens whose containers must carry the required label information. Such information will help preclude mix-ups in the subsequent handling of the specimens. Accession numbers or code numbers can be used for samples of specimens which are subjected to further analysis. For example, in histopathology the excised fixed tissue is a specimen which must carry all the label information. However, the blocks and slides prepared from that tissue can be identified by accession numbers. Similarly, in tissue residue analysis, the excised tissue is a specimen; whereas, tissue samples which are homogenized and otherwise prepared for further analysis are not specimens and need not carry full labeling.

SUBPART J: RECORDS AND REPORTS

58.185 REPORTING OF NONCLINICAL LABORATORY STUDY RESULTS

58.190 STORAGE AND RETRIEVAL OF RECORDS AND DATA

58.195 RETENTION OF RECORDS

1. What types of storage conditions are required for the storage of retained specimens?

    The Agency has not developed guidelines for storage conditions. The Agency does not expect heroic measures to be used, but conditions should be reasonable in light of the nature of the specimen. Storage conditions which foster accelerated deterioration should be avoided.

2. In section 58.185, it is stated that test and control article identification and characterization must appear in the final report signed by the study director. However, if the study director is affiliated with a contract laboratory, he/she has no need to know such details of a proprietary test article. Do you agree that such information can be appended to the final report by the sponsor rather than be provided by the study director?

    Yes.

3. Is the storage of archival material (tissues, slides, raw data) the responsibility of the testing laboratory or can this responsibility be assigned to the sponsor of the study?

    The GLPs permit these materials to be stored in the archives of either the testing laboratory or the sponsor. If they are stored in the sponsor's archives, the archives of the testing laboratory must identify the storage location.

4. If a sponsor agrees to characterize and store test articles submitted for study to a contractor, must the contractor also verify the characterization and provide storage for the test articles?

    No, but the contractor must identify the storage location.

5. What is the "completion date" of a nonclinical laboratory study?

    The completion date is the date that the study director signs the final report. Some discretion must be used however, since the protocol calls for a proposed "completion date." In this case, it would be adequate for the protocol to list a completion date for the in vivo phase and qualify it as such.

6. With respect to archival material, what is required to be listed as the date of the study?

    The study date would be the same as the completion date of the study.

7. Do all studies on a test article need to be submitted in support of an application for a research or marketing permit?

    All studies need be submitted, however, not all studies need be conducted in accord with the GLPs. The conforming amendments provide that a statement be included in the submission which identifies which studies have not been conducted in compliance with the GLPs and the extent of the non-compliance.

8. What should be included in the signed and dated reports of the individual scientists participating in the study?

    The final report prepared by the study director should have appended to it all reports written by other participating scientists. These reports should contain sufficient detail to enable the study director to write a final report which reflects the results of the study.

SUBPART K: DISQUALIFICATION OF TESTING FACILITIES

1. What can FDA do to force a laboratory to take corrective actions to achieve compliance with the GLPs? Are warnings given to the laboratory?

   FDA has a number of regulatory sanctions which can be brought to bear on a violative firm in order to bring about compliance with the law. These include rejection of studies, withdrawal of approval of marketed products if such products are supported by defective studies, prosecution and, after June 20, 1979, disqualification of the laboratory. FDA's present GLP enforcement policy is to provide adequate warning and to afford a reasonable opportunity to take corrective action.

2. Disqualifying a laboratory on the basis of failing to comply with one or more provisions of the GLPs raises the question of whether all violations are considered equally, are weighted, or are evaluated scientifically to consider the impact on the outcome of the study.

   A laboratory will not be considered for disqualification unless all of the following criteria are met:

   a. failure to comply with one or more provisions of the GLPs;

   b. the noncompliance adversely affected the validity of the studies;

   c. other lesser regulatory actions (warnings, rejection of individual studies) have not or will not be adequate to achieve compliance with the GLPs.

The violations of the various provisions of the GLPs are evaluated to assess their impact on the validity of the studies. It is impossible to assign weights to the various provisions of the GLPs. Noncompliance with the various provisions must be evaluated in the context of the entire laboratory operation and the kinds of studies being performed. Thus, a violation of a specific provision may be critical for one laboratory doing long-term studies and not for another laboratory engaged in short term studies.

3.   If a laboratory is disqualified, how long does the disqualification last? Under what conditions does reinstatement occur?

The disqualification will last until the laboratory submits in writing to the Commissioner, reasons for reinstatement including a detailed description of the corrective actions it has taken to assure that the violations which led to disqualification will not recur. Reinstatement will depend upon one or more inspections which show that the laboratory is in compliance with GLPs.

4.   Paragraph 231 of the preamble to the GLPs states: "The order of disqualification creates a rebuttable presumption that all studies previously conducted by the facility are unacceptable." Paragraph 226 states: "Studies conducted at facilities that are in substantial compliance will be presumed to be valid." Can we presume that studies conducted during a period when a lab is found to be substantially in compliance will be accepted by FDA as valid even if the laboratory is disqualified at a later date?

Yes, unless FDA develops information to the contrary.

5.   If a contract laboratory is disqualified because of a study performed for one sponsor, what effect does this have on other studies performed for other sponsors? What about studies underway at the time of disqualification?

FDA will not disqualify a laboratory on the basis of one invalid study. Disqualification is viewed as a most serious regulatory sanction by FDA and will only be imposed when the facts demonstrate that the laboratory is incapable of producing valid scientific data and will not take adequate corrective measures. In the event a laboratory is disqualified, all studies performed by the laboratory, including those in progress are presumed to be unacceptable unless the sponsors of those studies can establish to the satisfaction of FDA that the studies were not affected by the circumstances that led to disqualification.

6.   What steps must be taken by FDA prior to removal of a product from the market because of a rejected study which was pivotal to the assessment of safety?

If rejection of a study results in insufficient scientific data being available to support a decision on safety for a marketed product, FDA will initiate formal proceedings to withdraw the marketing approval of the product. These proceedings, for drugs, begin with a notice published in the FEDERAL REGISTER of FDA's proposal to withdraw approval setting forth the basis for the proposed action and affording affected parties an opportunity for a public hearing on the matter. If a hearing is requested, affected parties will have the opportunity to present additional facts at the hearing for the Agency to consider. The Commissioner's decision to withdraw or to continue the approval is based on the facts brought out at the hearing.

ENFORCEMENT STRATEGY

GENERAL POLICY

1. What is the regulatory basis for conducting GLP inspections? It would seem that by making the GLPs regulations instead of guidelines, that the attorneys and accountants are managing the studies. How does that produce good science?

    The GLP regulations are process-oriented; they are designed to assure that the data collected in a nonclinical laboratory study are valid and accurately reflect the responses of the test system. The GLP inspections are necessary to assess the degree of compliance with the GLPs. The science of a study depends on the appropriateness of the design selected to answer the questions raised in the use of the test article as well as the soundness of the conclusions drawn from the data collected in the study. The assessment of the scientific merit of a study is made by scientists.

2. Does FDA have the authority to audit an ongoing study of a product for which an application for a research or marketing permit has not yet been submitted to FDA?

    A distinction needs to be made between an audit of a study and a GLP inspection. An audit involves a comparison of raw data with completed reports to identify errors and discrepancies. A GLP inspection involves an assessment of the practices and procedures used to carry out the study and to record and store the data. FDA audits only studies which have been submitted or are intended to be submitted to the Agency. The FDA will, however, look at on-going studies whether or not they involve FDA regulated products for purposes of documenting the laboratory's adherence to GLPs; such an inspection does not, however, constitute a data audit of the study rather it is an audit of the "process."

3. What happens when a laboratory refuses to permit an inspection of its facilities?

    If the laboratory is actively conducting studies on investigational new drugs, investigational new animal drugs, or investigational devices, refusal to permit inspection is a violation of section 301(e) or (f) of the Act and the Agency will take whatever action is required to compel inspection.

    Where the Agency has reason to believe that the laboratory is in fact conducting nonclinical laboratory studies, a letter will issue to the laboratory stating that FDA will not accept any future studies performed by that laboratory in support of a research or marketing application. If the laboratory has not, or is not testing an FDA regulated product, it is also advised to contact the local FDA district office to arrange for an inspection should they anticipate engaging in such safety testing.

4. What happens if in the course of an inspection of a contract laboratory, the sponsor of the study selected for GLP inspection refuses to permit access to the study records?

The FDA investigator will select another study and proceed with the inspection. If the study originally selected for inspection involved an FDA regulated product, the Agency will pursue the matter directly with the sponsor.

5. If GLP regulations are not retroactive, will FDA audit pre-June 1979 studies? If so, will FDA investigators list non-conformance with GLPs on the FD-483 Notice of Observations associated with those studies?

    FDA will continue to audit pre-June 1979 studies for purposes of assessing not only the quality of a particular study, but also the general performance of the laboratory prior to the time when GLP regulations were first proposed in November 1976. This is necessary because many of the marketing applications pending before the Agency contain studies performed prior to 1976.

    While deviations from the GLPs will be noted in the FD-483 associated with these studies, the Agency will use this information only to make a judgment regarding the scientific acceptability of those studies and will not use the deviations to initiate regulatory action against the laboratory. After the June 1979 effective date, however, deviations from the GLPs could result in regulatory action against both the studies and the laboratories.

6. Will the GLPs apply to a study which has been completed to the June 20, 1979 effective date for which a final report will not be prepared until after?

    The GLP regulations became effective June 20, 1979, and those portions of studies underway, as of that date, even if only the final report, became subject to the regulations at that time.

7. Will a laboratory engaged in testing an FDA-regulated product be subject to a GLP inspection if a research or marketing application has not been submitted to the Agency, e.g., a new company developing its first products?

    Generally speaking, FDA inspects only those laboratories which have conducted studies submitted to the Agency. FDA strongly advises any laboratory which intends to engage in the safety testing of a regulated product, and which has not been previously inspected, to contact the local FDA district office and request a GLP Inspection.

8. Will FDA accept data from a study not conducted in accordance with GLPs for regulatory purposes?

    Even though a study has not been conducted totally in accordance with GLPs, FDA may accept the data from such a study if it can be demonstrated that the areas of non-compliance have not compromised the validity of that study. As a special corollary to this policy, FDA will take note of positive findings of toxicity in a study even though that study was not conducted in compliance with GLPs. While a technically bad study can never establish the absence of a safety risk, it may establish the presence of an unsuspected hazard or untoward effect.

9. Where can the Inflationary Impact Assessment Report of the GLPs be obtained?

    By writing to the: Hearing Clerk
                          Food and Drug Administration
                          5600 Fishers Lane
                          Rockville, MD 20857

10. How does FDA protect the confidentiality of valuable commercial or trade secret information given to an investigator during a GLP inspection?

    FDA employees are required by statute to protect the confidentiality of any trade secret or confidential commercial information which they may acquire in the performance of their duties. Thus any trade secret information which an FDA investigator may receive from a laboratory being inspected is exempt from public disclosure. Whenever the FDA receives a Freedom of Information Act request for a copy of the laboratory inspection report, all information which falls under the definition of trade secret or confidential commercial information will be purged from the report before it's released.

    From a practical standpoint, there is a "gray area" of information, which may or may not be privileged information. FDA personnel will make every effort to determine whether the rules of confidentiality apply in such cases. The final decision, however, will be FDA's.

11. Will FDA review non-GLP studies (range-finding, exploratory studies) in the course of conducting GLP inspections of studies intended to be submitted to the Agency? This is of particular concern in protecting proprietary research data. Will there be an opportunity for the inspected firm to do an FOI review before the final inspection report is written?

    FDA may review on-going non-GLP studies as described in question 23 on Subpart A and question 11 under "Inspections."

    The inspected firm may not review a draft inspection report for purposes of identifying what should not be released under FOI. Even if the Agency permitted this, which it does not, the fact that the report was made available to someone outside the Agency would immediately make that draft document available for public disclosure under the provisions of the FOI regulations.

12. Will foreign laboratories be inspected to determine their compliance with GLPs?

    Foreign laboratories which conduct studies submitted to the Agency will be inspected and held accountable to the same GLP requirements as U.S. laboratories. While FDA has no authority to inspect foreign labs, the Agency has adopted the policy of not accepting data from any laboratory (domestic or foreign) which refuses to permit an inspection of its facilities.

13. What accords have been made with foreign countries regarding GLPs and inspections?

FDA has signed a Memorandum of Understanding with Canada and Sweden which commit both countries to establish GLPs and an inspection system. Discussions which may lead to similar accords have been held with Great Britain and Switzerland. Informal expressions of interest have been received from other countries. The long-range objective of these bilateral agreements is reciprocal recognition of each country's GLP program.

14. Has FDA inspected its own animal research facilities for compliance with GLPs? Other Federal Laboratories?

Yes. To date, FDA has completed GLP inspections of all its animal research facilities and is taking steps to bring all its laboratories into compliance. FDA has also established contacts with the NH, DOD and USDA for purposes of scheduling inspections of laboratories performing safety studies intended to be submitted to the Agency.

15. Has FDA established liaisons with other Federal agencies regarding the GLP program?

Yes, liaisons have also been established with CPSC, EPA, and OSHA for purposes of furthering the objectives of the GLP program, scheduling inspections of Federal laboratories and sharing information resulting from the FDA program.

INSPECTIONS

1. Is it possible that an FDA investigator may take exception to a firm's definition of regulated and nonregulated laboratory studies? If such a difference of classification arises for a given study, how would you resolve the conflict with the FDA?

    Yes, it is possible. The testing facility may appeal any differences it has with the investigator first to the FDA district office and, if this is not satisfactory to FDA headquarters.

2. What is the estimated number of laboratories being inspected by FDA?

    FDA's inventory of laboratories subject to GLPs includes approximately 380 domestic laboratories and 110 foreign laboratories. The laboratories include sponsor laboratories, commercial contract laboratories and university laboratories.

3. Will the inspectional training course at the National Center for Toxicological Research be open to industry and academia?

    No. The training of industry and academic personnel to enable them to properly perform their duties is the responsibility of their employers. However, FDA is prepared to participate in any training courses which may be offered by industry associations or the academic community to the extent that resources will allow.

4. If the GLPs are Phase I of Bioresearch Monitoring, what other phases are anticipated by FDA?

    Other phases include new regulations on obligations of sponsors and monitors of clinical investigations, obligations of clinical investigators, and obligations of institutional review boards. Note that these regulations are directed towards efficacy data and the protection of human subjects whereas the GLPs are directed towards safety data.

5. Who makes the decision on whether or not a headquarters scientist participates in a GLP inspection? Why can't we have a headquarters scientist on each inspection?

    The scheduling bureau makes the decision. During the past two years, headquarters scientists have participated in about half of all GLP inspections and, with rare exception, the Bureau of Biologics assigns a headquarters scientist to each GLP inspection. Resources do not permit more extensive participation.

6. How are laboratories selected for inspection?

    Laboratories are selected for inspections bureaus within FDA. The criteria for selection are actual or potential involvement in studies associated with products regulated by FDA. Inspections will involve a specific study submitted to a bureau or a study selected from the firm's master list which is of interest to FDA.

7. How often can a laboratory expect to be inspected?

Routine surveillance inspections will occur at least once every two years or more frequently depending upon findings of previous inspections. However, more frequent inspections may occur when an audit of a specific study submitted to FDA or EPA in support of a marketing application is required.

Either type of inspection can result in more frequent visits if serious adverse findings are reported. These latter visits are considered compliance or follow-up inspections and are carried out to determine if correction of previous violative conditions have been made.

8. Will laboratories be notified in advance of an inspection?

Because of the comments received during the conferences and the experiences to date with this program, laboratories will generally be notified prior to inspection. However, compliance or special investigation inspections may not follow this procedure.

9. Can a laboratory postpone an inspection?

A facility may at the time of initial FDA contact request a postponement. Such a postponement may occur when personnel responsible for the conduct of the study to be audited will be unavailable at the anticipated inspection date. FDA expects to be reasonable in arranging for an inspection date. Unreasonable delays in scheduling the inspection will however be viewed by FDA as a refusal to permit an inspection.

10. Can a laboratory request an inspection? How?

A facility may request an inspection from either the local FDA district office or from FDA headquarters. However, an inspection will be initiated only with headquarters concurrence. Consideration will be given to the work schedules under which district management is operating.

11. If a laboratory is not performing a study on an FDA-regulated product at the time the investigator arrives, will the inspection still be carried out?

Routinely, GLP inspections are not scheduled unless the Agency has received a final report on a regulated product or has received submitted protocols, interim study reports, or knows that a study on a regulated product is underway. In the case of a laboratory that is not currently performing a study on a regulated product the laboratory will be asked to consent to an inspection. The FDA investigator will utilize an ongoing study, even though it is not associated with an FDA regulated product, to document the laboratory's compliance with GLPs. In such cases, the study will not be audited in terms of validating the raw data, and specifics of the study will not be included in the inspection report.

12. Will inspections cover other areas such as chemistry, physical testing, metallurgy, etc.?

To the extent that the protocol of a nonclinical laboratory study requires tests in the field of metallurgy, clinical chemistry, etc., we will examine and evaluate adherence to test specifications or protocol requirements.

13. Are firms notified of specific studies to be audited? Will sufficient time be allowed to seek authorization from the sponsor of the study to disclose the data to the FDA investigator? What happens if the sponsor of the study refuses to authorize the laboratory to disclose the records?

As stated with respect to prior notification of inspection, where FDA has an interest in auditing a study, ample time generally will be provided for the facility to seek authorization from the sponsor to disclose the data. In some cases, FDA investigators may begin inspecting the physical layout for the facilities while authorization to release the study records is being obtained. If the sponsor refuses to authorize disclosure of the records to the investigator, FDA will pursue the matter directly with the sponsor.

14. Can FDA investigators ask for records to which they are not legally entitled; can they engage in "fishing expeditions?"

It is not FDA policy to request documents during an inspection to which the Agency is not legally entitled. On occasion, the Agency may request such documents when pursuing an audit trail of a possible violation. Under these circumstances, it is the laboratory's prerogative to cooperate or refuse without fear of reprisal. The requests should be specific and pertinent to the inspection. The Agency discourages investigators from making vague requests to see documents with no specific purpose in mind.

15. Should the Form FD-483, Notice of Observations issued by the FDA investigator reflect current practices only; and should it include practices that were corrected during the course of the inspection?

The FD-483 can include historical practices which may affect the scientific validity of the nonclinical study in question even though subsequent correction may have occurred. Any corrective action taken by the facility will be noted by the investigator in the establishment inspection report.

16. What should a laboratory do when there is disagreement between the laboratory and the FDA investigator regarding the findings reflected in the FD-483 Notice of Observations?

At time of the observation, the management should discuss any differing opinions and attempt to clarify the investigator's perceptions or observations. The management may also, at the conclusion of the inspection, offer to explain what the management considers to be erroneous 483 observations. Should the matter in question remain unresolved, a written objection should be sent to the local FDA district director or a meeting with district personnel should be requested to attempt to resolve the issue.

17. What is the procedure for correcting errors in the FDA investigator's inspection report? Such errors can be damaging to the laboratories since the reports are ultimately available through FOI.

If in fact an error is made in an investigator's report, the matter should be immediately brought to the attention of FDA district management. If district management agrees with the complaint, the report will be amended and amended reports will be sent to all outside persons who may received the erroneous report. It should be stressed, however, that the time to change what a facility believes is an erroneous conclusion is when the FD-483 is discussed with laboratory management because as soon as the FD-483 is presented to management, it becomes available for public disclosure.

18. Does refusal to allow the FDA investigator access to certain information which the laboratory sincerely believes is not subject to FDA jurisdiction, constitute a refusal of inspection? How can disagreement of this kind be resolved?

Refusal to permit access to records which are associated with a study being audited or which preclude a judgment being made regarding compliance with GLPs, is considered a refusal of inspection with certain ensuing consequences. However, a facility may legitimately question FDA authority to review certain documents. Such objections and the reasons therefore, should be presented in writing or by telephone to the FDA district office management where the investigator is based. Each case will be individually reviewed both in the field and, if necessary at headquarters and a decision will be communicated to the inspected facility.

19. Will inspections and audits of foreign laboratories be carried out? Who pays for these inspections?

Inspections are being conducted of foreign facilities which have engaged in nonclinical studies which have been submitted to FDA in support of a marketing permit. FDA pays for travel and other expenses associated with such inspections.

20. In order for foreign laboratories to comply with the GLPs, do protocols, standard operating procedures, records, etc. have to be in English? Do FDA investigators bring interpreters with them to review records and data?

Submissions to FDA in support of a marketing application for a FDA-regulated product must be in English. Review of source documents at the site of the foreign facility may necessitate review of documents written in the language of the country of origin. FDA does not employ interpreters to accompany investigators on foreign inspections. It has been our experience that persons associated with the laboratory are normally fluent in the English language.

21. What kind of training does an FDA investigator have which qualifies him/her to conduct a GLP inspection or data audit? Does the investigator draw conclusions from his observations regarding the competence of the laboratory or quality of the studies?

Along with education in one of the natural or physical sciences, the individuals selected to conduce GLP inspections generally have had considerable experience inspecting facilities involved in drug manufacturing, biologics production, medical device assembly, food processing, and a range of other operations on products regulated by the Agency. In addition, the investigators conducting nonclinical laboratory inspections (GLPs) have undergone intensive training in the normal operating procedures of nonclinical testing facilities. This training which includes a full review of the Agency's policies and of the GLP regulations is accomplished at FDA's National Center for Toxicological Research located in Pine Bluff, Arkansas. Field investigators are encouraged to contact any resource within the Agency, i.e., scientists and other personnel of the various bureaus to resolve scientific questions that may arise during an inspection. Bureau scientists and not the investigators, draw conclusions regarding the competence of the laboratory of the quality of the study.

22. Does a laboratory manager have the right to ask for the FDA investigator's educational and experience qualifications prior to a GLP inspection?

Yes, questions regarding the formal training, educational experience, and on-the-job training of an individual investigator may be addressed to the investigator prior to a GLP inspection.

23. What can a laboratory manager do when he encounters an FDA investigator who is overly antagonistic or uncertain as to what he is looking for?

The Agency makes every effort to promote a professional attitude in its investigators including special training and selection of investigators for this program. However, if in the judgment of the laboratory manager there is a question as to the qualifications or attitude of the investigator, the local FDA district office director should be contacted.

24. What assurance does a firm have that confidential or trade secret information given to the FDA investigator will be safeguarded by the Agency? What happens when an FOI request for the inspection report is received by FDA?

Section 301(j) of the Food, Drug, and Cosmetic Act prohibits any employee from revealing for his/her advantage any information obtained in the course of carrying out his/her duties. Trade secrets and confidential commercial information are deleted from documents before they are released under FOI. Inspected firms may help by identifying information which they consider to be confidential when it is given to the investigator. FDA will however, exercise its own judgment, in accordance with the FOI regulations as to whether such information may properly be classified as confidential.

25. How can copies of inspection reports be obtained under FOI?

   Inspection reports may be obtained by making a request under FOI to:
   Freedom of Information Staff, HFI-35
   U.S. Food and Drug Administration
   5600 Fishers Lane
   Rockville, Maryland 20857

GOOD LABORATORY PRACTICE REGULATIONS, QUESTIONS AND ANSWERS

Since June 20, 1979, the agency has been asked many questions on the Good Laboratory Practice regulations (GLPs, 21 CFR 58). In accord with agency procedures, responses have been prepared and copies of the associated correspondence have been filed in the Dockets Management Branch (HFA-305). The responses have also been provided to the bioresearch monitoring program managers and to the district offices in order to ensure consistency of interpretation and equity of program operation. Unfortunately, the numerous filed correspondences contain many repeat questions that are not categorized to relate to the specific GLP subpart and section. On occasion, the answers appear to be somewhat cryptic. These disadvantages serve to limit the utility of the correspondences as advisories to our headquarters and field offices.

This document, therefore, consolidates all GLP questions answered by the agency during the past 2 years, clarifies the questions and answers as needed, and relates the questions and answers to the specific pertinent provisions of the GLPs. It represents a digest of some 30 letters, 160 memoranda of telephone conversations, 34 memoranda of meetings and 30 miscellaneous correspondences that have been issued by agency personnel. The document does not duplicate questions and answers that were dealt with in the August, 1979 Post Conference Report on the Good Laboratory Practice Regulations Management Briefings.

This document should be reviewed by field investigators prior to making GLP inspections and by headquarters personnel involved in the GLP program. Questions should be directed to:

> Dr. Paul D. Lepore*
> Bioresearch Monitoring Staff, HFC-30
> Food and Drug Administration
> 5600 Fishers Lane
> Rockville, MD 20857
> 301-443-2390
>
> *Retired, current contact is Stan W. Wollen

SUBPART A: GENERAL PROVISIONS

Section 58.1   Scope

1.   Do the GLPs apply to validation trials conducted to confirm the analytical methods used to determine the concentration of test article in animal tissues and drug dosage forms?

No.

2.   Do the GLPs apply to the following studies on animal health products: overdosage studies in the target species, animal safety studies in the target species, tissue residue accumulation and depletion studies, and udder irritation studies?

Yes.

3.   Do the GLPs apply to safety studies on cosmetic products?

No. Such studies are not carried out in support of a marketing permit. However, the GLPs represent good quality control; a goal that all testing facilities should strive to attain.

4.   Do safety studies done to determine the potential drug abuse characteristics of a test article have to be done under the GLPs?

Yes they do, but only when the studies are required to be submitted to the agency as part of an application for a research or marketing permit.

5.   Do the GLPs apply to the organoleptic evaluation of processed foods?

No.

6.   Do the GLPs apply to all of the analytical support work conducted to provide supplementary data to a safety study.

The GLPs apply to the chemical procedures used to characterize the test article, to determine the stability of the test article and its mixtures, and to determine the homogeneity and concentration of test article mixtures. Likewise, the GLPs apply to the chemical procedures used to analyze specimens (e.g., clinical chemistry, urinalysis). The GLPs do not apply to the work done to develop chemical methods of analysis or to establish the specifications of a test article.

7.   Is it possible to obtain an exemption from specific provisions of the GLPs for special non-clinical laboratory studies?

Yes. The GLPs were written with the aim of being applicable to a broad variety of studies, test articles and test systems. Nonetheless, the agency realizes that not all of the GLP provisions apply to all studies and, indeed, for some special studies, certain of the GLP provisions may compromise proper science. For this reason, laboratories may petition the agency for exemption for certain studies from some of the GLP provisions. The petition should contain sufficient facts to justify granting the exemption.

8. Are subcontractor laboratories that furnish a particular service such as ophthalmology exams, reading of animal ECGs, EEGs, EMGs, preparation of blocks and slides from tissues, statistical analysis and hematology covered by the GLPs?

   Yes, to the extent that they contribute to a study that is subject to the GLPs.

Section 58.3 Definitions

1. Are animal cage cards considered to be raw data?

   Raw data is defined as "any laboratory worksheets, records, memorandum, notes . . . that are the result of original observations and activities . . . and are necessary for the reconstruction and evaluation of the report of that study." Cage cards are not raw data if they contain information like animal number, study number, study dates, and cage number (information that is not the result of original observations and that is not necessary for study reconstruction). However, if an original observation is put on the cage cards, then all cards must be saved as raw data.

2. Are photo copies of raw data which are dated and verified by signature of the copier considered to be "exact" copies of the raw data?

   Yes.

3. Are records of quarantine, animal receipt, environmental monitoring, and instrument calibration considered to be raw data?

   Yes.

4. A laboratory conducts animal studies to establish a baseline set of data for a different test species/strain. No test article is administered but the toxicology laboratory facilities and procedures will be used and the resulting data may eventually be submitted to the agency as part of a research or marketing permit. Are the studies considered to be nonclinical laboratory studies that are covered by the GLPs?

   Generally, a nonclinical laboratory study involves a test article studied under laboratory conditions for the purpose of determining its safety. The cited example does not fit the definition so it would not be covered by the GLPs. Since the data from the baseline studies may be used to interpret the results of a nonclinical laboratory study, it is recommended, but not required, that the study be conducted in accord with GLPs in order to ensure valid baseline data.

5. The definition of "nonclinical laboratory study" excludes field trials in animals. What is a field trial in animals?

   A field trial in animals is similar to a human clinical trial. It is conducted for the purpose of obtaining data on animal drug efficacy and it is excluded from coverage under the GLPs.

6. Necropsies are done by prosectors trained by and working under the supervision of a pathologist. The necropsy data are recorded by the prosector, on data sheets, and when making the final report, the pathologist summarizes the data collected by the prosector as well as by him/herself. What constitutes the raw data in this example?

   Both the prosector's data sheets as well as the signed and dated report of the pathologist would be considered raw data.

7. Is a computer print-out derived from data transferred to computer media from laboratory data sheets considered to be raw data?

   No.

8. Are the assay plates used in the 10t1/2 mammalian cell transformation assay considered to be specimens?

   Yes.

9. If a firm uses parapathologists to screen tissue preparations, are the parapathologists' data sheets considered to be raw data?

   Yes.

Section 58.10   Applicability to Studies Performed Under Grants and Contracts

1. Certain contracts specify that a series of nonclinical laboratory studies be done on a single test article. Do the GLPs permit the designation of different study directors for each study under the contracts?

   Yes.

2. Do the GLPs require that a sponsor approve the study director for a contracted study?

   No. Testing facility management designates the study director.

3. A firm functions as a primary contractor for nonclinical laboratory studies. The actual studies are then subcontracted to nonclinical laboratories. Is the firm considered to be a "sponsor?"

The GLPs define "sponsor" as a person who initiates and supports a nonclinical laboratory. Sponsorship in the cited example would be determined by the specific provisions of the contract.

4. Who is responsible for test article characterization—the sponsor or the contractor?

The GLPs do not assign the responsibility in this area. The matter is a subject of the specific contractual arrangement between the sponsor and the contractor.

5. Do contract laboratories have to show the sponsor's name on the Master Schedule Sheet or can this information be coded?

The information can be coded but the code must be revealed to the FDA investigator on request.

6. A sponsor desires to contract for a nonclinical laboratory study to be conducted in a foreign laboratory. Must the sponsor notify the foreign laboratory that compliance with the U.S. GLPs is required?

Yes.

7. Must a contractor include in the final report information on test article characterization and stability when such information has been collected by the sponsor?

No. The contractor should identify in its final report which information will be subsequently supplied by the sponsor.

8. Must a sponsor reveal toxicology data already collected on a test article to a contract laboratory?

No. If use of the test article involves a potential danger to laboratory personnel, the contract laboratory should be advised so that appropriate precautions can be taken.

Section 58.15   Inspection of a Testing Facility

1. What is the usual procedure for the issuance of a Form FD-483?

The FD-483 is the written notice of objectionable practices or deviations from the regulations that is prepared by the FDA investigator at the end of the inspection. The items listed on the form serve as the basis for the exit discussion with laboratory management at which time management can either agree or disagree with the items and can offer possible corrective actions to be taken. Management may also respond to the district office in writing after it has had sufficient time to properly study the FD-483.

---

2. Will a laboratory subsequently be notified of GLP deviations not listed on the FD-483?

This does happen. The FDA investigator prepares an establishment inspection report (EIR) which summarizes the observations made at the laboratory and which contains exhibits concerning the studies audited (Protocols, SOPs, CV's, etc.) The EIR is then reviewed by District personnel as well as headquarters personnel. This review may reveal additional GLP deviations that should be and are communicated to laboratory management.

3. What kinds of domestic toxicology laboratory inspections does FDA perform and how frequently are they done?

FDA performs four kinds of inspections related to the GLPs and nonclinical laboratory studies. These include: A GLP inspection—an inspection undertaken as a periodic, routine determination of a laboratory's compliance with the GLPs, it includes examination of an ongoing study as well as a completed study; A data audit—an inspection made to verify that the information contained in a final report submitted to FDA is accurate and reflected by the raw data; A directed inspection—any of a series of inspections conducted for various compelling reasons (questionable data in a final report, tips from informers, etc.); a follow-up inspection—an inspection made sometime after a GLP inspection which revealed objectionable practices and conditions. The purpose of the follow-up inspection is to assure that proper corrective actions have been taken. GLP inspections are scheduled once every two years whereas the other kinds of inspections are scheduled as needed.

4. Should GLP investigators comment on the scientific merits of a protocol or the scientific interpretation given in the final report?

No. Their function is strictly a noting of observations and verification. Scientific judgments are made by the respective headquarters review units that deal with the test article.

5. Can a GLP EIR be reviewed by laboratory management prior to issuance?

No. The GLP EIR is an internal agency document which reflects the observations and findings of the FDA investigator. It cannot be released to anyone outside the agency until agency action has been completed and the released copy is purged of all trade secret information. Laboratories that disagree with portions of the EIR should write a letter which contains the area of disagreement to the local FDA District office. The laboratories can ask that their letters accompany the EIR whenever it is requested under the Freedom of Information Act.

6. Can FDA investigators take photographs of objectionable practices and conditions?

It is the agency position that photographs can be taken as a part of the inspection and this position has been sustained by a District Court decision.

7. The GLP Compliance Program requires the FDA investigator to select an ongoing study in order to inspect current laboratory operations. What criteria are used to select the study?

The studies are selected in accord with agency priorities, i.e., the longest term study on the most significant product.

8. Does FDA inspect international nonclinical laboratories once every two years?

No. Overseas laboratories are scheduled for inspection on the basis of having submitted to FDA the results of significant studies on important products.

9. What background materials are used by agency investigators to prepare for a GLP inspection?

Prior to inspection, the following materials are usually reviewed:

(a) The GLP regulations;

(b) The Management Briefings Post-Conference Report;

(c) Assorted memoranda and policy issuances;

(d) The GLP Compliance Program;

(e) The protocol of an ongoing study, if available;

(f) The final report of a completed study, if available;

(g) The inspection report of the most recent inspection.

10. How long does FDA allow a laboratory to effect corrective actions after an inspection has been made?

If the results of an inspection reveal that significant deviations from the GLPs exist, the laboratory will be sent a regulatory letter that lists the major deviations and that requests a response with 10 days. The response should describe those actions that the laboratory has taken or plans to take to effect correction. The response should also encompass items that were listed on the FD-483 and those that were discussed during the exit discussion with laboratory management. A specific time table should be given for accomplishing the planned actions. The reasonableness of the time table will be determined by FDA compliance staff, based on the needs of the particular situation.

For less significant deviations, the laboratory will be sent a Notice of Adverse Findings letter that also lists the deviations but that requests a response within 30 days. Again, the reasonableness of the response will be determined by FDA staff.

11. Does a laboratory's responsibility for corrective action listed on a FD-483 begin at the conclusion of an inspection or upon receipt of correspondence from the originating bureau in which corrective action is requested?

The FD-483 lists observations of violative conditions that have the capacity to adversely affect nonclinical laboratory studies. Corrective actions should be instituted as soon as possible.

12. Does FDA preannounce all GLP inspections?

Laboratory management is informed of all routine GLP inspections prior to the inspection, but special compliance or investigative inspections need not be preannounced.

SUBPART B: ORGANIZATION AND PERSONNEL

Section 58.29   Personnel

1.   For what sequence in the supervisory chain should position descriptions be available?

Position descriptions should be available for each individual engaged in or supervising the conducts of the study.

2.   Should current summaries of training and experience list attendance at scientific and technical meetings?

Yes. The agency considers such attendance as a valuable adjunct to the other kinds of training received by laboratory personnel.

3.   If certain specialists (pathologists, statisticians, opthalmologists, etc.) are contracted to conduct certain aspects of a study, need they be identified in the final report?

Yes.

4.   Does the QAU have to be composed of technical personnel?

No. Management is, however, responsible for assuring that "personnel clearly understand the functions they are to perform" (Section 58.31(f)) and that each individual engaged in the study has the appropriate combination of education, training and experience (Section 58.29(a)).

Section 58.31   Testing Facility Management

5.   Can the study director be the chief executive of a nonclinical laboratory?

No. The GLPs require that there be a separation of function between the study director and the QAU director. In the example, the QAU director would be reporting to the study director.

Section 58.33   Study Director

1.   The GLPs permit the designation of an "acting" or "deputy" study director to be responsible for a study when the study director is on leave. Should study records identify the designated "deputy" or "acting" study director?

Yes.

---

2. Is the study director responsible for adherence to the GLPs?

Yes.

Section 58.35 Quality Assurance Unit

1. As a QAU person, I have no expertise in the field of pathology. How do I audit pathology findings?

   The QAU is not expected to perform a scientific evaluation of a study nor to "second-guess" the scientific procedures that are used. QAU inspections are made to ensure that the GLPs, SOPs and protocols are being followed and that the data summarized in the final report accurately reflect the results of the study. A variety of procedures can be used to do this but certainly the procedures should include an examination and correlation of the raw data records.

2. Must the QAU keep copies of all protocols and amendments and SOPs and amendments?

   The QAU must keep copies of all protocols as currently amended. The only SOPs that the QAU are required to keep are those concerned with the operations and procedures of the QAU.

3. Does the QAU have to monitor compliance with regulations promulgated by other government agencies?

   The GLPs do not require this.

4. Can an individual who is involved in a nonclinical laboratory study perform QAU functions for portions of the study that the individual is not involved with?

   No. However, the individual can perform QAU functions for a study that he/she is not involved with.

5. Does the QAU review amendments to the final report?

   Yes.

6. What studies are required to be listed on the master schedule sheet?

   The master schedule sheet should list all nonclinical laboratory studies conducted on FDA regulated products and intended to support an application for a research or marketing permit.

7. May the QAU in its periodic reports to management and the study director recommend actions to solve existing problems?

Yes.

8. If raw data are transcribed and sent to the sponsor for (a) preparing the data in computer format or (b) performing a statistical analysis, what are the responsibilities of the QAU?

For (a) the QAU should assure that the computer formatted data accurately reflect the raw data. For (b) the statistical analyses would comprise a report from a participating scientist, therefore it should be checked by QAU and appended to the final report.

9. Can the QAU also be responsible for maintaining the laboratory archives?

Yes.

10. Can a QAU be constituted as a single person?

Yes, provided that the workload is not excessive and other duties do not prevent the person from doing an adequate job. It would be prudent to designate an alternate in case of disability/vacations/etc.

11. Who is responsible for defining study phases and designating critical study phases and can these be covered in the SOP?

The GLPs do not isolate this responsibility. Logically, the task should be done by the study director and the participating scientists working in concert with the QAU and laboratory management. It can be covered by an SOP.

SUBPART C: FACILITIES

Section 58.41   General

No questions were asked on the subject.

Section 58.43   Animal Care Facilities

1. Do the GLPs require clean/dirty separation for the animal care areas?

    No. They do require adequate separation of species and studies.

2. Do the GLPs require that separate animal rooms be used to house test systems and conduct different studies?

    No. The GLPs require separate areas adequate to assure proper separation of test systems, isolation of individual projects, animal quarantine and routine or specialized housing of animals, as necessary to achieve the study objectives.

3. Do the GLPs require that access to animal rooms be limited only to authorized individuals?

    No. However, undue stresses and potentially adverse influences on the test system should be minimized.

Section 58.45   Animal Supply Facilities

No questions were asked on the subject.

Section 58.47   Facilities for Handling Test and Control Articles

1. Do test and control articles have to be maintained in locked storage units?

    No. But accurate records of test and control article accountability must be maintained.

Section 58.49   Laboratory Operation Areas

No questions were asked on the subject.

Section 58.51   Specimen and Data Storage Facilities

1. What do the GLPs require with regard to facilities for the archives?

    Space should be provided for archives limited to access by authorized personnel. Storage conditions should minimize deterioration of documents and specimens.

Section 58.53   Administration and Personnel Facilities

No questions were asked on the subject.

SUBPART D: EQUIPMENT

Section 58.61   Equipment Design

No questions were asked on the subject.

Section 58.63   Maintenance and Calibration of Equipment

1. Has FDA established guidelines for the frequency of calibration of equipment (balances) used in nonclinical laboratory studies?

    The agency has not established guidelines for the frequency of calibration of balances used in nonclinical laboratory studies. This would be a large undertaking in part due to the wide variety of equipment that is available and to the differing workloads that would be imposed on the equipment. It is suggested that you work with the equipment manufacturers and your study directors to arrive at a suitable calibration schedule. The key point is that the calibration should be frequent enough to assure data validity. The maintenance and calibration schedules should be part of the SOPs for each instrument.

2. When an equipment manufacturer performs the routine equipment maintenance, do the equipment manufacturer's maintenance procedures have to be described in the facilities' SOPs?

    No. The facilities' SOPs would have to state that maintenance was being performed by the equipment manufacturer according to their own procedures.

SUBPART E: TESTING FACILITIES OPERATION

Section 58.51 Standard Operating Procedures

1. What amount of detail should be included in the standard operating procedures (SOPS)?

    The GLPs do not specify the amount of detail to be included in the SOPs. The SOPs are intended to minimize the introduction of systematic error into a study by ensuring that all personnel will be familiar with and use the same procedures. The adequacy of the SOPs is a key responsibility of management. A guideline of adequacy that could be used is to determine whether the SOPs are understood and can be followed by trained laboratory personnel.

2. Can the study director authorize changes in the SOPs?

    No. Approval of the SOPs and changes thereto is a function of laboratory management.

3. How many copies of the complete laboratory SOPs are needed?

    Each work station should have access to the SOPs applicable to the work performed at the station. A complete set of the SOPs, including authorized amendments, should be maintained in the archives.

4. Who approves the SOPs of the Quality Assurance Unit?

    Laboratory Management.

5. To what extent are computer programs to be documented as SOPs?

    The GLPs do not specify the contents of individual SOPs, but the SOP that deals with computerized data acquisition should include the purpose of the program, the specifications, the procedures, the end products, the language, the interactions with other programs, procedures for assuring authorized data entry and access, procedures for making and authorized changes to the program the source listing of the program and perhaps even a flow chart. The laboratory's computer specialists should determine what other characteristics need to be described in the SOP.

Section 58.83 Reagents and Solutions

1. What are the GLP requirements for labeling of reagents purchased directly from manufacturers?

    All reagents used in a nonclinical laboratory have to be labeled to indicate identify, titer or concentration, storage requirements, and expiration date. Purchased reagents usually carry all these items except for the expiration date, so the laboratory should label the reagent containers with an expiration date. The expiration date selected should be in line with laboratory experience and need not require specific stability testing.

2. How extensive should the procedures be for confirming the quality of incoming reagents used in nonclinical laboratory studies?

   Laboratory management should make this decision but the SOPs should document the actual procedures used.

3. Do the procedures used for preparing the S9 activator fraction (liver microsomal fraction from rats challenged with a toxin) have to be performed in accord with the GLPs?

   No. The GLPs consider the S9 activator fraction to be a reagent. Therefore, it must be labeled properly, stored properly, testing prior to use in accord with adequate SOPs, and it cannot be used if its potency is below established specifications.

4. Do the GLPs require the use of product accountability procedures for reagents and chemicals used in a nonclinical laboratory study?

   No.

Section 58.90   Animal Care

1. Can diseased animals received from a supplier be diagnosed, treated, certified "well" and then entered into a nonclinical laboratory study?

   The GLPs provide for this procedure by including provisions directed towards animal quarantine for isolation. The question of whether such animals can be entered into a study, however, is a scientific one that should be answered by the veterinarian-in-charge and the study director and other scientists involved in the study.

2. Do the GLPs prohibit the use of primates for multiple nonclinical laboratory studies?

   No. Again, the question is a scientific one and potential impact of multiple use on study interpretation should be carefully assessed.

3. Is a photocopy of an animal purchase order which has been signed and dated by the individual receiving the shipment sufficient proof of animal receipt?

   Yes, but actual shipping tickets are also acceptable.

4. Does FDA have guidelines for animal bedding?

   No, but the GLPs prohibit the use of bedding which can interfere with the objectives of the study.

GLP Audit Manual ©2000, Interpharm Press

5. Does FDA permit the sterilization of animal feed with ethylene oxide?

   No.

6. For certain test systems (timed-pregnant rodents), it is not possible to use long quarantine periods. Do the GLPs specify quarantine periods for each test system?

   No. The quarantine period can be established by the veterinary in charge of animal care and should be of sufficient length to permit evaluation of health status.

7. How are feed and water contaminants to be dealt with?

   The protocol should include a positive statement as to the need for conducting feed analysis for contaminants. If analysis is necessary, the identities and specifications for the contaminants should be listed. The need for analysis as well as the specifications should be determined by the study scientists. Water contaminants can be handled similarly.

8. How is the adequacy of bedding materials to be handled?

   This can be handled as are the analyses for possible contaminants in feed and water. The study director and associated scientists should consider the bedding and its possible impact on the study. The results of this consideration should appear in the protocol.

9. What do the GLPs require in regard to assuring the genetic quality of animals used in a nonclinical laboratory study?

   This is a scientific issue that is not specifically addressed by the GLPs. Suitability of the test system for use in a study is a protocol matter and any required testing procedures should be arrived at by the study scientists.

10. Do the GLPs require specific procedures for the microbiological monitoring of animals used in nonclinical laboratory studies?

    The procedures used should be in accord with acceptable veterinary medical practice.

11. The Japanese are preparing animal guidelines which are similar but not identical to the U.S. guidelines prepared by NIH. Would these be acceptable?

    Japanese guidelines that are similar, but no less stringent, in the important particulars with the NIH guidelines would be acceptable to FDA.

12. What is the frequency of feed contaminant analysis?

    If contaminant analyses are required by the protocol, then the GLPs require periodic analysis of the feed to ensure that the contaminant level is at or below that judged to be acceptable. Statistical procedures should be used to determine the frequency of analysis since this is dependent on the specific chemical characteristics of the interfering contaminant.

13. Is it necessary to use "official" methods of analysis to determine the levels of interfering contaminants?

No. The methods should be appropriate for the analysis and FDA reserves the right to examine the raw data supporting the analytical results.

14. Do the GLPs require production facilities to be dedicated to the manufacture of specific animal feeds used in nonclinical laboratory studies?

No.

15. Is a separate room required for animal necropsy?

No. the GLPs require separate areas and/or rooms as necessary to prevent any activity from having an adverse effect on the study. If the necropsy is done in an animal room, precautions should be taken to minimize disturbances that may interfere with the study.

SUBPART F: TEST AND CONTROL ARTICLES

Section 58.105   Test and Control Article Characterization

1.   Is it necessary to retain samples of feed from nonclinical laboratory studies in which the feed serves as the control article?

   Yes. It is not necessary, however, to retain reserve samples of feed from studies that involve test article administration by routes other than feed.

2.   What expiration date is placed on the label of test articles whose stability is being assessed concurrently with the conduct of the study?

   In this situation, the stability of the test article is unknown, but periodic analysis data exist. The label should contain a statement such as "see protocol" or "see" periodic analysis results" so that test article users will know that current analytical data should be examined prior to continued use of the test article.

3.   If analysis of the reserve samples is required by the Study Director or the QAU, is it permitted?

   Yes, but sufficient reserve sample should be retained so that the sample is not exhausted.

4.   Are physical and chemical tests conducted on test articles required to be done under the GLPs?

   According to section 58.105, such tests conducted to characterize the specific batch of test article used in the nonclinical laboratory study are covered.

Section 58.107   Test and Control Article Handling

1.   With regard to safety studies in large animals (cattle, horses, etc.), must test article accountability be maintained and can the animals be used for food purposes?

   Test article accountability must be maintained. For guidance on whether the treated animals can be used for food, you should contact the appropriate individuals in the Bureau of Veterinary Medicine.

Section 58.113   Mixtures of Articles with Carriers

1.   Do the GLPs require tests for homogeneity, concentration, and stability on mixtures of control articles used as positive controls?

   Yes.

---

2. Do test or control article concentration assay have to be performed on each batch of test or control article carrier mixture?

No. The GLPs require only periodic analysis of test or control article carrier mixtures.

3. What is the purpose of periodic analysis requirement for test or control article mixtures?

This requirement provides additional assurance that the test system is being exposed to protocol-specified quantities of test article. Whereas, in most instances proper assurance is obtained through adequate uniformity-of-mixing studies, adequate SOPs, and trained personnel, occasionally the mixing equipment can malfunction or other uncontrollable events can occur which lead to improper dosages. These events can be recognized through periodic analysis.

4. For acute studies, does the test article carrier mixture have to be analyzed (single dose studies)?

Yes, but the analysis need not be done prior to the study provided the mixture is stable in storage.

5. For liquid dosing studies where the test article mixture is made by dilution of the highest dose, which dose should be analyzed?

The lowest dose would be appropriate since it would confirm the efficacy of the dilution process, however, the GLPs do not prohibit the analysis of any of the other doses.

6. Do homogeneity studies need to be done on solutions and suspensions of test articles used in acute nonclinical laboratory studies?

The answers to these questions are yes for suspensions of test articles and no for true solutions of test articles.

7. The analysis of test article mixtures that are used in acute studies is problematic. Usually at the stage of product development, the analytical method is not fully developed. Also, getting the analytical department to schedule the analysis is difficult. Stability is not a problem since fresh solutions are used. In view of the fact that acute studies are not pivotal in gaining approval of a research or marketing permit, is it necessary to analyze test article mixtures?

Yes. Although acute studies may be of lesser importance in assessing the safety of human drugs, they are important for animal drugs, biological products and certain food additives. For this reason, there must be some assurance that the test system was dosed with protocol specified quantities of test article. The GLPs do not require that the analysis be done prior to the use of the test article mixture provided that the mixture is stable on storage.

 GLP Audit Manual ©2000, Interpharm Press

SUBPART G: PROTOCOL FOR THE CONDUCT
OF A NONCLINICAL LABORATORY STUDY

Section 58.120   Protocol

1.   What are the proposed starting and completion dates for a nonclinical laboratory study?

There is a good deal of confusion on these dates and proper interpretation impacts on several GLP areas. Accordingly, the following clarification is offered: At the time of protocol development, the study director is to propose to management the approximate time frame of the study. Section 58.120(a)(4), therefore, requires that the protocol contain the proposed starting and completion date of the study. These dates are somewhat discretionary provided that they are identified in the protocol. Suitable identification can be the date of first dosing of the test system to the date of last dosing, the date of allocation of the test system to the experimental units to the date of necropsy of the last animal on test, the date of receipt of the test system to the date of final histopathological examination, or any combination of these or any other logical starting and completion dates. After this, the protocol is signed by the study director and forwarded for approval to management. Management approves, if indicated, signs and dates and at this point the study becomes a regulated study and must be entered on the Master Schedule Sheet. The study is carried on the Master Scheduled Sheet until the study director submits a signed and dated final report. Thus, for Master Schedule Sheet purposes, the starting date of the study is the date of protocol approval by management and the completion date of the study is the date of signature of the final report by the study director. Neither of the foregoing time frames need to be used to define the study terms described in section 58.35(b)(3) and section 58.105(d). For these sections, the traditional terms found in the toxicology literature may be used.

2.   Must an analytical method be totally contained in the protocol?

No. The protocol must state the type and frequency of tests to be made. Type can be connoted by reference to literature citations or the SOPs as applicable.

3.   Does each nonclinical laboratory study require a sponsor-approved specific protocol?

Yes. However, the laboratory that conducts the study can also qualify as the sponsor of the study.

4.   Do unforeseen circumstances which occur during a study and which necessitate minor operational changes have to be reported as protocol amendments?

Unforeseen circumstances which have only a one time effect (different date of sample collection, animal weighings) need to be reported only in the raw data and the final report.

However, such circumstances which result in a systematic change, e.g., in the SOPs or in the protocol, should also be made by a protocol amendment. The protocol amendment need not be made in advance but should be made as rapidly as possible.

5. Pathologists at a firm would like to take tissues from animals in a nonclinical study which would be used to conduct exploratory research studies. The tissues would not be part of the nonclinical laboratory study design and the results would not necessarily pertain to the study objectives. What would the GLPs require in this case?

The protocol should state that tissues are to be taken from the experimental animals and that the tissues would be used for exploratory research purposes. If any effects were observed in the exploratory research studies which would influence the interpretation of the results of the nonclinical laboratory study, these effects must be reported in the final report.

6. Does the protocol have to list the SOPs used in a specific study?

The protocol must list the type and frequency of tests, analyses and measurements to be made in the study. Where these are covered by SOPs, they should be listed in the protocol.

7. Do the GLPs require that absorption studies be done on each test article?

No. The GLPs require that, if absorption studies are needed to achieve the scientific objectives of the study plan, the protocol should describe the methods to be used to determine absorption. Whether or not absorption studies are required is a scientific issue to be decided by the study scientists.

8. Who assesses protocol validity (No. of animals, test article dosage, test system, etc.)?

This is done by the study scientists using the scientific literature, published guidelines, advice from regulatory agencies, and prior experimental work.

Section 58.130   Conduct of a Nonclinical Laboratory Study

1. Does raw data collected in nonclinical laboratory studies have to be cosigned by a second individual?

No.

2. What are the GLP requirements that are applicable to computerized data-acquisition systems?

An acceptable system must satisfy the following criteria:

(1) Only authorized individuals can make data entries,

(2) data entries may not be deleted, but changes may be made in the form of dated amendments which provide the reason for data change,

(3) the data base must be made as tamperproof as possible,

(4) the SOPs should describe the procedures used for ensuring the validity of the data, and

(5) either the magnetic media or hard-copy printouts are considered to be raw data.

3. In Japan, employees do not sign raw data records but rather they use an official seal which is unique to the employee. Is this an acceptable procedure?

Yes.

4. Do tissue slides have to carry the complete sample labeling information stated in the GLPs?

No, accession numbers are permitted providing that these numbers can be translated into the information required under Section 58.130(c).

5. Is a positive notation (a statement of what was done in the raw data) required for routine laboratory operations such as:

(a) identifying animals,

(b) shaving or abrading rabbits,

(c) specific dosing procedures, and

(d) fasting of animals?

Yes.

6. Do the GLPs require the entry of raw data into bound notebooks?

No.

7. Is it acceptable to manually transcribe raw data into notebooks if it is verified accurate by signature and date?

Technically the GLPs do not preclude such an approach. It is not a preferred procedure, however, since the chance of transcription errors would exist. Accordingly, such an approach should be used only when necessary and in this event the raw data should also be retained.

SUBPART J: RECORDS AND REPORTS

Section 58.185   Reporting of Nonclinical Laboratory Study Results

1. Do contributing scientist's reports have to be prepared and appended to final reports or can the contributing scientist's report be included in the final report prepared by the study director and signed by each contributing scientist?

    The signed reports of contributing scientists should be appended to the final report.

2. Does Section 58.115(a) describe the format for submission of a final report?

    The cited section describes the information that has to be submitted in a final report but the specific format is left up to the laboratory.

3. Do all circumstances that may have affected the quality of the data have to be described in the final report?

    Yes.

4. Who approves the final report of a nonclinical laboratory study?

    The GLPs do not address the issue of approval of the final report. According to the GLPs, the final report is official when it is signed and dated by the study director. If persons reviewing the final report request changes, then such changes must be made by way of a formal amendment.

5. Can the chemistry information required by Section 58.185(a)(4) be located elsewhere in the application for a research marketing permit?

    Yes. The final report should, however, reference the location of the chemistry information.

6. Does everyone who participated in a study have to be identified in the final report?

    No. The final report need identify only the name of the study director, the names of other participating scientists, and the names of all supervisory personnel.

7. Does the phase of the study which has been inspected need to be identified in the QAU statement in the final report?

    No.

8. How are protocol deviations which are discovered after the completion of the study to be handled?

    The deviations should be described in the final report and in the study records.

9. How does the agency view interim reports of nonclinical laboratory studies?

    Interim reports are to be treated the same as final reports, i.e. they are to be reviewed by the QAU so that the summarized data accurately reflects the raw data.

Section 58.190 Storage and Retrieval of Records and Data

1. Certain raw data records are not study specific (pest control, instrument calibration). Must these be filed in the archives in each study file?

    No. These can be filed in a retrievable fashion such as chronological in the archive.

2. Where should the QAU records be retained?

    At the completion of a study, QAU records and inspection reports should be retained in the archives.

3. At the termination of a nonclinical laboratory study, can a contractor send all of the new raw data, study records, and specimens to the sponsor of the study?

    The regulations do not specifically address this issue. Section 58.195(g) requires contract laboratories that go out of business to transfer all raw data and records to the sponsor. Likewise, Section 58.190(b) permits raw data and study records to be stored elsewhere (other than the contract laboratory location) provided that the contract laboratory's archives have reference to the other locations and provided that the final study report identifies the other locations as directed by Section 58.195(a)(13).

    Consequently, it is permissible for the sponsor to retain all raw data and records from the date of termination of the nonclinical laboratory study. Common sense dictates, however, that the contract laboratory keep copies of the material that has been forwarded to the sponsor.

4. Can a study director or a pathologist be responsible for storing and retaining specimens and raw data?

    Yes, the GLPs permit multiple archival locations provided that these locations are identified in the central archives and that they provide adequate storage conditions and authorized access features.

1. With regard to blood and urine specimens which are analyzed for both labile and stable constituents, is it necessary to retain the specimen until the most stable constituent deteriorates?

    All specimens should be retained for the term required by the regulations or for as long as their quality permits meaningful reevaluation, whichever is shorter.

2. For a GLP regulated metabolism study, whole tissues are homogenized and aliquots thereof are used for analysis. Is it necessary to retain all of the remaining homogenate as a reserve sample?

    No, it is only necessary to retain a representative sample large enough to repeat the original measurements.

3. If animals used in acute studies are subjected to necropsy, is it necessary to retain the organs as study specimens?

    Yes.

CONFORMING AMENDMENTS

1. Do acute studies not done in conformity with the GLPs have to be identified in the conforming amendment statement?

   Yes.

2. How extensive should the conforming amendment statement be for preliminary exploratory studies that are exempt from GLP coverage?

   The statement should be brief and indicate the GLP-exempt status of the study.

3. For contracted nonclinical laboratory studies, who is responsible for preparing the GLP compliance statement required by the conforming amendments?

   The preparation of the conforming amendment statement is the responsibility of the product sponsor and the statement should be submitted as part of the application for a research or marketing permit. The contractor, however, should identify for the sponsor those non GLP practices which were used in each nonclinical laboratory study so that a proper conforming amendment statement can be prepared.

4. Who signs the conforming amendment statement?

   This can be done by the same individual in the firm who signs the official application for a research or marketing permit.

5. Is a specific conforming amendment statement as required by Part 314(f)(7) to be prepared for each nonclinical laboratory study?

   Yes. GLP deviations have to be identified for all nonclinical laboratory studies. This can be done by preparing a single comprehensive official filing. The conforming amendment statement in the official filing should be located in proximity to the animal safety studies section.

1. Have any nonclinical laboratories been disqualified since June 20, 1979?

   No.

2. Does FDA reject nonclinical laboratory studies that have not been conducted in full compliance with the GLPs?

   Not necessarily. The GLP Compliance Program provides guidance on the issue. For FDA to reject a study, it is necessary to find that there were deviations from the GLPs and that those deviations were of such a nature as to compromise the quality and integrity of the study covered by the agency inspection.

3. Must copies of the SOPs be submitted along with an application for a research or marketing permit?

   No.

4. What should be done about nonclinical laboratory studies that are stopped prior to completion?

   The agency recognizes that a variety of circumstances (disease outbreak, power failures, etc.) can lead to the premature termination of a nonclinical laboratory study. In these cases, a short final report should be prepared that describes the reasons for study termination.

5. Has the agency established permissible limits for environmental controls (temperature, humidity and lighting) for the animal facilities?

   No, these are scientific matters that should be described in the protocol and/or the SOPs. Of course, accurate records should be maintained.

FOOD AND DRUG ADMINISTRATION
COMPLIANCE PROGRAM GUIDANCE MANUAL

CHAPTER 48 — BIORESEARCH MONITORING

SUBJECT:	IMPLEMENTATION DATE
GOOD LABORATORY PRACTICE (Nonclinical Laboratories)	OCTOBER 1, 1997
	COMPLETION DATE
	SEPTEMBER 30, 2000

DATA REPORTING	
PRODUCT CODES	PRODUCT/ASSIGNMENT CODES
51Z OR 52Z 45Z, 46Z 57Z, 99Z	04808 Chemical Contaminants (EPA) 09808 Food Additives 41808 Biologics (Therapeutics) 42808 Biologics (Blood) 45808 Biologics (Vaccines)
60Z, 61Z 68Z, 69Z 73Z, 74Z 94Z OR 95Z	48808 Human Drugs 68808 Animal Drugs 83808 Medical and Radiological Devices

SAMPLES — Appropriate 7 digit code for product collected

* Current Change

DATE OF ISSUANCE:
TRANSMITTAL NO.   TN 98-22   08/17/98
FORM FDA 2438 (9/98)

FIELD REPORTING REQUIREMENTS

A. **To Headquarters:** All establishment inspection reports (EIRs), complete with attachments, exhibits, and any related post-inspection correspondence should be submitted promptly to the assigning Center.

A copy of an EIR containing findings of such a serious nature that they raise the possibility of one or more violations of the FD&C Act or other federal statutes should be forwarded to the district compliance branch at the time the EIR is sent to the Center.

The District should immediately notify the responsible Center program office via electronic mail system, fax, or by phone of any significant adverse inspectional or analytical findings or other information which may affect the agency's new product approval decisions. *In these instances, a copy of the FDA 483 shall be faxed to the Center contact identified in the assignment.*

B. **To Other Districts:** The analyzing laboratory will submit copies of the Collection Report and the Analytical Work sheets to the assignment's originating district office and the assigning Center.

PART I—BACKGROUND

Under the laws administered by the Food and Drug Administration (FDA), industry is required to submit evidence to ensure the safety of a broad spectrum of FDA-regulated products prior to their marketing and/or clinical testing in humans and animals. Products that require evidence of safety include the following: human and animal drugs, human biological products, medical devices and diagnostic products, food additives, color additives, and electronic products.

Extensive animal and other types of nonclinical testing are required to provide evidence of safety. The product sponsor is responsible for conducting and reporting the safety tests necessary for demonstrating the safety of its product.

FDA uses the data from these nonclinical safety tests to answer the following types of questions:

- What risks may be associated with further testing in clinical studies involving humans or animals?

- What adverse effects are likely?

- What is the fundamental toxicity profile?

- What are the potential teratogenic, carcinogenic, or other degenerative effects?

- What is the no-effect dose in the test system?

- What level of use in foods can be safely supported?

The importance of nonclinical laboratory studies to FDA decisions demands that all safety studies be conducted according to scientifically-sound protocols and with detailed attention to quality control. In the mid 1970s FDA inspections of several nonclinical safety testing laboratories and audits of study data revealed serious problems with the conduct of safety studies and the quality and integrity of the data derived from them. As a result of these findings, FDA published the Good Laboratory Practice (GLP) Regulations, 21 CFR 58, which became effective June 1979. The regulations establish basic standards for conducting and reporting nonclinical safety testing and are intended to assure the quality and integrity of safety data submitted to FDA.

FDA relies heavily on documented adherence to GLP requirements by nonclinical testing laboratories in judging the acceptability of safety data submitted in support of research and/or marketing permits. FDA has implemented this program of regular inspections and data audits to monitor laboratory compliance with the GLP requirements.

* Current Change

DATE OF ISSUANCE:
TRANSMITTAL NO.   TN 98-22   08/17/98
FORM FDA 2438 (9/98)

PART II—IMPLEMENTATION

OBJECTIVES

- To verify the quality and integrity of submitted data;

- To inspect nonclinical laboratories engaging in safety studies that are intended to support applications for research or marketing permits for regulated products, approximately every two years; and

- To audit ongoing and completed laboratory safety studies and determine degree of compliance with GLP regulations (21 CFR 58).

A. Program Management Instructions

This program provides for the inspection of public, private, and government nonclinical laboratories which may be performing tests on food and color additives, generally recognized as safe substances, human and animal drugs, human medical devices, biological products, pesticides, disinfectants, and electronic products.

Assignments for the initial GLP inspection of a specific university department or government facility are to be initiated only after the facility has been informed, by a letter from the *Bioresearch Monitoring Program Coordinator, Division of Compliance Policy (HFC-230),* Office of Enforcement, of our intent to inspect. These assignments will include a data audit of a completed study which has been submitted to the agency. Notification of subsequent inspections of these laboratory facilities is not necessary.

Subcontractor and off-site laboratories are covered by the GLPs to the degree that they provide data for a nonclinical laboratory study. They are only required to be in compliance with those parts of the GLPs that are appropriate to the nature of their contributions to a study.

B. Types of Inspections

1. Surveillance Inspections

Surveillance inspections are periodic, routine determinations of a laboratory's compliance with GLP regulations utilizing studies in progress and/or studies recently completed.

Current Change

DATE OF ISSUANCE:
TRANSMITTAL NO.   TN 98-22   08/17/98
FORM FDA 2438 (9/98)

2. Directed Inspections

Directed inspections are assigned by the FDA Centers to achieve a specific purpose such as the following:

a. verifying the reliability, integrity, and compliance of important or critical safety studies currently being reviewed in support of pending applications for product research or pre-marketing approval;

b. investigating issues involving potentially unreliable safety data and/or violative studies brought to FDA's attention in reports from whistle-blowers, news media, industry complaints, FDA reviewers, other government contacts, or from other sources;

c. reinspecting laboratories classified OAI, usually within 6 months after the firm responds to a Warning Letter; and

d. verifying the results from third party audits or sponsor audits submitted to FDA for consideration in deciding whether to accept or reject questionable or suspect studies.

C. Inspection Teams (FDA Personnel)

Team inspections may be conducted.

1. Team Leader

A field investigator will serve as team leader and is operationally responsible for the conduct of the inspection. Responsibilities of the team leader are explained fully in Investigations Operations Manual (IOM)502.4.

2. Field Analyst/Headquarters Participant

Field analyst/headquarters personnel will serve as scientific or technical support to the team leader and shall participate in the inspection by:

a. attending pre-inspection conferences when/if scheduled;

b. participating in the entire on-site inspection if permitted by agency priorities; and

c. providing support, when requested by the team leader, in the preparation of specific sections of the inspection report, including portions of the report where the headquarters personnel's expertise is especially useful.

---

* Current Change

DATE OF ISSUANCE:
TRANSMITTAL NO.   TN 98-22   08/17/98
FORM FDA 2438 (9/98)

Any difficulties involving field analysts or headquarters participation in the inspection should be discussed with District management and, if not resolved, immediately referred to the Office of Regional Operations (ORO) contact for this program.

D. Joint Inspections with Other Government Agency Personnel

Joint inspections with other government agency personnel may be conducted under this program. In all instances, the FDA field representative will function as the team leader. Specific instructions regarding inter-agency coordination, reporting procedures, etc., are contained in the program circular associated with this program.

E. Confirmation of Schedule

All inspections of commercial laboratories are to be conducted without prior notification unless otherwise instructed by the assigning Center. Arrangements to conduct the initial inspection of university and government laboratories should be made soon after the laboratory has received the Bioresearch Monitoring Program Coordinator's notification letter.

F. Field Responsibilities

1. Alert Center contact to facilities that are becoming operational under GLPs.

2. Notify the Center contact, identified in the assignment, of any significant delay or postponement of an inspection assignment, or when an inspection has been scheduled.*If possible, this notification should be made a week or more prior to the start of the inspection.*

Current Change

DATE OF ISSUANCE:
TRANSMITTAL NO.   TN 98-22   08/17/98
FORM FDA 2438 (9/98)

PART III—INSPECTIONAL

A. Operations

The investigator will determine the current state of GLP compliance by evaluating the laboratory facilities, operations, and study performance as outlined in Parts III, C and D of this program.

If the assignment does not identify a study for coverage or if the referenced study is not suitable to assess all portions of current GLP compliance, the investigator will select studies as necessary to evaluate all areas of laboratory operations. NOTE: Inspection and audit of studies performed for submission to other government agencies, Environmental Protection Agency, National Toxicology Program, National Cancer Institute, etc., will not be conducted without specific headquarters (HFC-230,)authorization. However, this authorization is not necessary to briefly look at one of these studies to assess the ongoing operations of a portion of the facility. When additional studies are selected, first priority should be given to FDA studies for submission to the assigning Center.

Examine and collect the firm's master schedule for all studies listed since the last GLP inspection or last two years and select studies as defined in 21 CFR 58.3(d). If this is the initial inspection of the firm, review the entire master schedule. If there are studies identified as non-GLP by the firm, determine the nature of several of these studies to verify the accuracy of this designation. See 21 CFR 58.1 and 58.3(d). In contract laboratories determine who decides if a study is a GLP study.

Ongoing and completed studies should be selected using the following criteria:

- Safety studies conducted on FDA-regulated products that have been initiated or completed since the last GLP inspection; *or*

- Safety studies that encompass the full scope of laboratory operations; *or*

- Studies that are significant to safety assessment, e.g. carcinogenicity, reproduction, chronic toxicity studies; *or*

- *Studies that encompass operations for several species of animals.*

The investigator is encouraged to contact the Center for guidance on study selection.

1. **Ongoing Studies** — Obtain a copy of the study protocol and determine the schedule of activities that will be underway during the inspection. This information should be used to schedule inspections of ongoing laboratory operations, as well as equipment and facilities associated

* Current Change

with the study. If there are no activities underway in a given area for the study selected, evaluate the area based on ongoing activities. Ongoing studies are good sources for obtaining samples of test article/carrier mixtures.

2. Completed Studies — The data audit should be carried out as outlined in Part III, D. If possible, accompany laboratory personnel when they retrieve the study data to assess the adequacy of data retention, storage, and retrieval as described in Part III, C 10.

B. Reporting

The EIR should contain the headings as prescribed in the IOM in addition to the specific headings outlined in Parts III, C and D below. In areas where no deficiencies were observed, the operations inspected must be identified but need not be described in detail. Any adverse findings should be fully explained and documented in the EIR.

Because the Centers are solely responsible for issuing post- inspection correspondence, each EIR must include the full name, title, academic degree(s), and mailing address of the most responsible person *(For universities the most responsible individual would typically be a member of theadministration, e.g., department chair, dean)*. An FDA 483 listing adverse inspectional observations will be issued under this program. Findings should not be listed on the FDA 483 under the following circumstances:

1. Findings are deviations that have been observed and corrected by the firm through its internal audit procedures; and

2. Findings are minor and constitute one-time occurrences which have no impact on the firm's operations and/or study conduct.

Findings which are not considered significant enough to be listed on the FDA 483 may be discussed with the firm's management. Such discussions must be reported in the EIR.

Documentation necessary to support the observations listed on the FDA 483 should be collected, submitted, and referenced in the EIR narrative. Suspected violations under Title 21 must be documented sufficiently to form the basis for legal or administrative action. Violations under Title 18 require extensive documentation. Discuss the situation with your supervisor prior to embarking on this type of coverage.

C. Establishment Inspections

The facility inspection should be guided by the GLP regulations. The following areas should be evaluated and described as appropriate:

1. **Organization and Personnel (21 CFR 58.29, 58.31, 58.33)**

   Purpose: To determine whether the organizational structure is appropriate to ensure that studies are conducted in compliance with GLP regulations, and to determine whether management, study directors, and laboratory personnel are fulfilling their responsibilities under the GLPs. Determine Study Director workload through review of the Master Schedule.

   a. **Management Responsibilities (21 CFR 58.31)** — Identify the various organizational units, their role in carrying out GLP study activities, and the intermediate and top level management responsible for these organizational units. This includes identifying personnel who are in a location other than that of the facility and identifying their line of authority. If there is a need for inspection of outside or contract facilities, the investigator should contact the assigning Center for guidance.

   Determine if management has procedures for assuring that the responsibilities in 58.31 can be carried out. Look for evidence of management involvement or lack thereof, in the areas listed:

   – Assignment of study directors;

   – Establishment and support of the Quality Assurance Unit (QAU), including assuring that deficiencies reported by the QAU are communicated to the study directors and acted upon;

   – *Assuring that test and control articles or mixtures are appropriately tested for identity, strength, purity, stability, and uniformity;

   – Assuring that all study personnel are informed of and follow any specific or special test and control article handling and storage procedures;*

   – Providing required study personnel, resources, facilities, equipment, and materials;

   – Providing QAU oversight;

   – Reviewing and approving protocols and standard operating procedures (SOPs); and

   – Providing GLP or technical training as needed.

   b. **Personnel (21 CFR 58.29)** — Review personnel records, policies, and operations to determine if:

---

* Current Change

- Summaries of training and job descriptions are maintained and current for selected employees engaged in the study;

- Personnel have been adequately trained to carry out the study functions that they perform;

- Personnel have been trained in GLPs; and

- Practices are in place to ensure that employees take necessary health precautions, wear appropriate clothing, and report illnesses so as to avoid contamination of the test and control articles and test systems.

The investigator should identify key laboratory and management personnel, including any outside consultants used, and review their curriculum vitae (CV), training records, and position descriptions. Facility SOPs covering training and personnel practices should also be reviewed.

If the firm has computerized operations, determine the following (See also attachment A):

- Who was involved in the design, development, and validation of the computer system;

- Who is responsible for the operation of the computer system, including inputs, processing, and output of data; and

- If computer system personnel have training commensurate with their responsibilities, including professional training and training in GLPs.

Interview and observe key study personnel to assess their training and performance of assigned duties.

c. **Study director (21 CFR 58.33)** — *Assess the extent of the study director's actual study involvement and participation. In those instances when the study director is located off-site, review any correspondence/records between the testing facility and the off-site study director that ensures that the study director is being kept immediately apprised of any problems that may affect the quality and integrity of the study.

In addition, assess the procedures by which the study director:*

- Develops and amends protocols and distributes them to all involved parties;

- Records, observes, and verifies the accuracy of study observations;

- Assures that data are collected according to protocol and SOPs;

- Documents unforeseen circumstances that may affect the quality and integrity of the study and implements corrective action;

Current Change

– Assures that study personnel are familiar with and adhere to the study protocol and SOPs; and

– Assures that study data are transferred to the archives.

d. **EIR Documentation and Reporting** — Collect exhibits to document deficiencies. This may include SOPs, organizational charts, position descriptions, and CVs, as well as study-related memos, records, and reports for the studies selected for review.

2. **QAU (21 CFR 58.35)**

Purpose: To determine if the firm has an effective, independent QAU which monitors significant study events and overall facility operations, reviews records and reports, and assures management of GLP compliance.

a. **QAU Operations** — **(21 CFR 58.35(b-d))** — Review QAU SOPs to ensure that they cover all methods and procedures for carrying out the required QAU functions, and confirm that they are being followed. Verify that SOPs exist and are being followed for QAU activities including, but not limited to, the following:

– Maintenance of a master schedule sheet;

– Maintenance of copies of all protocols and amendments;

– Scheduling of its in-process inspections and audits;

– Inspection of each nonclinical laboratory study at intervals adequate to assure the integrity of the study, and maintenance of records of each inspection;

– Notification to the study director and management of any problems which are likely to affect the integrity of the study;

– Submission of periodic status reports on each study to the study director and management;

– Review of the final study report;

– Preparation of a statement to be included in the final report which specifies the dates inspections were made and findings reported to management and to the study director; and

– Inspection of computer operations.

Verify that, for any given study, the QAU is entirely separate from and independent of the personnel engaged in the direction and conduct of that study.

---

* Current Change

Estimate the time QAU personnel spend in performing in-process inspection and final report audits. Determine if the time spent is sufficient to detect problems in critical study phases. Determine if there are an adequate number of personnel to perform the required functions.

NOTE: The investigator may request the firm's management to certify in writing that inspections are being implemented, performed, documented, and followed-up in accordance with this section [see 58.35(d)].

b. **EIR Documentation and Reporting** — Obtain a copy of the master schedule sheet dating from the last routine GLP inspection or covering the past two years.

If too voluminous, representative pages to allow for headquarters review may be obtained. When master schedule entries are coded, obtain the code key.

Deficiencies should be fully reported and documented in the EIR. Documentation to support deviations may include copies of QAU SOPs, list of QAU personnel, their CVs or position descriptions, study-related records, protocols, and final reports.

3. **Facilities (21 CFR 58.41 – 51)**

Purpose: Assess whether the facilities are of adequate size and design.

a. **Facility Inspection**

- Review environmental controls and monitoring procedures for critical areas (i.e., animal rooms, test article storage areas, laboratory areas, handling of bio-hazardous material, etc.) and determine if they appear adequate and are being followed.

- Review the SOPs that identify materials used for cleaning critical areas and equipment, and assess the facility's current cleanliness.

- Determine whether there are appropriate areas for the receipt, storage, mixing, and handling of the test and control articles.

- Determine whether separation is maintained in rooms where two or more functions requiring separation are performed.

- Determine that computerized operations and archived computer data are housed under appropriate environmental conditions.

b. **EIR Documentation and Reporting** — Identify which facilities, operations, SOPs, etc., were inspected. Only significant changes in the facility from previous inspections need to be described. Facility floor plans may be collected to illustrate problems or changes.

Current Change

Specify and document any overcrowding conditions which would lead to contamination of test articles or to unusual stress of test systems.

4. Equipment (21 CFR 58.61 – 63)

Purpose: To assess whether equipment is maintained and operated in a manner which ensures valid results.

a. **Equipment Inspection** — Check the following:

 – The general condition, cleanliness, and ease of maintenance of equipment in various parts of the facility;

 – The heating, ventilation, and air conditioning system design and maintenance, including documentation of filter changes and temperature/humidity monitoring in critical areas;

 – Whether equipment is located where it is used and, where required, that it is located in a controlled environment;

 – Records for cleaning and decontamination of non-dedicated equipment for preparation of different test and control article carrier mixtures, to verify that procedures are followed to prevent cross contamination;

 – For representative pieces of equipment check the availability of the following:

   o SOPS and/or operating manuals;

   o Maintenance schedule and log;

   o Standardization/calibration procedure, schedule, and log; and

   o Standards used for calibration and standardization.

   o For computer systems, check that the following procedures exist and are documented (See also attachment A):

 – Validation study, including validation plan and documentation of the plan's completion;

   o Maintenance of equipment, including storage capacity and back-up procedures;

   o Control measures over changes made to the computer system, which include the evaluation of the change, necessary test design, test data, and final acceptance of the change;

Current Change

DATE OF ISSUANCE:
TRANSMITTAL NO.   TN 98-22   08/17/98
FORM FDA 2438 (9/98)

      ○ Evaluation of test data to assure that data is accurately transmitted and handled properly when analytical equipment is directly interfaced to the computer; and

      ○ Procedures for emergency back-up of the computer system (e.g., back-up battery system and data forms for recording data in the event of a computer failure or power outage).

b. **EIR Documentation and Reporting** — The EIR should list which equipment, records, and procedures were inspected and the covered studies that they relate to. Detail any deficiencies which might result in contamination of test articles, uncontrolled stress to test systems, and/or erroneous test results.

5. **Testing Facility Operations (21 CFR 58.81)**

Purpose: To determine if the facility has established and follows written SOPs necessary to carry out study operations in a manner designed to ensure the quality and integrity of the data.

a. **SOP Evaluation**

– Review the SOP index and representative samples of SOPs to ensure that written procedures exist to cover at least all of the areas identified in 58.81(b).

– Verify that only current SOPs are available at the personnel work stations.

– Review key SOPs in detail and check for proper authorization signatures and dates, and general adequacy with respect to the content (i.e., SOPs are clear, complete, and can be followed by a trained individual).

– Verify that changes to SOPs are properly authorized and dated and that a historical file of SOPs is maintained.

– Ensure that there are procedures for familiarizing employees with SOPs.

– Determine that there are SOPs to ensure the quality and integrity of data, including input (data checking and verification), output (data control), and an audit trail covering all data changes.

– Verify that a historical file of outdated or modified computer programs is maintained. If the firm does not maintain old programs in digital form, ensure that a hard copy of all programs and data has been made and stored.

– Verify that SOPs are periodically reviewed for current applicability and that they are representative of the actual procedures in use.

Current Change

    – Review select SOPs and observe employees performing the operation to evaluate SOP adherence and familiarity.

  b. **EIR Documentation and Reporting** — Submit SOPs, data collection forms, and raw data records as exhibits where necessary to support and illustrate deficiencies.

6. **Reagents and Solutions (21 CFR 58.83)**

Purpose: To determine that the facility ensures the quality of reagents at the time of receipt and subsequent use.

    – Review the procedures used to purchase, receive, label, and determine the acceptability of reagents and solutions for use in the studies.

    – Verify that reagents and solutions are labeled to indicate identity, titer or concentration, storage requirements, and expiration date.

    – Verify that the chemistry profile data accompanying control reagents are used.

    – Check that storage requirements are being followed.

7. **Animal Care (21 CFR 58.90)**

Purpose: To assess whether animal care and housing is adequate to preclude stress and uncontrolled influences which could alter the response of test *system to the test article.*

  a. **General Inspection** — Inspect the animal room(s) housing the study to observe operations, protocol and SOP adherence, and study records. Refer to *IOM 145.2* prior to inspecting sub-human primate facilities.

    – Determine that there are adequate SOPs covering environment, housing, feeding, handling, and care of laboratory animals, and that the SOPs and the protocol instructions are being followed.

    – Review pest-control procedures and identify individuals responsible for the program. If pest control is done by an outside firm, does someone from the laboratory accompany the exterminator at all times? Does the laboratory know what chemicals are being used?

    – Determine whether the facility has an Institutional Animal Care and Use Committee (IACUC). Review a copy of the Committee's Standard Operating Procedures and the most recent committee minutes to verify committee operation.

    – Determine that all newly received animals are appropriately isolated, identified, and their health status evaluated.

Current Change

DATE OF ISSUANCE:
TRANSMITTAL NO.   TN 98-22   08/17/98
FORM FDA 2438 (9/98)

- Verify that treatment given to animals that become diseased is authorized and documented.

- Randomly compare individual animal identification against corresponding housingunit identification and dose group designations to assure that animals are appropriately identified.

- Review daily observation logs and randomly verify their accuracy for animals reported as dead or having external gross lesions or masses.

- Ensure that animals of different species, or animals of the same species on different projects, are separated as necessary.

- Verify that cages, racks, and accessory equipment are cleaned and sanitized, and that appropriate bedding is used.

- Determine that feed and water samples are collected at appropriate sources, analyzed periodically, and that analytical documentation is maintained.

b. **EIR Documentation and Reporting** — The EIR should identify which areas, operations, SOPs, studies, etc., were covered.

8. **Test and Control Articles (21 CFR 58.105 – 113)**

Purpose: To determine that procedures exist to assure that test and control articles meet protocol specifications throughout the course of the study, and that accountability is maintained.

a. **Characterization and Stability (21 CFR 58.105)** — The responsibility for carrying out appropriate characterization and stability testing may be assumed by the facility performing the study or the study sponsor.

Verify that procedures are in place to ensure that:

- The acquisition, receipt and storage of test articles, and means used to prevent deterioration and contamination are as specified;

- The identity, strength, purity, etc., to define the test and control *articles* are determined for each batch and are documented;

- The stability of test and control articles is documented;

- The transfer of samples from the collecting facility to the analytical laboratory is documented;

- Storage containers are appropriately labeled and assigned for the duration of the study; and

Current Change

— Reserve samples of test and control articles for each batch are retained for studies lasting more than four weeks.

b. **Test and Control Article Handling (21 CFR 58.107)**

— Determine that there are adequate procedures for handling test and control articles in the following areas:

○ Proper identification and storage;

○ Distribution to preclude contamination, deterioration, or damage; and

○ Documentation for receipt and distribution.

— Inspect test and control article storage areas to verify that environmental controls, container labeling, and storage are adequate.

— Observe test and control article handling and identification during the distribution and administration to the test system.

— Review accountability records for proper completion and, if possible, randomly verify their accuracy by comparing actual amounts in inventory against amounts reported. For completed studies verify documentation of final test and *control* article reconciliation.

c. **Mixtures of Articles with Carriers (58.113)** — If possible, observe the preparation, sampling, testing, storage, and administration of mixtures. Verify that analytical tests are conducted as appropriate to:

— Determine uniformity of mixtures and to determine periodically the concentration of the test or control article in the mixture; and

— Determine stability as required under study conditions.

d. **EIR Documentation and Reporting** — Identify the test articles, SOPs, facilities, equipment, operations, etc., covered. *Review the analytical raw data versus the reported results for accuracy of reporting and overall integrity of the data that is being collected. Where deficiencies or other information suggest a problem with the purity, identity, strength, etc., of the test article or the concentration of test article mixtures, document and obtain copies of the analytical raw data and any related reports. Consideration should be given to proceed with collecting a sample as described Part III, G.*

9. **Protocol and Conduct of Nonclinical Study (21 CFR 58.120 – 130)**

   Purpose: To determine if study protocols are properly written and authorized, and that studies are conducted in accord with the protocol.

* Current Change

DATE OF ISSUANCE:
TRANSMITTAL NO.   TN 98-22   08/17/98
FORM FDA 2438 (9/98)

a. **Study Protocol (21 CFR 58.120)**

- Review SOP for protocol preparation and approval and determine that it is followed.

- Review protocol to determine if it contains required elements.

- Review all changes, revisions, amendments, etc., to the protocol to ensure that they are properly authorized, signed, and dated.

- Verify that all copies of the protocol contain any applicable amendments.

b. **Conduct of Nonclinical Study (21 CFR 58.130)** — Evaluate the following laboratory operations, facilities, and equipment to verify conformity with the protocol and SOP requirements:

- Test system monitoring;

- Storage and identification of specimens;

- Recording of data, both manual and automated;

- Raw data corrections do not obscure the original entry and the reason for change is documented;

- *Verify that animals are randomized according to protocol and SOPs;*

- Procedures exist for the collection and identification of specimens and data handling, storage, and retrieval; and

- Procedures exist for limiting access to data and programs to authorized personnel.

c. **EIR Reporting and Documentation** — Fully identify the study(ies) examined. Report in detail any deficiencies observed and collect exhibits to document adverse observations. Submit, as exhibits, a copy of all protocols and amendments that were reviewed.

10. **Records and Reports (21 CFR 58.185 – 195)**

Purpose: To assess the firm's ability to store and retrieve study data, reports, slides, and specimens in a manner which maximizes their integrity and utility.

a. **Reporting of Study Results (21 CFR 58.185)** — Determine if the facility generates a final report for each study conducted. For selected studies, obtain the final report, and verify that it contains the following:

- The required elements in 21 CFR 58.185(a)(1-14), including a description of any computer program changes;

Current Change

- Dated signature of the study director;

- Amendments to the final report clearly identify that part of the report that is being amended and the reasons therefore; and

- The identity of any subcontractors, by name, address, and portion of the study contracted.

b. **Storage and Retrieval of Records and Data (21 CFR 58.190)** — Verify that all raw data, documentation, protocols, specimens, and final reports are retained. For randomly selected individual test systems, determine that all raw data, specimens, slides, etc., have been retained.

- Review SOPs covering collection and identification of specimens and data handling, storage, and retrieval.

- Check archive facilities for degree of controlled access and adequacy of environmental controls, including computer media storage conditions. It is not necessary that all data and specimens be in the same physical archive. For raw data and specimens retained elsewhere, the archives must make specific reference to those other locations.

- Determine that records are indexed in a way to allow access to data stored on computer media.

- Ensure that there are controlled procedures for adding or removing material.

- Review archive records for the removal and return of data and specimens. Check for any unexplained or prolonged removals.

- Determine how and where original computer data and backup copies are stored.

c. **EIR Documentation and Reporting** — Identify the individual(s) responsible for the archives. Describe and document deficiencies.

D. **Data Audit**

In addition to the procedures outlined above for evaluating the overall GLP compliance of a firm, the inspection should include the audit of at least one completed study. Studies for audit may be assigned by the Center or selected by the investigator as described in Part III, A. The audit will include a comparison of the protocol and amendments, raw data, records, and specimens against the final report to substantiate that protocol requirements were met and that findings were fully and accurately reported.

For each study audited, the study records should be reviewed for quality, i.e., to ensure that data:

* Current Change

DATE OF ISSUANCE:
TRANSMITTAL NO.   TN 98-22   08/17/98
FORM FDA 2438 (9/98)

- Are recorded legibly in ink promptly at the time of observation or data collection;

- Are signed or initialed and dated by the individual entering the data;

- Changes to original entries shall:

  o Not obscure the original entry,

  o Indicate the reason for change, and

  o Be signed or initialed and dated by the person making the change;

- Are completed for all multiple data entry forms. A written explanation of any entry blanks should be retained with the study records; and

- Are accurately recorded, transcribed, and represented in the final report.

1. **General**

   - Determine if there were any significant changes in the facilities, operations, and QAU function other than those currently reported.

   - Determine whether the equipment used was inspected, standardized, and calibrated prior to, during, and after use in the study. If equipment malfunctioned, review the remedial action taken, and ensure that the final report addresses whether the malfunction *affected* the study.

   - Determine if approved SOPs existed during the conduct of the study.

   - Review the protocol and check against requirements in 21 CFR 58.120.

   - Review the final report for the study director's dated signature and the QAU statement as required in 21 CFR 58.35(b)(7).

2. **Protocol vs. Final Report** — Study methods described in the final report should be compared against the protocol and the SOPs to confirm that requirements were met in all essential areas. Examples include, but are not limited to, the following:

   - Selection and acquisition of the test system, i.e., species, source, weight range, number, age, date ordered, date received;

   - Procedure for receipt, examination, and isolation of newly received animals;

   - Methods of test system identification, housing, and assignment to study;

   - Types and occurrences of diseases and clinical observations prior to and during the study, as well as any treatments administered;

   - Limits for, frequency of, and methods of sampling and analysis for contamination in feed, water, and bedding;

* Current Change

- Preparation and administration of test and/or control articles;

- Analysis of test article/test article carrier mixture;

- Observation of the test system response to test article;

- Handling of dead or moribund animals;

- Collection and analysis of specimens and data;

- Necropsy;

- Histopathology; and

- Pathology.

3. **Final Report vs. Raw Data** — The audit should include a detailed review of study records and raw data to confirm that findings in the final report are substantiated in the raw data and completely and accurately reflect the raw data. The study files, memos, and raw data associated with the study should be examined. Representative samples of raw data should be audited against what is reported in the final report. Examples of types of data to be checked include, but are not limited to, the following:

- Animal weight records;

- Food consumption and body weight gain records;

- Test system observation records and dosing records;

- Protocol-required analyses, e.g., urinalysis, hematology, blood chemistry, ophthalmologic exams, etc.;

- Necropsy and gross pathology observations records; and

- Histopathology, including records for tissue processing and slide preparation.

The investigator should include, by random selection from the dose groups, a representative number of animals to be traced from receipt through final histopathologic exam of the specimens. Check for the following:

- Consistent test system identification;

- Consistent trends in each parameter measured, including the upper and lower limits established as acceptable values;

- Explanations for data outliers;

- Correlation between in vivo test system observations and gross pathology;

- Correlation between gross pathology observations and histopathologic examinations;

- Correlation between what is reported in "interim" reports as compared to the "final" reported values and observations; and

- Documentation of unforeseen circumstances, when they occurred, and whether the circumstance had an impact on the study.

4. **Specimen Retention vs. Final Report** — The audit should include random examination of selected specimens in the archives for confirmation of the number and identity of specimens reported in the final report.

5. **EIR Documentation and Reporting** — All studies utilized should be fully identified: sponsor's name, study title, study director's name, test article name and code, test system, initiation and completion dates, final report date, and if available, to which IND/NDA/NADA/IDE or petition submitted. For major personnel engaged in the study, obtain names, position descriptions, and summaries of training and experience,

Fully describe any significant deficiencies in the conduct or reporting of the study reviewed. A copy of the narrative study report, protocol, and study records should be collected and submitted as necessary.

E. **Refusal to Permit Inspection**

Upon refusal to permit inspection, the facility should be advised by the FDA investigator that it is the policy of FDA not to accept studies submitted to the agency in support of any research or marketing permit if those studies were performed or conducted by a facility that has refused to permit inspection.

REFUSALS TO ALLOW INSPECTION OF FACILITIES OR STUDIES SHOULD BE REPORTED IMMEDIATELY BY TELEPHONE TO HFC-230, *(301- 827-0425), and followed-up by FAX to HFC-230 (301-827-0482)* with an information copy to the assigning Center.

Partial refusals, including refusal to permit access to, and copying of the master schedule (including any code sheets), SOPs, and other documents pertaining to the inspection will be treated as a "total" refusal to permit inspection.

F. **Sealing of Research Data**

Whenever an investigator encounters questionable or suspicious records under any Bioresearch Monitoring Program, and is unable to copy or review them immediately, and has

Current Change

reason for preserving the integrity of those records, the investigator is to immediately contact by telephone HFC-230 for instructions. Refer to IOM 453.3 for more information.

G. **Samples**

Collection of samples should be considered when the situation under audit or surveillance suggests that the facility had, or is having, problems in the area of characterization, stability, storage, contamination, or dosage preparation. The investigator should contact the assigning Center and the *Division of Field Science (HFC-140)* before samples are collected. If the field investigator collects a sample, a copy of the methodology of the sponsor or the testing facility *must also* be obtained. The copy will be sent to HFC-140 for designation of a laboratory to perform the sample analysis according to instrumental capabilities and the availability of the selected district laboratory. The investigator should contact HFC-140 at *(301) 443-7103* for specific disposition/handling instructions.

The investigator must provide complete documentation with each sample for possible legal/administrative actions.

Samples may include the following:

Physical samples of carrier, test article, control article, or test and control article carrier mixtures for identity, strength, potency, purity, composition or other characteristics which will appropriately define the test or control articles; and

Physical samples of specimens, including wet tissues, tissue blocks, and slides.

H. **Report Format**

1. **Abbreviated Reporting** — The districts may use abbreviated reporting (except for initial inspections) for the following types of assignments depending upon the nature and severity of any adverse findings:

   GLP surveillance inspections of a facility when it is apparent from the findings that the inspection may result in an NAI or VAI classification. These reports must include enough documented information to support the respective classification.

   Directed inspections and data audits provided the report fully covers all aspects of the specific topic of the inspection (i.e., operations, past deficiencies, assigned studies, etc.) and documents significant adverse findings to support the anticipated classification.

* Current Change

DATE OF ISSUANCE:
TRANSMITTAL NO.   TN 98-22   08/17/98
FORM FDA 2438 (9/98)

FORMAT: The investigator, with supervisory approval, may prepare an abbreviated report*. A sentence shall be included that clearly states that the report is an abbreviated format. In accordance with IOM 593.1 the report should include the following:*

   a.   Summary of findings;

   b.   Description of any significant changes in the laboratory management, facilities, or operations since the previous inspection;

   c.   Identity of the studies selected for audit and review; and

   d.   Identification of the areas, operations, and amount of data that were inspected (e.g., identity and percent of records reviewed in a specific area, number of animals tracked, percent of slides checked for accountability, etc.).

   All adverse observations will be fully discussed in the EIR and documented to permit assessment of their impact on the quality and integrity of the involved study. An abbreviated report need not address every heading under Part III, C. It should cover only those headings pertinent to the areas outlined above.

   If the Center review reveals that an abbreviated report is inadequate, the Center will notify HFC-132 that additional information is required.

2.   **Full Reporting** — The EIR should contain the headings as prescribed in the IOM, in addition to the specific headings outlined in Part III, C and Part III, D. The report must always include sufficient information and documentation to support the recommended classification. A complete report will be prepared and submitted in the following situations:

   a.   The initial GLP inspection of a facility;

   b.   All inspections that may result in an OAI classification; and/or

   c.   Any assignment specifically requesting a full report.

Current Change

PART IV—ANALYTICAL

A. ANALYZING LABORATORIES

All field and headquarters laboratories.

B. ANALYSIS

Samples may be submitted for a determination for identity, strength, potency, purity, or composition. Sample analysis will follow the exact methodology of the sponsor or the testing facility. The methodology must accompany the samples.

In those cases where the nonclinical laboratory does not have a methodology, due to the fact that the stability of the test and control articles in the mixture has not been determined, the established standard operating procedures of the nonclinical laboratory shall be used as described under Section 58.113(a)(2) of the Regulations.

C. METHOD INQUIRIES AND ASSIGNMENT OF APPROPRIATE DISTRICT LABORATORIES

Call ORO, Division of Field Science (HFC-140) at *(301) 443-7103.*

D. REPORTING

The analyzing laboratory will submit copies of the Collection Report and the Analytical Worksheets to the originating district office and to the Center with reference to the related EIR.

* Current Change

DATE OF ISSUANCE:
TRANSMITTAL NO.   TN 98-22   08/17/98
FORM FDA 2438 (9/98)

PART V—REGULATORY/ADMINISTRATIVE

A. District EIR Classification Authority

The District is encouraged to review and initially classify inspection reports generated under this compliance program, including those containing data audits.

The Districts will not classify EIRs generated in inspections that are conducted for other government agencies.

B. Center EIR Classification Authority

The Center has the final classification authority for all bioresearch monitoring inspection reports. Instances may arise when a Center review results in a reclassification of an EIR reviewed and classified by the District. The Center will provide to the appropriate District copies of all final classifications, including any reason for changes.

C. EIR Classifications

The following guidance is to be used in conjunction with the instructions in FMD-86 for District and Center classification of EIRs generated under this compliance program:

1. NAI — No objectionable conditions or practices were found during the inspection *(or the objectionable conditions found do not justify further regulatory action).*

2. VAI — Objectionable conditions or practices were found * but the District is not prepared to take or recommend any administrative or regulatory action.*

3. OAI —* Regulatory and/or Administrative actions will be recommended.*

D. Post-Inspectional Correspondence

Because the Centers are solely responsible for issuing post-inspection correspondence, the District should advise the Center of any response to the FDA 483 or any other communication (written or oral) with the facility concerning the inspection. Similarly, if the headquarters unit has communication (including any written correspondence) with the facility following the inspection, including any judicial/administrative action, the District will be advised of such communication and will be provided a copy of memorandum of contact.

Current Change

DATE OF ISSUANCE:
TRANSMITTAL NO.   TN 98-22   08/17/98
FORM FDA 2438 (9/98)

E. District Follow-up

All District follow-up action, including reinspection, will usually be done on headquarters assignment. On occasion, District compliance branches may initiate case development activities and may issue investigative assignments whenever review of the inspection report raises the possibility of severe violations of the FD&C Act or other federal statutes. This intention is to be immediately communicated to the involved Center and to the Office of Enforcement, Division of Compliance Policy(HFC-230).

F. Regulatory/Administrative Actions

The regulatory/administrative actions that can be used under this compliance program are not mutually exclusive. Follow-up of an OAI inspection may involve the use of one or more of the following:

- Warning Letter

- Reinspection

- Informal Conference

- Third Party Validation of a Nonclinical Study or Studies

- Rejection of a Nonclinical Study or Studies

- Disqualification of the Facility

- Injunction/Prosecution

- Withholding or Revocation of Marketing Permit

- Termination of IND or INAD

- Application Integrity Policy

* Current Change

DATE OF ISSUANCE:
TRANSMITTAL NO.   TN 98-22   08/17/98
FORM FDA 2438 (9/98)

GLP Audit Manual ©2000, Interpharm Press

399

PART VI—REFERENCES AND PROGRAM CONTACTS

References

Guide for the Care and Use of Laboratory Animals, DHHS Publication No. (NIH) 96-23

The Animal Welfare Act, 9 CFR Parts 1, 2, 3

21 CFR 11 - Electronic Records; Electronic Signatures Regulation effective August 1997.

21 CFR 58.1 - 58.219 Good Laboratory Practice Regulations effective June 1979, and amended effective October 1987

Good Laboratory Practice Regulations, Management Briefings, Post Conference Report, August 1979

Good Laboratory Practice Regulations, Questions and Answers, June 1981

"Toxicological Principles for the Safety Assessment of Direct Food Additives and Color Additives used in Food," FDA, Bureau of Foods

"Guide to Inspection of Computerized Systems in Drug Processing," February 1983

"Technical Reference on Software Development Activities," July 1987

"Guide For Detecting Fraud in Bioresearch Monitoring Inspections," April 1993

Copies of the listed references may be obtained from *HFC-130* or HFC-140.

Program Contacts

When technical questions arise on a specific assignment, or when additional information or guidance is required, contact the assigning Center. Operational questions should be addressed to *HFC-130*.

Specific Contacts

Office of the Associate Commissioner for Regulatory Affairs

Office of Enforcement (OE), Division of Compliance Policy: *Dr. James F. McCormack,* HFC-230, 301-827-0425, FAX 301-827-0482.

Current Change

DATE OF ISSUANCE:
TRANSMITTAL NO.   TN 98-22   08/17/98
FORM FDA 2438 (9/98)

Office of Regional Operations (ORO), Division of Emergency and Investigational Operations, Drugs Group (HFC-132): Dr. Thaddeus Sze, *HFC-130*, 301-827-5649, FAX 301-443-6919.

Center for Drug Evaluation and Research (CDER) Division of Scientific Investigations, *Good Laboratory Practices and Bioequivalence Investigations Branch: Mr. Ty Fujiwara,* HFD-345, 301-594-1023, FAX 301-594-1204.

Center for Biologics Evaluation and Research (CBER) *Bioresearch Monitoring Staff: Ms. Pat Holobaugh, HFM-650, 301-594-1077, FAX 301-827-6221.*

Center for Veterinary Medicine (CVM) Biioresearch Monitoring Staff: Ms. Dorothy Pocurull, HFV-234, *301-594-1785,* FAX 301-594-1812.

Center for Devices and Radiological Health (CDRH) Division of Bioresearch Monitoring:*Mr. Rodney Allnutt, HFZ-312, 301-594-4723,* FAX 301-594-4731.

Center for Food Safety and Applied Nutrition (CFSAN) *Division of Product Policy: Dr. John Welsh, HFS-207, 202-418-3057,* FAX 202-418-3126.

* Current Change

DATE OF ISSUANCE:
TRANSMITTAL NO.   TN 98-22   08/17/98
FORM FDA 2438 (9/98)

PART VII—HEADQUARTERS' RESPONSIBILITIES

A. Centers

1. Review all EIRS and regulatory/administrative recommendations forwarded by the districts.

2. Notify OE when the EIR involves other government agencies.

3. Provide OE with a list of firms on a quarterly basis, or as other priorities dictate, to be scheduled for inspection.

4. Select studies to be audited and provide necessary support documents.

5. Prepare letters for issuance by the Center, ORA, or Commissioner as appropriate.

6. Issue all assignments to the field.

7. Recommend compliance program changes to OE.

8. Distribute copies of EIR correspondence to districts and OE*(OE should receive only copies of correspondence regarding OAI classifications and instances where the final center classification differs from the initial district classification)*.

9. Provide inspectional support to the field by direct participation and in consultant capacity.

10. Copy all Centers whenever a warning letter has issued to a specific laboratory.

B. Office of Regulatory Affairs, Office of Enforcement (OE), Division of Compliance Policy

1. *Is the* liaison with other federal agencies and foreign governments with whom FDA has Memoranda of Agreement or Memoranda of Understanding.

2. Coordinates the distribution and review of multi-center and inter-agency EIRs including the planning, coordination, and designation of a lead Center for *EIR classification, issuance of correspondence, and follow-up assignments*.

3. Concurs with Center on recommended administrative and regulatory actions.

4. Coordinates modifications and future issuances of this compliance program.

Current Change

DATE OF ISSUANCE:
TRANSMITTAL NO.   TN 98-22   08/17/98
FORM FDA 2438 (9/98)

5. Resolves issues involving compliance or enforcement policy.

7. Coordinates Center and other federal agency inspection assignments.

8. Prepares notice of first inspection letters to university and government laboratories.

C. Office of Regional Operations (ORO), Division of *Emergency and Investigational Operations, (DEIO)*, and Division of Field Science (DFS)

1. DEIO:

    a. Provides inspection quality assurance, training of field personnel, and operational guidance.

    b. Maintains liaison with Centers and Field Offices and resolves operational questions.

    c. Coordinates and schedules joint Center and multi- District inspections.

2. DFS:

Assigns laboratories for sample analysis and responds to method inquiries.

---

* Current Change

DATE OF ISSUANCE:
TRANSMITTAL NO.   TN 98-22   08/17/98
FORM FDA 2438 (9/98)

ATTACHMENT A

COMPUTER SYSTEMS

The intent of this attachment is to collect, in one place, references to computer systems found throughout Part III . Computer systems and operations should be thoroughly covered during inspection of any facility. *FDA published Electronic Records; Electronic signatures; Final Rule [21 CFR 11] on March 20, 1997, The rule became effective on August 20, 1997. Records in electronic form that are created, modified, maintained, archived, retrieved, or transmitted under any records requirement set forth in agency regulations must comply with 21 CFR 11.* No additional reporting is required under this Attachment.

Personnel — (21 CFR 58.29) (PART III C.1.b.)

Determine the following:

- Who was involved in the design, development, and validation of the computer system;

- Who is responsible for the operation of the computer system, including inputs, processing, and output of data;

- If computer system personnel have training commensurate with their responsibilities, including professional training and training in GLPs; and

QAU Operations — (21 CFR 58.35(b-d)) (PART III C.2.a.)

Verify SOPs exist and are being followed for QAU inspections of computer operations.

Facilities (21 CFR 58.41 - 51) (PART III C.3.a.)

Determine that computerized operations and archived computer data are housed under appropriate environmental conditions.

Equipment (21 CFR 58.61 - 63) (PART III C.4.a.)

For computer systems, check that the following procedures exist and are documented:

- Validation study, including validation plan and documentation of the plan's completion;

Current Change

– Maintenance of equipment, including storage capacity and back-up procedures;

– Control measures over changes made to the computer system, which include the evaluation of the change, necessary test design, test data, and final acceptance of the change;

– Evaluation of test data to assure that data is accurately transmitted and handled properly when analytical equipment is directly interfaced to the computer; and

– Procedures for emergency back-up of the computer system, (e.g., back-up battery system and data forms for recording data in the event of a computer failure or power outage).

Testing Facility Operations (21 CFR 58.81) (PART III C.5.a.)

Verify that a historical file of outdated or modified computer programs is maintained.

Records and Reports (21 CFR 58.185 - 195) (PART III C.10.a.)

Verify that the final report contains the required elements in 58.185(a)(1-14), including a description of any computer program changes.

Storage and Retrieval of Records and Data (21 CFR 58.190) (PART III C.10.b.)

Check archive facilities for degree of controlled access and adequacy of environmental controls with respect to computer media storage conditions.

Determine that records are indexed in a way to allow for access to data stored on computer media.

Determine how and where original computer data and backup copies are stored.

☆U.S. Government Printing Office: 1998 — 433-061/80024

* Current Change

DATE OF ISSUANCE:
TRANSMITTAL NO.   TN 98-22   08/17/98
FORM FDA 2438 (9/98)

PART 58—GOOD LABORATORY PRACTICE FOR NONCLINICAL LABORATORY STUDIES

AUTHORITY: 21 U.S.C. 342, 346, 346a, 348, 351, 352, 353, 355, 360, 360b–360f, 360h–360j, 371, 379e, 381; 42 U.S.C. 216, 262, 263b–263n.

SOURCE: 43 FR 60013, Dec. 22, 1978, unless otherwise noted.

Subpart A—General Provisions

58.1 Scope.

(a) This part prescribes good laboratory practices for conducting nonclinical laboratory studies that support or are intended to support applications for research or marketing permits for products regulated by the Food and Drug Administration, including food and color additives, animal food additives, human and animal drugs, medical devices for human use, biological products, and electronic products. Compliance with this part is intended to assure the quality and integrity of the safety data filed pursuant to sections 406, 408, 409, 502, 503, 505, 506, 510, 512–516, 518–520, 721, and 801 of the Federal Food, Drug, and Cosmetic Act and sections 351 and 354–360F of the Public Health Service Act.

(b) References in this part to regulatory sections of the Code of Federal Regulations are to chapter I of title 21, unless otherwise noted.

[43 FR 60013, Dec. 22, 1978, as amended at 52 FR 33779, Sept. 4, 1987; 64 FR 399, Jan. 5, 1999]

Effective Date Note: At 64 FR 399, Jan. 5, 1999, in § 58.1, paragraph (a) was amended by removing "507,", effective May 20, 1999.

§ 58.3 Definitions.

As used in this part, the following terms shall have the meanings specified:

(a) *Act* means the Federal Food, Drug, and Cosmetic Act, as amended (secs. 201–902, 52 Stat. 1040 *et seq.,* as amended (21 U.S.C. 321–392)).

(b) *Test article* means any food additive, color additive, drug, biological product, electronic product, medical device for human use, or any other article subject to regulation under the act or under sections 351 and 354–360F of the Public Health Service Act.

(c) *Control article* means any food additive, color additive, drug, biological product, electronic product, medical device for human use, or any article other than a test article, feed, or water that is adminis-

tered to the test system in the course of a nonclinical laboratory study for the purpose of establishing a basis for comparison with the test article.

(d) *Nonclinical laboratory study* means in vivo or in vitro experiments in which test articles are studied prospectively in test systems under laboratory conditions to determine their safety. The term does not include studies utilizing human subjects or clinical studies or field trials in animals. The term does not include basic exploratory studies carried out to determine whether a test article has any potential utility or to determine physical or chemical characteristics of a test article.

(e) *Application for research or marketing permit* includes:

(1) A color additive petition, described in part 71.

(2) A food additive petition, described in parts 171 and 571.

(3) Data and information regarding a substance submitted as part of the procedures for establishing that a substance is generally recognized as safe for use, which use results or may reasonably be expected to result, directly or indirectly, in its becoming a component or otherwise affecting the characteristics of any food, described in §§ 170.35 and 570.35.

(4) Data and information regarding a food additive submitted as part of the procedures regarding food additives permitted to be used on an interim basis pending additional study, described in § 180.1.

(5) *An investigational new drug application,* described in part 312 of this chapter.

(6) A *new drug application,* described in part 314.

(7) Data and information regarding an over-the-counter drug for human use, submitted as part of the procedures for

classifying such drugs as generally recognized as safe and effective and not misbranded, described in part 330.

(8) Data and information about a substance submitted as part of the procedures for establishing a tolerance for unavoidable contaminants in food and food-packaging materials, described in parts 109 and 509.

(9) Data and information regarding an antibiotic drug submitted as part of the procedures for issuing, amending, or repealing regulations for such drugs, described in § 314.300 of this chapter.

(10) *A Notice of Claimed Investigational Exemption for a New Animal Drug*, described in part 511.

(11) A *new animal drug application*, described in part 514.

(12) [Reserved]

(13) An *application for a biologics license*, described in part 601 of this chapter.

(14) An *application for an investigational device exemption*, described in part 812.

(15) An *Application for Premarket Approval of a Medical Device*, described in section 515 of the act.

(16) A *Product Development Protocol for a Medical Device*, described in section 515 of the act.

(17) Data and information regarding a medical device submitted as part of the procedures for classifying such devices, described in part 860.

(18) Data and information regarding a medical device submitted as part of the procedures for establishing, amending, or repealing a performance standard for such devices, described in part 861.

(19) Data and information regarding an electronic product submitted as part of the procedures for obtaining an exemption from notification of a radiation safety defect or failure of compliance with a radiation safety performance standard, described in subpart D of part 1003.

(20) Data and information regarding an electronic product submitted as part of the procedures for establishing, amending, or repealing a standard for such product, described in section 358 of the Public Health Service Act.

(21) Data and information regarding an electronic product submitted as part of the procedures for obtaining a variance from any electronic product performance standard as described in § 1010.4.

(22) Data and information regarding an electronic product submitted as part of the procedures for granting, amending, or extending an exemption from any electronic product performance standard, as described in § 1010.5.

(f) *Sponsor* means:

(1) A person who initiates and supports, by provision of financial or other resources, a nonclinical laboratory study;

(2) A person who submits a nonclinical study to the Food and Drug Administration in support of an application for a research or marketing permit; or

(3) A testing facility, if it both initiates and actually conducts the study.

(g) *Testing facility* means a person who actually conducts a nonclinical laboratory study, i.e., actually uses the test article in a test system. *Testing facility* includes any establishment required to register under section 510 of the act that conducts nonclinical laboratory studies and any consulting laboratory described in section 704 of the act that conducts such studies. *Testing facility* encompasses only those operational units that are being or have been used to conduct nonclinical laboratory studies.

(h) *Person* includes an individual, partnership, corporation, association, scientific or academic establishment, government agency, or organizational unit thereof, and any other legal entity.

(i) *Test system* means any animal, plant, microorganism, or subparts thereof to which the test or control article is administered or added for study. *Test system* also includes appropriate groups or components of the system not treated with the test or control articles.

(j) *Specimen* means any material derived from a test system for examination or analysis.

(k) *Raw data* means any laboratory worksheets, records, memoranda, notes, or exact copies thereof, that are the result of original observations and activities of a nonclinical laboratory study and are necessary for the reconstruction and evaluation of the report of that study. In the event that exact transcripts of raw data have been prepared (e.g., tapes which have been transcribed verbatim, dated, and verified accurate by signature), the exact copy or exact transcript may be substituted for the original source as raw data. *Raw data* may include photographs, microfilm or microfiche copies, computer printouts, magnetic media, including dictated observations, and recorded data from automated instruments.

(l) *Quality assurance unit* means any person or organizational element, except the study director, designated by testing facility management to perform the duties relating to quality assurance of nonclinical laboratory studies.

(m) *Study director* means the individual responsible for the overall conduct of a nonclinical laboratory study.

(n) *Batch* means a specific quantity or lot of a test or control article that has been characterized according to § 58.105(a).

(o) *Study initiation date* means the date the protocol is signed by the study director.

(p) *Study completion date* means the date the final report is signed by the study director.

[43 FR 60013, Dec. 22, 1978, as amended at 52 FR 33779, Sept. 4, 1987; 54 FR 9039, Mar. 3, 1989; 64 FR 56448, Oct. 20, 1999]

EFFECTIVE DATE NOTE: At 64 FR 399, Jan. 5, 1999, § 58.3 was amended by removing and reserving paragraph (e)(9), effective May 20, 1999.

§ 58.10 Applicability to studies performed under grants and contracts.

When a sponsor conducting a nonclinical laboratory study intended to be submitted to or reviewed by the Food and Drug Administration utilizes the services of a consulting laboratory, contractor, or grantee to perform an analysis or other service, it shall notify the consulting laboratory, contractor, or grantee that the service is part of a nonclinical laboratory study that must be conducted in compliance with the provisions of this part.

§ 58.15 Inspection of a testing facility.

(a) A testing facility shall permit an authorized employee of the Food and Drug Administration, at reasonable times and in a reasonable manner, to inspect the facility and to inspect (and in the case of records also to copy) all records and specimens required to be maintained regarding studies within the scope of this part. The records inspection and copying requirements shall not apply to quality assurance unit records of findings and problems, or to actions recommended and taken.

(b) The Food and Drug Administration will not consider a nonclinical laboratory study in support of an application for a research or marketing permit if the testing facility refuses to permit inspection. The determination that a nonclinical laboratory study will not be considered in support of an application for a research or marketing permit does not, however, relieve the applicant for such a permit of any obligation under any applicable statute or regulation to submit the results of the study to the Food and Drug Administration.

Subpart B—Organization and Personnel

§ 58.29 Personnel.

(a) Each individual engaged in the conduct of or responsible for the supervision of a nonclinical laboratory study shall have education, training, and experience, or combination thereof, to enable that individual to perform the assigned functions.

(b) Each testing facility shall maintain a current summary of training and experience and job description for each individual engaged in or supervising the conduct of a nonclinical laboratory study.

(c) There shall be a sufficient number of personnel for the timely and proper conduct of the study according to the protocol.

(d) Personnel shall take necessary personal sanitation and health precautions designed to avoid contamination of test and control articles and test systems.

(e) Personnel engaged in a nonclinical laboratory study shall wear clothing appropriate for the duties they perform. Such clothing shall be changed as often as necessary to prevent microbiological, radiological, or chemical contamination of test systems and test and control articles.

(f) Any individual found at any time to have an illness that may adversely affect the quality and integrity of the nonclinical laboratory study shall be excluded from direct contact with test systems, test and control articles and any other operation or function that may adversely affect the study until the condition is corrected. All personnel shall be instructed to report to their immediate supervisors any health or medical conditions that may reasonably be considered to have an adverse effect on a nonclinical laboratory study.

§ 58.31 Testing facility management.

For each nonclinical laboratory study, testing facility management shall:

(a) Designate a study director as described in § 58.33, before the study is initiated.

(b) Replace the study director promptly if it becomes necessary to do so during the conduct of a study.

(c) Assure that there is a quality assurance unit as described in § 58.35.

(d) Assure that test and control articles or mixtures have been appropriately tested for identity, strength, purity, stability, and uniformity, as applicable.

(e) Assure that personnel, resources, facilities, equipment, materials, and methodologies are available as scheduled.

(f) Assure that personnel clearly understand the functions they are to perform.

(g) Assure that any deviations from these regulations reported by the quality assurance unit are communicated to the study director and corrective actions are taken and documented.

[43 FR 60013, Dec. 22, 1978, as amended at 52 FR 33780, Sept. 4, 1987]

§ 58.33 Study director.

For each nonclinical laboratory study, a scientist or other professional of appropriate education, training, and experience, or combination thereof, shall be identified as the study director. The study director has overall responsibility for the technical conduct of the study, as well as for the interpretation, analysis, documentation and reporting of results, and represents the single point of study control. The study director shall assure that:

(a) The protocol, including any change, is approved as provided by § 58.120 and is followed.

(b) All experimental data, including observations of unanticipated responses of the test system are accurately recorded and verified.

(c) Unforeseen circumstances that may affect the quality and integrity of the nonclinical laboratory study are noted when they occur, and corrective action is taken and documented.

(d) Test systems are as specified in the protocol.

(e) All applicable good laboratory practice regulations are followed.

(f) All raw data, documentation, protocols, specimens, and final reports are transferred to the archives during or at the close of the study.

[43 FR 60013, Dec. 22, 1978; 44 FR 17657, Mar. 23, 1979]

§ 58.35 Quality assurance unit.

(a) A testing facility shall have a quality assurance unit which shall be responsible for monitoring each study to assure management that the facilities, equipment, personnel, methods, practices, records, and controls are in conformance with the regulations in this part. For any given study, the quality assurance unit shall be entirely separate from and independent of the personnel engaged in the direction and conduct of that study.

(b) The quality assurance unit shall:

(1) Maintain a copy of a master schedule sheet of all nonclinical laboratory studies conducted at the testing facility indexed by test article and containing the test system, nature of study, date study was initiated, current status of each study, identity of the sponsor, and name of the study director.

(2) Maintain copies of all protocols pertaining to all nonclinical laboratory studies for which the unit is responsible.

(3) Inspect each nonclinical laboratory study at intervals adequate to assure the integrity of the study and maintain written and properly signed records of each periodic inspection showing the date of the inspection, the study inspected, the phase or segment of the study inspected, the person performing the inspection, findings and problems, action recommended and taken to resolve existing problems, and any scheduled date for reinspection. Any problems found during the course of an inspection which are likely to affect study integrity shall be brought to the attention of the study director and management immediately.

(4) Periodically submit to management and the study director written status reports on each study, noting any problems and the corrective actions taken.

(5) Determine that no deviations from approved protocols or standard operating procedures were made without proper authorization and documentation.

(6) Review the final study report to assure that such report accurately describes the methods and standard operating procedures, and that the reported results accurately reflect the raw data of the nonclinical laboratory study.

(7) Prepare and sign a statement to be included with the final study report which shall specify the dates inspections were

made and findings reported to management and to the study director.

(c) The responsibilities and procedures applicable to the quality assurance unit, the records maintained by the quality assurance unit, and the method of indexing such records shall be in writing and shall be maintained. These items including inspection dates, the study inspected, the phase or segment of the study inspected, and the name of the individual performing the inspection shall be made available for inspection to authorized employees of the Food and Drug Administration.

(d) A designated representative of the Food and Drug Administration shall have access to the written procedures established for the inspection and may request testing facility management to certify that inspections are being implemented, performed, documented, and followed-up in accordance with this paragraph.

(Information collection requirements approved by the Office of Management and Budget under control number 0910–0203)

[43 FR 60013, Dec. 22, 1978, as amended at 52 FR 33780, Sept. 4, 1987]

Subpart C—Facilities

§ 58.41 General.

Each testing facility shall be of suitable size and construction to facilitate the proper conduct of nonclinical laboratory studies. It shall be designed so that there is a degree of separation that will prevent any function or activity from having an adverse effect on the study.

[52 FR 33780, Sept. 4, 1987]

§ 58.43 Animal care facilities.

(a) A testing facility shall have a sufficient number of animal rooms or areas, as needed, to assure proper: (1) Separation of species or test systems, (2) isolation of individual projects, (3) quarantine of animals, and (4) routine or specialized housing of animals.

(b) A testing facility shall have a number of animal rooms or areas separate from those described in paragraph (a) of this section to ensure isolation of studies being done with test systems or test and control articles known to be biohazardous, including volatile substances, aerosols, radioactive materials, and infectious agents.

(c) Separate areas shall be provided, as appropriate, for the diagnosis, treatment, and control of laboratory animal diseases. These areas shall provide effective isolation for the housing of animals either known or suspected of being diseased, or of being carriers of disease, from other animals.

(d) When animals are housed, facilities shall exist for the collection and disposal of all animal waste and refuse or for safe sanitary storage of waste before removal from the testing facility. Disposal facilities shall be so provided and operated as to minimize vermin infestation, odors, disease hazards, and environmental contamination.

[43 FR 60013, Dec. 22, 1978, as amended at 52 FR 33780, Sept. 4, 1987]

§ 58.45 Animal supply facilities.

There shall be storage areas, as needed, for feed, bedding, supplies, and equipment. Storage areas for feed and bedding shall be separated from areas housing the test systems and shall be protected against infestation or contamination. Perishable supplies shall be preserved by appropriate means.

[43 FR 60013, Dec. 22, 1978, as amended at 52 FR 33780, Sept. 4, 1987]

§ 58.47 Facilities for handling test and control articles.

(a) As necessary to prevent contamination or mixups, there shall be separate areas for:

(1) Receipt and storage of the test and control articles.

(2) Mixing of the test and control articles with a carrier, e.g., feed.

(3) Storage of the test and control article mixtures.

(b) Storage areas for the test and/or control article and test and control mixtures shall be separate from areas housing the test systems and shall be adequate to preserve the identity, strength, purity, and stability of the articles and mixtures.

§ 58.49 Laboratory operation areas.

Separate laboratory space shall be provided, as needed, for the performance of the routine and specialized procedures required by nonclinical laboratory studies.

[52 FR 33780, Sept. 4, 1987]

§ 58.51 Specimen and data storage facilities.

Space shall be provided for archives, limited to access by authorized personnel only, for the storage and retrieval of all raw data and specimens from completed studies.

Subpart D—Equipment

§ 58.61 Equipment design.

Equipment used in the generation, measurement, or assessment of data and equipment used for facility environmental control shall be of appropriate design and adequate capacity to function according to the protocol and shall be suitably located for operation, inspection, cleaning, and maintenance.

[52 FR 33780, Sept. 4, 1987]

§ 58.63 Maintenance and calibration of equipment.

(a) Equipment shall be adequately inspected, cleaned, and maintained. Equipment used for the generation, measurement, or assessment of data shall be adequately tested, calibrated and/or standardized.

(b) The written standard operating procedures required under § 58.81(b)(11) shall set forth in sufficient detail the methods, materials, and schedules to be used in the routine inspection, cleaning, maintenance, testing, calibration, and/or standardization of equipment, and shall specify, when appropriate, remedial action to be taken in the event of failure or malfunction of equipment. The written standard operating procedures shall designate the person responsible for the performance of each operation.

(c) Written records shall be maintained of all inspection, maintenance, testing, calibrating and/or standardizing operations. These records, containing the date of the operation, shall describe whether the maintenance operations were routine and followed the written standard operating procedures. Written records shall be kept of nonroutine repairs performed on equipment as a result of failure and malfunction. Such records shall document the nature of the defect, how and when the defect was discovered, and any remedial action taken in response to the defect.

(Information collection requirements approved by the Office of Management and Budget under control number 0910–0203)

[43 FR 60013, Dec. 22, 1978, as amended at 52 FR 33780, Sept. 4, 1987]

Subpart E—Testing Facilities Operation

§ 58.81 Standard operating procedures.

(a) A testing facility shall have standard operating procedures in writing setting forth nonclinical laboratory study methods that management is satisfied are adequate to insure the quality and integrity of the data generated in the course of a study. All deviations in a study from standard operating procedures shall be authorized by the study director and shall be documented in the raw data. Significant changes in established standard operating procedures shall be properly authorized in writing by management.

(b) Standard operating procedures shall be established for, but not limited to, the following:

(1) Animal room preparation.

(2) Animal care.

(3) Receipt, identification, storage, handling, mixing, and method of sampling of the test and control articles.

(4) Test system observations.

(5) Laboratory tests.

(6) Handling of animals found moribund or dead during study.

(7) Necropsy of animals or postmortem examination of animals.

(8) Collection and identification of specimens.

(9) Histopathology.

(10) Data handling, storage, and retrieval.

(11) Maintenance and calibration of equipment.

(12) Transfer, proper placement, and identification of animals.

(c) Each laboratory area shall have immediately available laboratory manuals and standard operating procedures relative to the laboratory procedures being performed. Published literature may be used as a supplement to standard operating procedures.

(d) A historical file of standard operating procedures, and all revisions thereof, including the dates of such revisions, shall be maintained.

[43 FR 60013, Dec. 22, 1978, as amended at 52 FR 33780, Sept. 4, 1987]

§ 58.83 Reagents and solutions.

All reagents and solutions in the laboratory areas shall be labeled to indicate identity, titer or concentration, storage requirements, and expiration date. Deteriorated or outdated reagents and solutions shall not be used.

§ 58.90 Animal care.

(a) There shall be standard operating procedures for the housing, feeding, handling, and care of animals.

(b) All newly received animals from outside sources shall be isolated and their health status shall be evaluated in accordance with acceptable veterinary medical practice.

(c) At the initiation of a nonclinical laboratory study, animals shall be free of any disease or condition that might interfere with the purpose or conduct of the study. If, during the course of the study, the animals contract such a disease or condition, the diseased animals shall be isolated, if necessary. These animals may be treated for disease or signs of disease provided that such treatment does not interfere with the study. The diagnosis, authorizations of treatment, description of treatment, and each date of treatment shall be documented and shall be retained.

(d) Warm-blooded animals, excluding suckling rodents, used in laboratory procedures that require manipulations and observations over an extended period of time or in studies that require the animals to be removed from and returned to their home cages for any reason (e.g., cage cleaning, treatment, etc.), shall receive appropriate identification. All information needed to specifically identify each animal within an animal-housing unit shall appear on the outside of that unit.

(e) Animals of different species shall be housed in separate rooms when necessary. Animals of the same species, but used in different studies, should not ordinarily be housed in the same room when inadvertent exposure to control or test articles or animal mixup could affect the outcome of either study. If such mixed housing is necessary, adequate differentiation by space and identification shall be made.

(f) Animal cages, racks and accessory equipment shall be cleaned and sanitized at appropriate intervals.

(g) Feed and water used for the animals shall be analyzed periodically to ensure that contaminants known to be capable of interfering with the study and reasonably expected to be present in such feed or water are not present at levels above those specified in the protocol. Documentation of such analyses shall be maintained as raw data.

(h) Bedding used in animal cages or pens shall not interfere with the purpose or conduct of the study and shall be changed as often as necessary to keep the animals dry and clean.

(i) If any pest control materials are used, the use shall be documented. Cleaning and pest control materials that interfere with the study shall not be used.

(Information collection requirements approved by the Office of Management and Budget under control number 0910–0203)

[43 FR 60013, Dec. 22, 1978, as amended at 52 FR 33780, Sept. 4, 1987; 54 FR 15924, Apr. 20, 1989; 56 FR 32088, July 15, 1991]

Subpart F—Test and Control Articles

§ 58.105 Test and control article characterization.

(a) The identity, strength, purity, and composition or other characteristics which will appropriately define the test or control article shall be determined for each batch and shall be documented. Methods of synthesis, fabrication, or derivation of the test and control articles shall be documented by the sponsor or the testing facility. In those cases where marketed products are used as control articles, such products will be characterized by their labeling.

(b) The stability of each test or control article shall be determined by the testing facility or by the sponsor either: (1) Before study initiation, or (2) concomitantly according to written standard operating procedures, which provide for periodic analysis of each batch.

(c) Each storage container for a test or control article shall be labeled by name, chemical abstract number or code number, batch number, expiration date, if any, and, where appropriate, storage conditions necessary to maintain the identity, strength, purity, and composition of the test or control article. Storage containers shall be assigned to a particular test article for the duration of the study.

(d) For studies of more than 4 weeks' duration, reserve samples from each batch of test and control articles shall be retained for the period of time provided by § 58.195.

(Information collection requirements approved by the Office of Management and Budget under control number 0910–0203)

[43 FR 60013, Dec. 22, 1978, as amended at 52 FR 33781, Sept. 4, 1987]

§ 58.107 Test and control article handling.

Procedures shall be established for a system for the handling of the test and control articles to ensure that:

(a) There is proper storage.

(b) Distribution is made in a manner designed to preclude the possibility of contamination, deterioration, or damage.

(c) Proper identification is maintained throughout the distribution process.

(d) The receipt and distribution of each batch is documented. Such documentation shall include the date and quantity of each batch distributed or returned.

§ 58.113 Mixtures of articles with carriers.

(a) For each test or control article that is mixed with a carrier, tests by appropriate analytical methods shall be conducted:

(1) To determine the uniformity of the mixture and to determine, periodically, the concentration of the test or control article in the mixture.

(2) To determine the stability of the test and control articles in the mixture as required by the conditions of the study either:

(i) Before study initiation, or

(ii) Concomitantly according to written standard operating procedures which provide for periodic analysis of the test and control articles in the mixture.

(b) [Reserved]

(c) Where any of the components of the test or control article carrier mixture has an expiration date, that date shall be clearly shown on the container. If more than one component has an expiration date, the earliest date shall be shown.

[43 FR 60013, Dec. 22, 1978, as amended at 45 FR 24865, Apr. 11, 1980; 52 FR 33781, Sept. 4, 1987]

Subpart G—Protocol for and Conduct of a Nonclinical Laboratory Study

§ 58.120 Protocol.

(a) Each study shall have an approved written protocol that clearly indicates the objectives and all methods for the conduct of the study. The protocol shall contain, as applicable, the following information:

(1) A descriptive title and statement of the purpose of the study.

(2) Identification of the test and control articles by name, chemical abstract number, or code number.

(3) The name of the sponsor and the name and address of the testing facility at which the study is being conducted.

(4) The number, body weight range, sex, source of supply, species, strain, substrain, and age of the test system.

(5) The procedure for identification of the test system.

(6) A description of the experimental design, including the methods for the control of bias.

(7) A description and/or identification of

the diet used in the study as well as solvents, emulsifiers, and/or other materials used to solubilize or suspend the test or control articles before mixing with the carrier. The description shall include specifications for acceptable levels of contaminants that are reasonably expected to be present in the dietary materials and are known to be capable of interfering with the purpose or conduct of the study if present at levels greater than established by the specifications.

(8) Each dosage level, expressed in milligrams per kilogram of body weight or other appropriate units, of the test or control article to be administered and the method and frequency of administration.

(9) The type and frequency of tests, analyses, and measurements to be made.

(10) The records to be maintained.

(11) The date of approval of the protocol by the sponsor and the dated signature of the study director.

(12) A statement of the proposed statistical methods to be used.

(b) All changes in or revisions of an approved protocol and the reasons therefor shall be documented, signed by the study director, dated, and maintained with the protocol.

(Information collection requirements approved by the Office of Management and Budget under control number 0910–0203)

[43 FR 60013, Dec. 22, 1978, as amended at 52 FR 33781, Sept. 4, 1987]

§ 58.130 Conduct of a nonclinical laboratory study.

(a) The nonclinical laboratory study shall be conducted in accordance with the protocol.

(b) The test systems shall be monitored in conformity with the protocol.

(c) Specimens shall be identified by test system, study, nature, and date of collection. This information shall be located on the specimen container or shall accompany the specimen in a manner that precludes error in the recording and storage of data.

(d) Records of gross findings for a specimen from postmortem observations should be available to a pathologist when examining that specimen histopathologically.

(e) All data generated during the conduct of a nonclinical laboratory study, except those that are generated by automated data collection systems, shall be recorded directly, promptly, and legibly in ink. All data entries shall be dated on the date of entry and signed or initialed by the person entering the data. Any change in entries shall be made so as not to obscure the original entry, shall indicate the reason for such change, and shall be dated and signed or identified at the time of the change. In automated data collection systems, the individual responsible for direct data input shall be identified at the time of data input. Any change in automated data entries shall be made so as not to obscure the original entry, shall indicate the reason for change, shall be dated, and the responsible individual shall be identified.

(Information collection requirements approved by the Office of Management and Budget under control number 0910–0203)

[43 FR 60013, Dec. 22, 1978, as amended at 52 FR 33781, Sept. 4, 1987]

Subparts H–I [Reserved]

Subpart J—Records and Reports

§ 58.185 Reporting of nonclinical laboratory study results.

(a) A final report shall be prepared for each nonclinical laboratory study and shall include, but not necessarily be limited to, the following:

(1) Name and address of the facility performing the study and the dates on which the study was initiated and completed.

(2) Objectives and procedures stated in the approved protocol, including any changes in the original protocol.

(3) Statistical methods employed for analyzing the data.

(4) The test and control articles identified by name, chemical abstracts number or code number, strength, purity, and composition or other appropriate characteristics.

(5) Stability of the test and control articles under the conditions of administration.

(6) A description of the methods used.

(7) A description of the test system used. Where applicable, the final report shall include the number of animals used, sex, body weight range, source of supply, species, strain and substrain, age, and procedure used for identification.

(8) A description of the dosage, dosage regimen, route of administration, and duration.

(9) A description of all cirmcumstances that may have affected the quality or integrity of the data.

(10) The name of the study director, the names of other scientists or professionals, and the names of all supervisory personnel, involved in the study.

(11) A description of the transformations, calculations, or operations performed on the data, a summary and analysis of the data, and a statement of the conclusions drawn from the analysis.

(12) The signed and dated reports of each of the individual scientists or other professionals involved in the study.

(13) The locations where all specimens, raw data, and the final report are to be stored.

(14) The statement prepared and signed by the quality assurance unit as described in § 58.35(b)(7).

(b) The final report shall be signed and dated by the study director.

(c) Corrections or additions to a final report shall be in the form of an amendment by the study director. The amendment shall clearly identify that part of the final report that is being added to or corrected and the reasons for the correction or addition, and shall be signed and dated by the person responsible.

[43 FR 60013, Dec. 22, 1978, as amended at 52 FR 33781, Sept. 4, 1987]

§ 58.190 Storage and retrieval of records and data.

(a) All raw data, documentation, protocols, final reports, and specimens (except those specimens obtained from mutagenicity tests and wet specimens of blood, urine, feces, and biological fluids) generated as a result of a nonclinical laboratory study shall be retained.

(b) There shall be archives for orderly storage and expedient retrieval of all raw data, documentation, protocols, specimens, and interim and final reports. Conditions of storage shall minimize deterioration of the documents or specimens in accordance with the requirements for the time period of their retention and the nature of the documents or specimens. A testing facility may contract with commercial archives to provide a repository for all material to be retained. Raw data and specimens may be retained elsewhere provided that the archives have specific reference to those other locations.

(c) An individual shall be identified as responsible for the archives.

(d) Only authorized personnel shall enter the archives.

(e) Material retained or referred to in the archives shall be indexed to permit expedient retrieval.

(Information collection requirements approved by the Office of Management and Budget under control number 0910–0203)

[43 FR 60013, Dec. 22, 1978, as amended at 52 FR 33781, Sept. 4, 1987]

§ 58.195 Retention of records.

(a) Record retention requirements set forth in this section do not supersede the record retention requirements of any other regulations in this chapter.

(b) Except as provided in paragraph (c) of this section, documentation records, raw data and specimens pertaining to a nonclinical laboratory study and required to be made by this part shall be retained in the archive(s) for whichever of the following periods is shortest:

(1) A period of at least 2 years following the date on which an application for a research or marketing permit, in support of which the results of the nonclinical laboratory study were submitted, is approved by the Food and Drug Administration. This requirement does not apply to studies supporting investigational new drug applications (IND's) or applications for investigational device exemptions (IDE's), records of which shall be governed by the provisions of paragraph (b)(2) of this section.

(2) A period of at least 5 years following the date on which the results of the nonclinical laboratory study are submitted to the Food and Drug Administration in support of an application for a research or marketing permit.

(3) In other situations (e.g., where the nonclinical laboratory study does not result in the submission of the study in support of an application for a research or marketing permit), a period of at least 2 years following the date on which the study is completed, terminated, or discontinued.

(c) Wet specimens (except those specimens obtained from mutagenicity tests and wet specimens of blood, urine, feces, and biological fluids), samples of test or control articles, and specially prepared material, which are relatively fragile and differ markedly in stability and quality during storage, shall be retained only as long as the quality of the preparation affords evaluation. In no case shall retention be required for longer periods than those set forth in paragraphs (a) and (b) of this section.

(d) The master schedule sheet, copies of protocols, and records of quality assurance inspections, as required by § 58.35(c) shall be maintained by the quality assurance unit as an easily accessible system of records for the period of time specified in paragraphs (a) and (b) of this section.

(e) Summaries of training and experience and job descriptions required to be maintained by § 58.29(b) may be retained along with all other testing facility employment records for the length of time specified in paragraphs (a) and (b) of this section.

(f) Records and reports of the maintenance and calibration and inspection of equipment, as required by § 58.63(b) and (c), shall be retained for the length of time specified in paragraph (b) of this section.

(g) Records required by this part may be retained either as original records or as true copies such as photocopies, microfilm, microfiche, or other accurate reproductions of the original records.

(h) If a facility conducting nonclinical testing goes out of business, all raw data, documentation, and other material specified in this section shall be transferred to the archives of the sponsor of the study. The Food and Drug Administration shall be notified in writing of such a transfer.

[43 FR 60013, Dec. 22, 1978, as amended at 52 FR 33781, Sept. 4, 1987; 54 FR 9039, Mar. 3, 1989]

Subpart K—Disqualification of Testing Facilities

§ 58.200 Purpose.

(a) The purposes of disqualification are:

(1) To permit the exclusion from consideration of completed studies that were conducted by a testing facility which has failed to comply with the requirements of the good laboratory practice regulations until it can be adequately demonstrated that such noncompliance did not occur during, or did not affect the validity or acceptability of data generated by, a particular study; and

(2) To exclude from consideration all studies completed after the date of disqualification until the facility can satisfy the Commissioner that it will conduct studies in compliance with such regulations.

(b) The determination that a nonclinical laboratory study may not be considered in support of an application for a research or marketing permit does not, however, relieve the applicant for such a permit of any obligation under any other applicable regulation to submit the results of the study to the Food and Drug Administration.

§ 58.202 Grounds for disqualification.

The Commissioner may disqualify a testing facility upon finding all of the following:

(a) The testing facility failed to comply with one or more of the regulations set forth in this part (or any other regulations regarding such facilities in this chapter);

(b) The noncompliance adversely affected the validity of the nonclinical laboratory studies; and

(c) Other lesser regulatory actions (e.g., warnings or rejection of individual studies) have not been or will probably not be adequate to achieve compliance with the good laboratory practice regulations.

§ 58.204 Notice of and opportunity for hearing on proposed disqualification.

(a) Whenever the Commissioner has information indicating that grounds exist under § 58.202 which in his opinion justify disqualification of a testing facility, he may issue to the testing facility a written notice proposing that the facility be disqualified.

(b) A hearing on the disqualification shall be conducted in accordance with the requirements for a regulatory hearing set forth in part 16 of this chapter.

§ 58.206 Final order on disqualification.

(a) If the Commissioner, after the regulatory hearing, or after the time for requesting a hearing expires without a request being made, upon an evaluation of the administrative record of the disqualification proceeding, makes the findings required in § 58.202, he shall issue a final order disqualifying the facility. Such order shall include a statement of the basis for that determination. Upon issuing a final order, the Commissioner shall notify (with a copy of the order) the testing facility of the action.

(b) If the Commissioner, after a regulatory hearing or after the time for requesting a hearing expires without a request being made, upon an evaluation of the administrative record of the disqualification proceeding, does not make the findings required in § 58.202, he shall issue a final order terminating the disqualification proceeding. Such order shall include a statement of the basis for that determination. Upon issuing a final order the Commissioner shall notify the testing facility and provide a copy of the order.

§ 58.210 Actions upon disqualification.

(a) Once a testing facility has been disqualified, each application for a research or marketing permit, whether approved or not, containing or relying upon any nonclinical laboratory study conducted by the disqualified testing facility may be examined to determine whether such study was or would be essential to a decision. If it is determined that a study was or would be essential, the Food and Drug Administration shall also determine whether the study is acceptable, notwithstanding the disqualification of the facility. Any study done by a testing facility before or after disqualification may be presumed to be unacceptable, and the person relying on the study may be required to establish that the study was not affected by the circumstances that led to the disqualification, e.g., by submitting validating information. If the study is then determined to be unacceptable, such data will be eliminated from consideration in support of the application; and such elimination may serve as new information justifying the termination or withdrawal of approval of

the application.

(b) No nonclinical laboratory study begun by a testing facility after the date of the facility's disqualification shall be considered in support of any application for a research or marketing permit, unless the facility has been reinstated under § 58.219. The determination that a study may not be considered in support of an application for a research or marketing permit does not, however, relieve the applicant for such a permit of any obligation under any other applicable regulation to submit the results of the study to the Food and Drug Administration.

[43 FR 60013, Dec. 22, 1978, as amended at 59 FR 13200, Mar. 21, 1994]

§ 58.213 Public disclosure of information regarding disqualification.

(a) Upon issuance of a final order disqualifying a testing facility under § 58.206(a), the Commissioner may notify all or any interested persons. Such notice may be given at the discretion of the Commissioner whenever he believes that such disclosure would further the public interest or would promote compliance with the good laboratory practice regulations set forth in this part. Such notice, if given, shall include a copy of the final order issued under § 58.206(a) and shall state that the disqualification constitutes a determination by the Food and Drug Administration that nonclinical laboratory studies performed by the facility will not be considered by the Food and Drug Administration in support of any application for a research or marketing permit. If such notice is sent to another Federal Government agency, the Food and Drug Administration will recommend that the agency also consider whether or not it should accept nonclinical laboratory studies performed by the testing facility. If such notice is sent to any other person, it shall state that it is given because of the relationship between the testing facility and the person being notified and that the Food and Drug Administration is not advising or recommending that any action be taken by the person notified.

(b) A determination that a testing facility has been disqualified and the administrative record regarding such determination are disclosable to the public under part 20 of this chapter.

§ 58.215 Alternative or additional actions to disqualification.

(a) Disqualification of a testing facility under this subpart is independent of, and neither in lieu of nor a precondition to, other proceedings or actions authorized by the act. The Food and Drug Administration may, at any time, institute against a testing facility and/or against the sponsor of a nonclinical laboratory study that has been submitted to the Food and Drug Administration any appropriate judicial proceedings (civil or criminal) and any other appropriate regulatory action, in addition to or in lieu of, and prior to, simultaneously with, or subsequent to, disqualification. The Food and Drug Administration may also refer the matter to another Federal, State, or local government law enforcement or regulatory agency for such action as that agency deems appropriate.

(b) The Food and Drug Administration may refuse to consider any particular nonclinical laboratory study in support of an application for a research or marketing permit, if it finds that the study was not conducted in accordance with the good laboratory practice regulations set forth in this part, without disqualifying the testing facility that conducted the study or undertaking other regulatory action.

§ 58.217 Suspension or termination of a testing facility by a sponsor.

Termination of a testing facility by a sponsor is independent of, and neither in lieu of nor a precondition to, proceedings or actions authorized by this subpart. If a sponsor terminates or suspends a testing facility from further participation in a nonclinical laboratory study that is being conducted as part of any application for a research or marketing permit that has been submitted to any Center of the Food and Drug Administration (whether approved or not), it shall notify that Center in writing within 15 working days of the action; the notice shall include a statement of the reasons for such action. Suspension or termination of a testing facility by a sponsor does not relieve it of any obligation under any other applicable regulation to submit the results of the study to the Food and Drug Administration.

[43 FR FR 60013, Dec. 22, 1978, as amended at 50 FR 8995, Mar. 6, 1985]

§ 58.219 Reinstatement of a disqualified testing facility.

A testing facility that has been disqualified may be reinstated as an acceptable source of nonclinical laboratory studies to be submitted to the Food and Drug Administration if the Commissioner determines, upon an evaluation of the submission of the testing facility, that the facility can adequately assure that it will conduct future nonclinical laboratory studies in compliance with the good laboratory practice regulations set forth in this part and, if any studies are currently being conducted, that the quality and integrity of such studies have not been seriously compromised. A disqualified testing facility that wishes to be so reinstated shall present in writing to the Commissioner reasons why it believes it should be reinstated and a detailed description of the corrective actions it has taken or intends to take to assure that the acts or omissions which led to its disqualification will not recur. The Commissioner may condition reinstatement upon the testing facility being found in compliance with the good laboratory practice regulations upon an inspection. If a testing facility is reinstated, the Commissioner shall so notify the testing facility and all organizations and persons who were notified, under § 58.213 of the disqualification of the testing facility. A determination that a testing facility has been reinstated is disclosable to the public under part 20 of this chapter.

Printed and bound by CPI Group (UK) Ltd, Croydon, CR0 4YY

23/10/2024

01778259-0019